D1613802

DIVERSITY OF INSECT FAUNAS

Previous Symposia of the Royal Entomological Society

NO. 1. INSECT POLYMORPHISM edited by J.S. Kennedy
London: 1961

NO. 2. INSECT REPRODUCTION edited by K.C. Highnam
London: 1964

NO. 3. INSECT BEHAVIOUR edited by P.T. Haskell
London: 1966

NO. 4. INSECT ABUNDANCE edited by T.R.E. Southwood
Blackwell Scientific Publications, Oxford: 1968

NO. 5. INSECT ULTRASTRUCTURE edited by A.C. Neville
Blackwell Scientific Publications, Oxford: 1970

NO. 6. INSECT/PLANT RELATIONSHIPS edited by H.F. van Emden
Blackwell Scientific Publications, Oxford: 1973

NO. 7. INSECT FLIGHT edited by R.C. Rainey
Blackwell Scientific Publications, Oxford: 1976

NO. 8. INSECT DEVELOPMENT edited by P.A. Lawrence
Blackwell Scientific Publications, Oxford: 1976

SYMPOSIA OF THE ROYAL ENTOMOLOGICAL SOCIETY OF LONDON: NUMBER NINE

Diversity of Insect Faunas

EDITED ON BEHALF OF THE SOCIETY BY

L. A. MOUND & N. WALOFF

PUBLISHED FOR
THE ROYAL ENTOMOLOGICAL SOCIETY
41 QUEEN'S GATE, LONDON SW7
BY
BLACKWELL SCIENTIFIC PUBLICATIONS
OXFORD LONDON EDINBURGH MELBOURNE

Osney Mead, Oxford OX2 0EL
8 John Street, London WC1N 2ES
9 Forrest Road, Edinburgh EH1 2QH
P.O. Box 9, North Balwyn, Victoria, Australia

ISBN 0 632 00352 9

First published 1978

British Library Cataloguing in Publication Data

Diversity of insect faunas.
– (Royal Entomological Society of London. Symposia; no. 9).
1. Insects – Ecology – Congresses
I. Mound, Laurence Alfred II. Waloff, N
III. Series
595.7'05'24 QL461

ISBN 0-632-00352-9

Distributed in the USA by
Halsted Press, a Division of
John Wiley & Sons Inc
New York

Set by Preface Ltd, Salisbury, Wiltshire
Printed and bound in Great Britain by
Butler & Tanner Ltd, Frome

Foreword

Variability of living organisms, and one of its results the diversity of faunas, has always occupied the minds of biologists. Recently, this theme has received a fresh impetus and has been reinvigorated by the critical examination of the relevant mathematical models by theoretical biologists, by the experimental approach of ecologists and by the new data obtained by palaeoecologists. This volume presents the different approaches of an international group of scientists, from a variety of biological disciplines, to this fundamental and rapidly expanding theme.

The taxonomist is concerned with the constituents of diversity and their origin, whereas the ecologist is concerned with understanding the present day complexity of the systems of living organisms. Insects lend themselves well to these studies as their small size and the intricacies of their behavioural and physiological responses enable them to utilise a multitude of microhabitats within the different ecosystems. This, and their great motility and aerial dispersal, enables them to respond to environmental changes in space and time, and it is these attributes that may lie at the root of their diversification and diversity.

January 1978

LAURENCE MOUND
NADIA WALOFF

Acknowledgements

This volume comprises the twelve papers which were read at the Ninth Symposium of the Royal Entomological Society held in Imperial College, London, on 22–23 September 1977. Chairmen at the three sessions were Professor Robert M. May, Dr L.R. Taylor and Professor T.R.E. Southwood.

The Society is grateful to the following organisations for substantial financial contributions towards the cost of the meeting.

May & Baker Ltd
Shell Research Ltd
The Wellcome Foundation Ltd

The costs of the colour plates in this volume were met by a grant from the British Museum (Natural History).

Contents

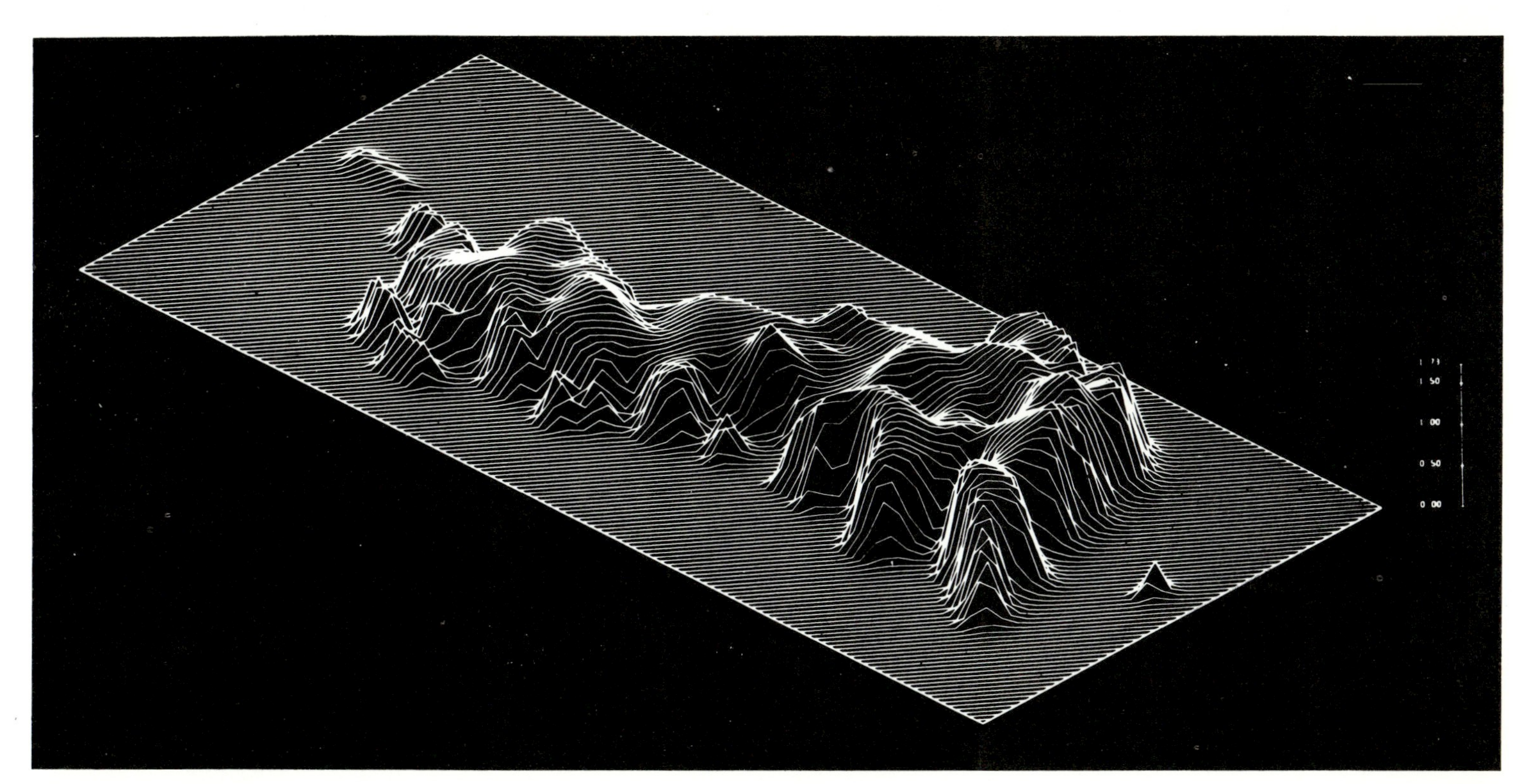

Frontispiece. Isometric projection of α-diversity surface of United Kingdom Macrolepidoptera viewed by an observer at high altitude S.W. of Ireland. Samples were taken by the Rothamsted Insect Survey daily at about 140 sites from 1968 to 1974. The figure was created using the SYMVU package at University of London Computer Centre by R.A.J. Taylor.
Parameters are as follows

Azimuth – direction from which the view is taken – 45° from Greenwich.
Altitude – angle of view from horizontal – 30°.
Scale, 1 = 30α

1 • Bates, Williams, Hutchinson — a variety of diversities

L. R. TAYLOR

Rothamsted Experimental Station,
Harpenden, Hertfordshire, England

Introducing this Symposium on the Diversity of Insect Faunas, I am conscious of two aspects of the subject that excite considerable comment. One of them is the proliferation of measurements for diversity during the last two decades. The resulting confusion has raised doubts about the validity of the property being measured, and I will return to this later. The other striking aspect is the great variety of approaches to the subject and this is reflected in the papers in this Symposium.

There are papers on diversity in time and space, within and between species, within and between habitats, and on a range of scales from hours to eons and centimetres to continents. 'Diversity' so pervades every aspect of biology that each author may safely interpret the word as he wishes and there is consequently no central theme to the subject. We cannot be sure if this flexibility is healthy or due to lack of discipline, but it can be traced back to the beginnings of interest in biological diversity.

The original inspiration came from the impact made on the receptive imaginations of the great Victorian naturalist-travellers by their first contact with the profusion of tropical life. They had been prepared by a new awareness of the unity of living organisms and their responses were intuitive. Each individual responded differently to the stimulus received, and this we still do today.

In the young Charles Darwin, biological diversity evoked a sensation of delight that lasted vividly until the end of his life (Darwin, F., 1887, p. 64). This emotion aroused his curiosity about the diversity of, amongst other things, 'a most singular group of finches' and brought him to consider that 'mystery of mysteries' which became evolutionary biology (Darwin, C. 1890, pp. 362–3).

Alfred Russel Wallace equated the endless diversity of living things with life itself and made it a criterion to distinguish the living from the non-living. It was he who first questioned the relationship between diversity and the stability of populations, not from a study of numbers but from the behaviour of plants observed by an obscure Lincolnshire clergyman (Wallace, 1910). Wallace's numerical approach to the work of De Candolle and the species-area relationship founded the science of bio-geography, inspired by his experience of diversity on the Amazons (Wallace, 1853) and later in the Malay Archipelago (Wallace, 1894). As with Darwin, it was a major factor influencing his discovery of evolution.

Whilst Darwin dealt with evolutionary time and Wallace with global space, Henry Walter Bates (1892) was concerned with more modest ecological scales of time and space, changes from place to place and over periods of a few generations. It is this scale of diversity that I propose to develop here.

Batesian diversity

Bates saw diversity's dual nature clearly; individuals and species. He associated it most strongly with insects, especially with the adult, day-flying Lepidoptera. Like Darwin and Wallace, his interest was intuitive and largely aesthetic. Although it later became the basis of the study of within-species diversity in mimicry and polymorphism, there is another aspect to which he frequently returned. In *The Naturalist on the River Amazons* (1892 edn. p. 351) he wrote, 'No description can convey an adequate notion of the beauty and diversity in form of this class of insects'. 'The species and varieties are of almost endless diversity' (p. 127); some, being excessively common, 'assembled in densely-packed masses, so that the beach looked as though variegated with beds of crocuses'; at these times three-fourths of the individuals belonged to a single species *Callidryas statira*† two other *Callidryas* species being rather less numerous; but sometimes an 'infinite number of curious and rare species may be taken' (p. 52). 'It will convey some idea of the diversity of butterflies when I mention that about 700 species are found within about an hour's walk . . . when the total number, in the British Isles does not exceed 66'.

Here we have all the components for a modern view of high diversity; a large number of species, a few of which are very common whilst many are rare at the same place and time, and this is contrasted with the paucity of fauna in another place or time. In fact, the property can be described by a hollow curve, defining the relative commonness and rarity of species at a place, derived from the numbers of species (S) and the numbers of individuals (N) in a sample. It is not concerned with diversity within a species and hence does not deal with immatures or morphs, but concentrates on the number of reproductive adults, and it is the overall multi-species comparison that is of interest. Bates did not quantify this property, subsequently referred to here as Batesian diversity, but it is easy for an entomologist to visualise the Batesian diversity of Lepidoptera from these descriptions because he can conceive, and even obtain, a sample of thousands of individuals in hundreds of species. For the primatologist, for example, it is more difficult.

Williams' index

After several unsuccessful attempts to manipulate diversity data, the statistical approach was effected at last when Corbet returned from Malaya in 1940 with a collection of 9000 butterflies in 316 species and tried to describe the frequency distribution with a hyperbolic series (Corbet, 1941). Like all museum collections, Corbet's butterfly sample was strongly biassed by deliberate selection of the rare species, and Williams then made the crucial advance of obtaining a sample of 15 200 individual moths of 234 species in which the selection was made at night by a trap instead of by day with a net and was therefore not biassed for rarity because the selection was independent of abundance. He succeeded in interesting Fisher in the statistical problem and Fisher resolved it using the two-parameter Log-series distribution with a sampling parameter X and a diversity parameter α. In his introduction to the classical paper by Fisher *et al* (1943), Corbet drew attention to the non-linearity of the relationship between N and S, (Fig. 1.1) and the statistical nature of the solution, drawing a parallel with the inadequacy of arithmetic means to represent some sets

†Callidryas is a synonym of *Phoebis*.

of collective biological measurements. In other words, 'diversity' was seen as a property of the multi-species population that is equivalent to 'density' in a single-species population. Fisher's development of the solution was based on the assumption that a Gamma distribution describes species abundance; it did not imply a functional model of diversity itself, nor of the underlying species interactions. The resulting diversity statistic, α, is purely empirical, to be justified only by its usefulness in describing what Bates saw and Williams sampled. It is unfortunate that it acquired the epithet 'index', since it is a straightforward parametric statistic, and subsequent reference to α-diversity here means Batesian diversity measured by the statistic α.

The picture drawn by Bates was fairly static. He described diversity at a place and time and made no comment about the dynamics of the species' populations involved. When sampling active adults, diversity changes hourly and seasonally (Fig. 1.1). Williams summed his nightly samples over a year to eliminate these diel and phenological components. He treated α-diversity as a stabilising property of multi-species populations over-riding the instabilities of each component species. In replicate samples from the same place in successive years, the position occupied by a named species in the sequence of commonness and rarity changes. With better data, we can now appreciate that this is because the numbers of all species respond to environmental change by fluctuating constantly in space as well as in time (Fig. 1.2). In contrast, the α-diversity of a multi-species population of adult moths, summed over a year, produces a quasi-stable map of Batesian diversity (Fig. 1.3) that changes more slowly, at the average rate of environmental change experienced by all the included species. This creates problems of weighting the statistic by species, a difficulty which will be returned to later.

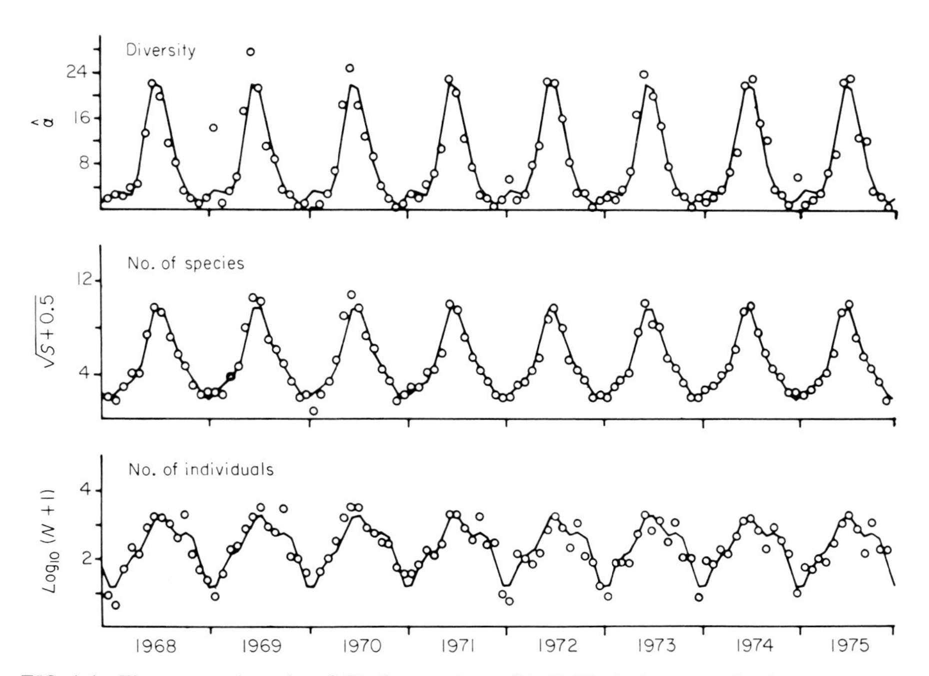

FIG. 1.1. The seasonal cycle of N, the number of individuals in a sample, is appropriately transformed by logs, that of S, the number of species, by square roots. α-diversity combines these and the resulting cycle is more or less skewed depending on the site and taxonomic group. Data for moths at Geescroft, Rothamsted.

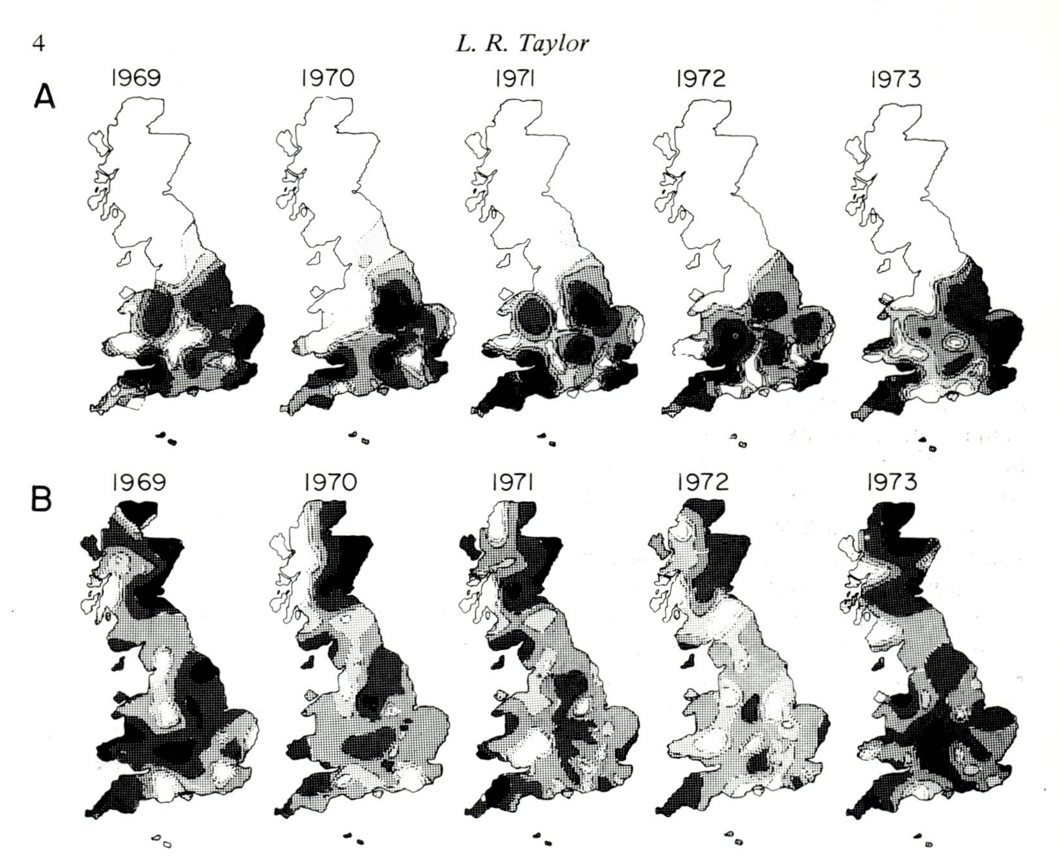

FIG. 1.2. The spatial distribution of moths changes in each generation with a highly specific behavioural component that contributes to the diversity structure and makes random process models inappropriate. Density layers 0, 1–2, 3–9, 10–31, 32–99, 100–315, 316–999. **A** *Agrochola lychnidis.* **B** *Xanthorhoe fluctuata*. (after Taylor & Taylor, 1977).

FIG. 1.3. The geographical distribution of α-diversity is highly complex but, unlike density for single species (Fig. 1.2), it is spatially stable. Contour intervals < 15; 15–; 20–; 25–; 30–; 35–; 40–; 45–; > 50. (after Taylor *et al* 1978).

Hutchinson's question

Williams posed many questions in his penetrating study of the implications and applications of α-diversity, published in a series of over twenty papers (see Williams, 1964), including questions about the evolutionary significance of the data he examined. It was left to Hutchinson, however, to restate the most appealing question at the right time and place and so elicit a flood of replies. Although the issue is of fundamental evolutionary significance and the question had been asked many times before in different ways, by Williams and others, the form of Hutchinson's (1962) question was disarmingly simple; 'Why are there so many kinds of animals?' The simplicity is, of course deceptive. As it is phrased, it could be answered 'Because there are 5×10^9 square kilometers of the Earth's surface,' and this would explain nothing. It does, however, bring out the duality in the question; 'What is the relationship between number of species (S^*) and area?' and 'Why is this so?'.

The answer can now be seen to depend on biomass, because a given area can support only a limited amount of life, so size of individuals determines the number in the multi-species population per unit area, i.e. density (ρ_m). The answer also introduces coordinate space, latitude and longitude, because the ice-cap can support less biomass than the tropics and, at a finer scale, a sand-dune less than a meadow; in other words 'Batesian-scale' place must be considered because the environment is heterogeneous. It also involves Batesian-scale time, because none of these sites or regions are permanent.

The question may, therefore, be rephrased, 'How is Batesian diversity (S corrected for ρ and hence for place and time) related to biomass (corrected for size)' and, 'Why is this so?' It reflects Wallace's question about diversity and stability with a different emphasis. The relationship can be examined if Batesian diversity is measured by $\hat{\alpha}$ and size is eliminated from biomass by using N, the number of adult individuals of a taxonomically related group of species of roughly the same order of size, in a reasonably small standard area. An adequate experimental species complex, based on Williams' own experience and technique, is the nocturnal Macro-lepidoptera in a set of light trap samples throughout an island the size of Great Britain (Fig. 1.4). The relationship is not well defined because so many other

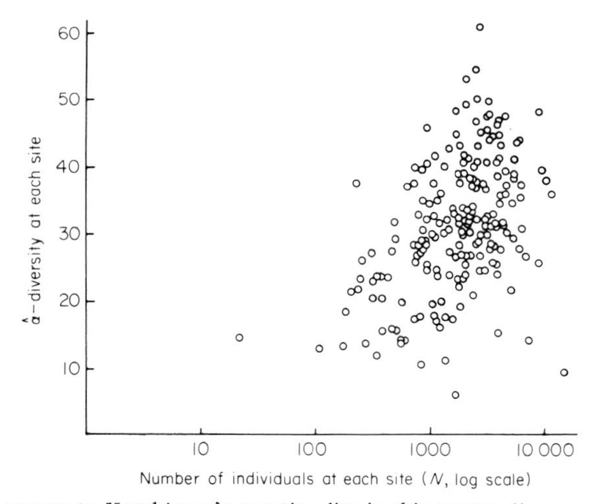

FIG. 1.4. The answer to Hutchinson's question lies in this scatter diagram relating $\hat{\alpha}$-diversity to total numbers, N, at sites throughout Great Britain, but it is obscured by the effects of many other environmental determinants. The problem is to identify and account for these and so reduce the scatter.

environmental factors affect the variance of diversity. To remove their effects and answer Hutchinson's question requires an accurate understanding of the statistics' behaviour and the discovery and definition of these other environmental determinants. Some idea of the complexity of the question can be obtained from the scale of the data required to answer it. More than two million individuals in over six hundred species are included in Fig. 1.4 and it is still not certain that the data are adequate.

This paper is thus an attempt to rephrase Hutchinson's question more precisely in terms of Batesian diversity and to indicate the magnitude of the problem of answering it with Williams' technique. It does not reach a conclusion because the project is not yet finished, but suggests that a solution may be possible.

The requirements of a diversity statistic

Hutchinson's revised question has two parts, 'How is diversity related to density in a multi-species population?' and 'Why?' Clearly the second part cannot be answered before the first, and the answer to that depends on the analytical techniques used. In other words, there must be a well-established statistic for measuring diversity before its relationships can be investigated. Coming at a time of great theoretical excitement in ecology, Hutchinson's question had the unfortunate side-effect of proliferating indices for diversity. Many of these were labelled 'theoretical', being based on hypotheses from population dynamics, which in turn depend on improbable, random, spatial behaviour of individuals. Hence they have no more biological claim to serious consideration than the professedly empirical ones, unless they function more efficiently. Unfortunately, few such indices have been systematically examined in relation to real data and many have been abandoned, but, as mentioned earlier, this has damaged the credibility of the subject (Hurlbert, 1971). Part of the problem has been to collect adequate data and it will become evident that the quantity required inhibits rapid progress. The other part of the problem is that, no matter how theoretically alluring a statistic may be, it is of little value unless it fulfils its purpose, but this is not easy to specify precisely until the statistic is available.

We are here concerned with factors affecting the frequency distribution of relative abundance. The available statistics, however derived and whether parametric or not, emphasise different aspects of that distribution. Unlike a model based on mathematical or population dynamics theory, a descriptive distribution model is chosen only to graduate the data and so to emphasise that part of the distribution which responds to the relevant independent variables. Such a model will also support the statistic for those sets of data that are, for unknown reasons, erratic and which always occur (see later). They cannot simply be discarded, for it is the behaviour of the statistic that is of interest, not the fit of the model. The choice of a statistic can only be justified if it is more effective than others in discriminating the effects of the relevant independent variates, i.e. more of its variance can be accounted for. However, only experience in using the statistic can decide which those variables are. The technique and the analysis must, therefore, be developed together.

Batesian diversity is concerned with the distribution of observable units (individuals) within categories (species) in relatively small areas, and its measurement presents certain problems. These have been approached differently by botanists and zoologists because they originate in the fundamental concept of a sample and its relationship to area. Fisher *et al* (1943) avoided the issues. Using individuals caught in a trap, they treated sample size as being N, the number of individuals. From a consideration of plants, it is clear that sample size cannot so easily be divorced from area (see for example Whittaker, 1972).

Dimensional properties of a sample

Individuals occupy space so that the ecological property actually measured by a sample is always density, ρ, number of individuals (or mass of plant material) per unit area, that is, a mean in space and/or time that smooths out the detail, within the sampling area, of a conceptual density surface (Taylor, 1963) whose shape is always changing (Fig. 1.2). The larger the area required to yield a statistically useful number of individuals (N), the more smoothing of the density has already occurred during its collection. Plainly a very large N can only be accumulated over a considerable area, and/or time, and much of the intrinsic information at the Batesian scale will then be lost. This is not a problem in biogeography because larger areas are being considered and Batesian scale density can usefully be smoothed out.

The same surface concept applies to the density of species, S, or whatever other classifying categories may be used. It is usually assumed that the unit area is common to both N and S and they can then be used without the qualifying dimension. However, when all the species in an area have been recorded ($S = S^*$), further sampling will continue to increase N. The concept 'total species', S^*, seems to imply that S is a sample from some larger, definable, population. This appears to me to be suspect without further qualification, but in any event N and S do not increase linearly with area and time.

The conceptual surface we call diversity (see frontispiece) combines the two density surfaces, for N and S, and its measurement suffers from their faults. If, for example, individuals are so thinly spread ($\rho \to 0$) that N falls to unity in the sample, although diversity may be high, it cannot be measured. The same applies to species. It then becomes necessary to increase sample area or time to obtain a better estimate of N and S, but, since their inter-relationship with respect to space and time is not linear, this becomes a questionable exercise unless the behaviour of the statistical model relating N and S is well understood. Interpreting the behaviour of the statistics is only practicable with large taxa with many common species, so that the sample areas can be small and replication adequate.

Preston's veil-line

Common logic suggests that the frequency distribution for N individuals (which vary logarithmically) in S species (which is a Poisson variate) (Taylor *et al*, 1976) is a Log-normal (Preston, 1948). The logarithmic abscissa raises some problems in converting a continuous distribution to a discrete one (Bliss, 1963) but these can now be overcome with better computing facilities by using the Poisson-log-normal (Kempton & Taylor, 1974). A more serious and fundamental problem lies in the nature of the categories themselves. When expressed as frequency distributions, samples commonly reflect in miniature the frequency distribution of the parent population because each unit has a prospect, however small, of being taken from the highest density category. In samples taken to measure diversity, the categories are themselves being sampled at the same time as their contents, so the larger the sample the more new categories are included from the supposed parent population, thereby changing the shape of the distribution; i.e. if the whole sample contains only 25 individuals, there cannot be a species represented by more than 25, but this is not so for a larger sample of, say, 250 individuals. If a large sample yields the full form of the Log-normal distribution, reducing the sample progressively cuts off the left-hand side of the distribution by moving the origin to the right, and hiding part of the distribution behind what Preston called the 'veil-line' (Fig. 1.5). Eventually the mode is passed and only the right-hand tail, a hollow curve, remains in the small sample.

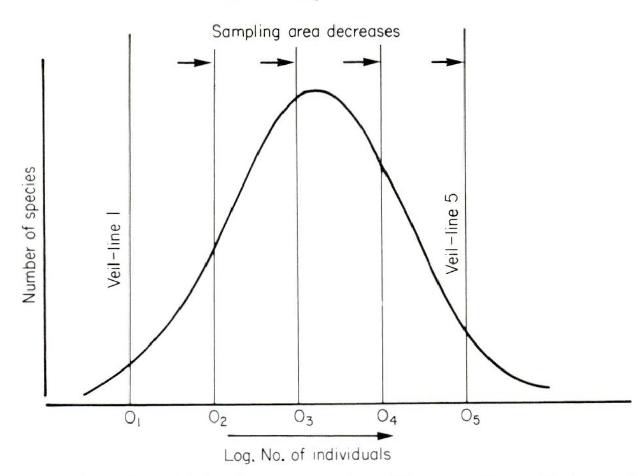

FIG. 1.5. Applied to data in which the categories (S) as well as their contents (N) are dependent on the size of the sample, the Log-normal distribution has a mobile origin that moves across the distribution to conceal progressively more of it behind Preston's veil-line as sampling area decreases.

It is this size of sample yielding a hollow curve (Fig. 1.6B) that Bates described and Fisher defined and which is commonly obtained in trap catches of moths.

When a number of samples like those in Fig. 1.6B are summed over a large area of biogeographical dimensions, the familiar bell-shaped distribution is restored (Fig. 1.6A). It can no longer be described by the Log-series but by the Log-normal, defined by the three parameters S^*, the total number of species under the curve, μ and σ, the mean and standard

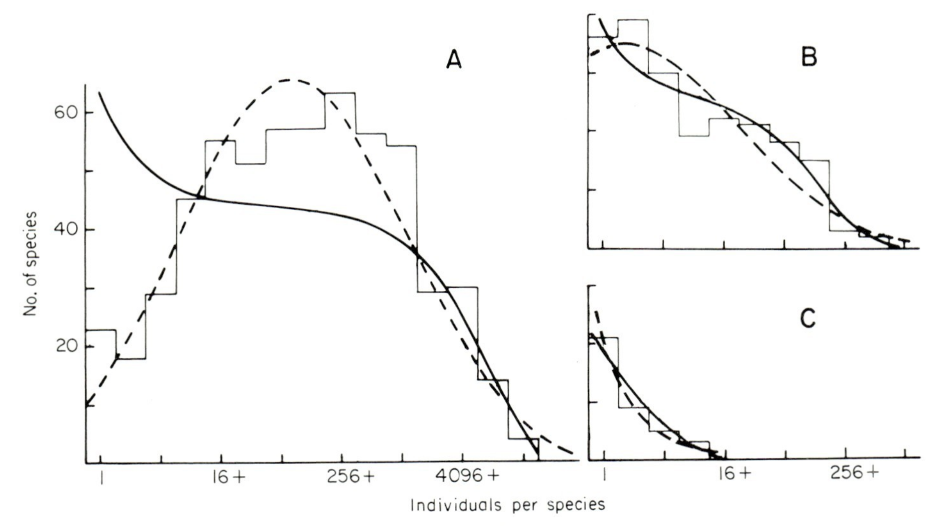

FIG. 1.6.A Summed over 225 sites throughout Great Britain, a large light trap sample of $N = 656\ 943$ moths in $S = 585$ species yields a full bell-shaped Log-normal curve (– – – – –) and is not fitted by the Log-series (———). **B** A typical annual sample from a single stable site is one-tailed and fits equally well to the Log-series and the Log-normal. Data from Geescroft, 1970; $N = 10\ 705$; $S = 205$. **C** When the site is impoverished (urban), the frequency distribution becomes even more concave and the Log-normal responds to this (but see Fig. 1.9). Data from Isleworth, 1970; $N = 153$; $S = 38$.

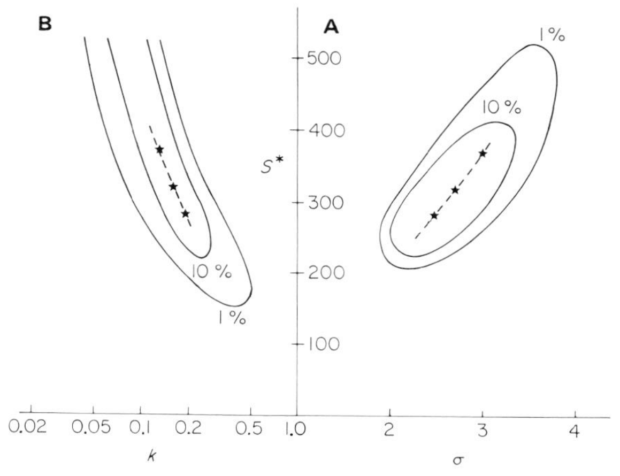

FIG. 1.7.A The likelihood surface for the Log-normal model, showing contours for values of S^* and σ, that differ at the 10% and 1% level from the maximum (★). Along the ridge (– – – – –) pairs of values of S^* and σ cannot be differentiated, but the ratio S^*/σ is almost constant. Data from Geescroft 1970 (see Fig. 1.6B). **B** The Gamma model yields a similar surface in which only the product S^*k can be fixed from the data, in this instance a smaller sample. Data from Barnfield, 1971; $N = 1496$; $S = 130$, (11.7B after Kempton & Taylor, 1974).

deviation. Thus, to estimate the whole distribution from the right-hand tail only, is inappropriate and Fig. 1.7A shows that the parameters S and σ cannot be separately estimated from a sample like those in Fig. 1.6B. In other words, species richness (S^*) and the property 'evenness' (σ), which is a measure of uniformity of density of the species at the site, are inextricably confounded in samples at Batesian scales of space and time. This has led to much misunderstanding of diversity as an ecological, as distinct from a biogeographical concept, and to the role of the Log-normal distribution which has been used less than its ease of comprehension would seem to justify. However, the ratio S^*/σ can be more accurately estimated even from these small samples (see Fig. 1.7). This means that a Batesian diversity statistic must combine richness and evenness to have any practical value, because they cannot be measured separately and because S^* itself is of doubtful validity unless it relates to a clearly circumscribed area.

The sampling dilemma

The sampling dilemma presented by the veil-line which demands a large sample, and the variable density surface which may cause a large sample to obscure the information, can only be resolved by developing a diversity statistic that is efficient for small, hollow-curved, samples.

At large sample sizes there is little to choose between the Log-normal and the Gamma distributions for descriptive power (Fig. 1.8). The Gamma distribution, however, has an advantage over the Log-normal. The defining parameters are S^*, μ, and k, where k is an inverse measure of proportional variance, fulfilling a function similar to σ in the Log-normal. S^* and k in the Gamma are highly correlated in small samples (Fig. 1.7B), as are S^* and σ in the Log-normal (Fig. 1.7A), and, because k is a reciprocal, only the product kS^* can be estimated efficiently, like S^*/σ. However, at low values of k, that is in small samples, the Gamma distribution reaches a limiting condition defined by the Log-series in which $\alpha = kS^*$.

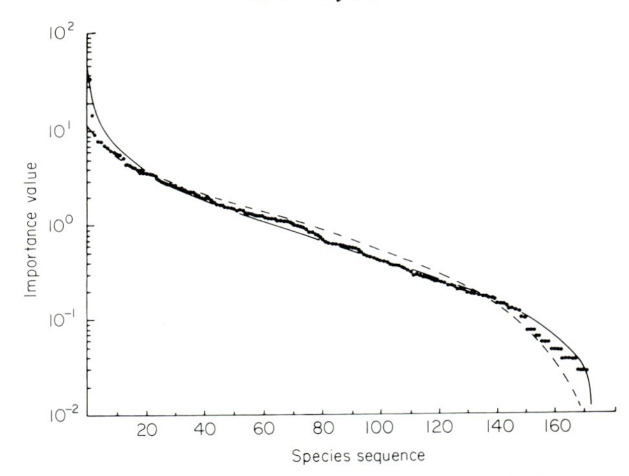

FIG. 1.8. Large samples, in this instance of plants, are equally well fitted by the Log-normal and the Gamma distributions (data from Lamont *et al*, 1977; after Kempton & Taylor, 1978). Importance value is a measure of biomass. ——— Log-normal; – – – – – Gamma.

Thus the rationale of the Log-series is that it represents the 'small sample' condition of the Gamma distribution and this in turn is not usually distinguishable, in practice, from the Log-normal. Corbet's parallel is thus even more apposite than he realised. The use of the log-series $\hat{\alpha}$ is exactly analogous to the use of arithmetic means for biological data which nearly always derive from multiplicative processes where a longer series of observations would show geometrical variation; the arithmetic mean is then an approximation to the geometrical mean justified by the small sample.

Weighting for species

In accumulated annual samples of moths from light traps used by the Rothamsted Insect Survey (Taylor, 1974), $\alpha = kS^*$ from the Log-series and $\lambda = S^*/\sigma$ from the Log-normal are

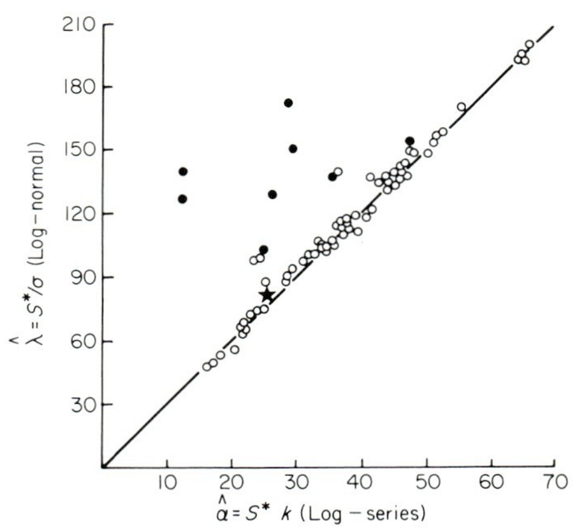

FIG. 1.9. $\hat{\lambda} = S^*/\sigma$, from the Log-normal, and $\hat{\alpha} = S^*k$, from the Log-series, yield similar values for stable sites where both distributions fit (○). The more flexible Log-normal fits better in impoverished sites (see Fig. 1.6C) where singletons predominate (●), but yields improbably high diversity values. (★) fits neither distribution, being inexplicably bimodal, but both yield reasonable parameter values (after Kempton & Taylor, 1974).

almost indistinguishable when the Log-series fits (Fig. 1.9). The Log-normal, being more flexible fits all the sets of data in Fig. 1.9 (except one which was bimodal and fitted by neither distribution) whilst the Log-series does not and the deviant points disclose another interesting property of the Log-series.

As mentioned earlier, the choice of a diversity statistic is based on assumptions about which species' are most relevant to the purpose in hand. Because of the movement inherent in all species, samples from impoverished sites, where the resident population is at very low density, may be flooded occasionally by numbers of a passing migrant species or, more likely, samples may be overweighted by single immigrant individuals from many vagrant species bordering the impoverished zone (see Fig. 1.6C). The resulting distribution then has a very high proportion of singletons and is strongly concave. These curves tend to project unrealistic values for S^*/σ using the Log-normal, and the wild points in Fig. 1.9 are from such sites. In this instance the less flexible Log-series gives a diversity value less subject to the vagaries of the non-resident species, whose mobility provides the constantly changing succession of singletons, and is more dependent on the mid-range species' resident at, and therefore more representative of, the site.

In a similar way, $\hat{\alpha}$ is also less influenced by the other tail of the distribution, the single most common species in the sample. Because of the logarithmic variations in levels of density, the commonest species contributes disproportionately to the fluctuations in N. Sampled repeatedly at the same site, the commonest species may represent from 10 to 25% of the sample in different years (Fig. 1.10). To be effective, a Batesian diversity statistic should not be appreciably affected by this. The Shannon–Weaver information statistic, I, (Shannon & Weaver, 1949) is adversely influenced by the vagaries of the commonest species and, M, the inverse Simpson–Yule statistic (Simpson, 1949; Yule, 1944) even more so. By contrast, $\hat{\alpha}$ is only marginally affected.

The value of a parametric statistic is that the parameter estimate is stable against random or unwanted deviations in the data, providing the model is fitted to weight the appropriate part of the distribution. The Log-series heavily accentuates the median range of commonness, being less sensitive to either the sparse or the abundant species. It is not a particularly good fit to most sets of data (Fig. 1.11) being less sensitive than, say, the

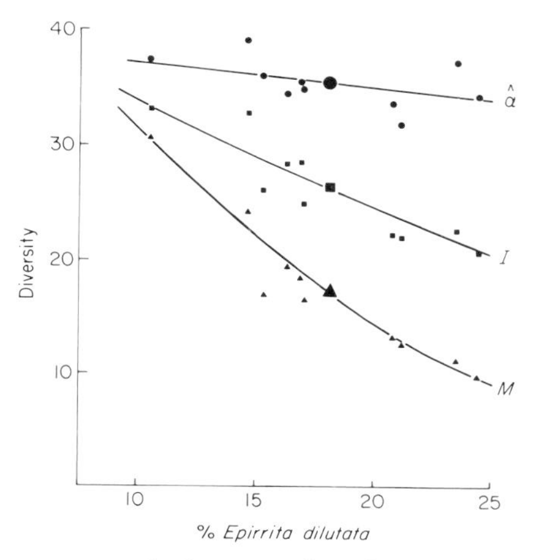

FIG. 1.10. Change from year to year in the proportion of the commonest species in the sample profoundly affects M (the inverse Simpson–Yule statistic), and less so I (the Shannon–Weaver statistic), but hardly affects $\hat{\alpha}$. Data from Geescroft, 1965–1974 (after Taylor *et al*, 1976).

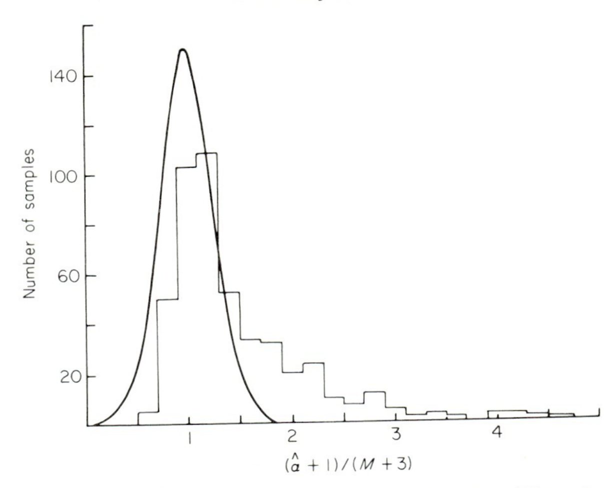

FIG. 1.11. Distribution of the test statistic for goodness-of-fit of the Log-series $[(\hat{\alpha} + 1)/(M + 3)]$ for all available sites indicates that the measured distributions are systematically more long-tailed than the Log-series expectation, given by the smooth curve (after Taylor *et al*, 1976).

Log-normal to the tails of extremely skewed samples, but the measured statistic $\hat{\alpha}$ is normally distributed in samples (Fig. 1.12) and this is of greater concern for its use in subsequent analysis.

Discriminant ability

In the final analysis, the criterion for selecting a diversity statistic must be that it performs the function required of it. To analyse the components of diversity in relation to the controlling environmental variables, the first essential is to discriminate between sites, as Bates did. The seasonal cycle of diversity, however measured, is very marked (Fig. 1.1) and differentially skewed at different sites so that a whole year's sample is a minimum requirement for use in differentiating between the spatial effects of the environment.

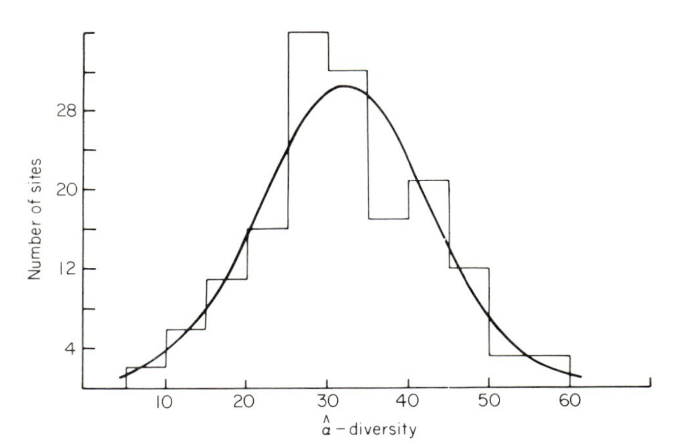

FIG. 1.12. In spite of the failure of the log-series to yield a perfect fit to the data, the diversity statistic $\hat{\alpha}$ is Normally distributed between sites (after Taylor *et al*, 1976).

TABLE 1.1. The ability of various population statistics to discriminate between sites, measured by the spatial variance relative to the within-site variance in time

	α	I	S	M	$\log N$	λ	S^*	σ
F-ratio	201.53	103.89	85.25	57.87	32.58	76.16	9.42	3.94

Based on total annual moth samples from 9 sites for 4 replicate years.

Provided other trends on an ecological time-scale are not detectable, the year to year fluctuations in diversity at a site can then be used as replicates for a within-site variance against which to assess between-site variances for different diversity statistics (Table 1.1).

Neither the number of replicates (four years) nor of sites (nine) are adequate to regard the variance ratios in Table 1 1 as more than a provisional indication of general priorities, but these are clear. Just as it is impossible to discriminate between richness and evenness within a site with these kinds of samples, so S^* and σ are useless to discriminate between sites. Combined together as $\lambda = S^*/\sigma$, their discriminant ability is at least better than mere biomass ($\log N$) but no better than S, the number of species in the sample or M, the inverse Simpson–Yule statistic which is so dependent on the commonest species. The information statistic, I, may be slightly better, but none approach $\hat{\alpha}$ in discriminant ability. Fisher *et al* (1943) are entirely vindicated; $\hat{\alpha}$, in practice, provides a completely efficient statistic for moth diversity in these samples.

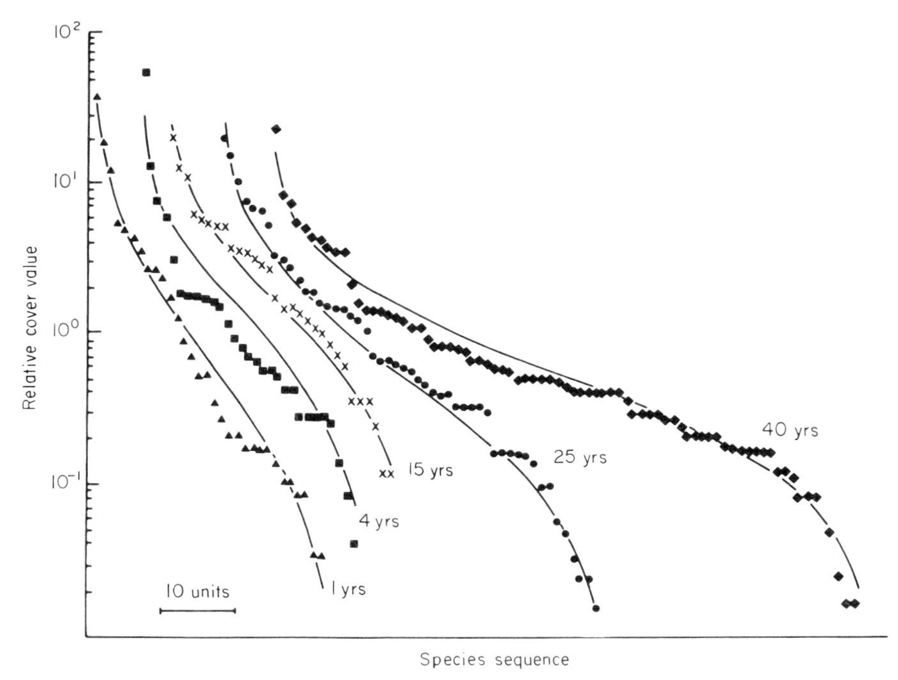

FIG. 1.13. These sequential samples from a reverting flora can be fitted by different models at different stages, but the Log-normal fits all and yields a single parameter (0.371 λ) to relate to time (data from Bazzaz 1975; after Kempton & Taylor, 1976). Relative cover value is another measure of biomass.

It is not, however, desirable to change parameters during a sampling sequence; otherwise comparison between samples are not profitable, and this raises difficulties of comparison within some repeated sampling systems where S changes markedly with time. It has been suggested, for example, that the series of sets of data, from the reversion-to-type of flora after agriculture disturbance, shown in Fig. 1.13, justify a change of model from the Geometric to the Log-normal, within the series (Bazzaz, 1975; May, 1976). However, in this instance, the Poisson-log-normal serves equally well at all stages in the series and provides a parameter, $0.371\ S^*/\sigma$, which reflects accurately the progression in the curves (Kempton & Taylor, 1976). In such instances switching models seems only to add confusion unless the theoretical considerations are over-riding. In the present state of the art, when theory is so inadequate, it may be better to take the simpler course. When there is a prospect of samples extending above and below the median of the log-normal, i.e. both large and small samples are possible, it may be best to use the Q-statistic, the slope of the cumulative species curve in the interval between the quartiles (Kempton & Wedderburn, 1978).

The variance of $\hat{\alpha}$

As mentioned earlier, $\hat{\alpha}$ is normally distributed in samples from Great Britain and so it may be analysed untransformed in relation to environmental determinants in an attempt to answer Hutchinson's question.

The first expectation is that latitude should account for some of the variance of $\hat{\alpha}$ and so it does (Fig. 1.14), but only for 5% of the variance between sites of all kinds. When the urban sites are removed, latitude accounts for 10% of the variance of the remaining, less disturbed, sites. Site disturbance, measured by land use within about 100 m of each site, given arbitrary scoring for areas in five categories (arrived at by iteration), can account for 25% of the variance between sites (Taylor *et al*, 1978), (Fig. 1.15). Combined with latitude we may expect 30–40% of the variance to be explained.

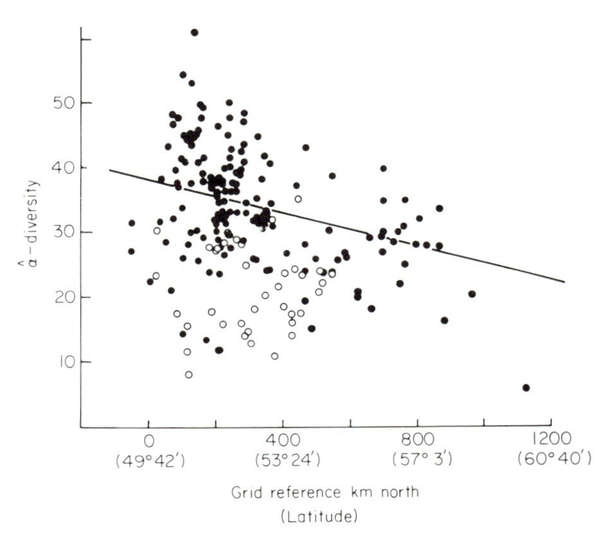

FIG. 1.14. Latitude limits the maximal potential $\hat{\alpha}$-diversity at a site, but much of the variance at low latitudes is accountable to urbanisation. ○ = urban sites, not included in regression; $\hat{\alpha} = 37.9 - 0.0013\ x$ where x is the grid reference.

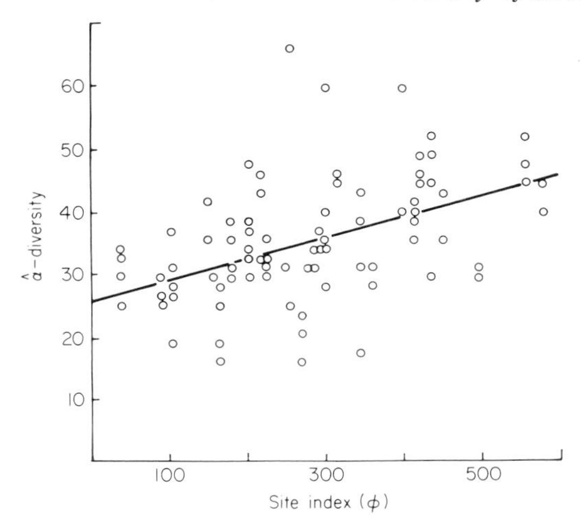

FIG. 1.15. Land use measured by an area-weighted index (ϕ) for proportion of woodland, hedgerow, gardens, grass, arable and buildings, in the immediate vicinity of the trap, derived by iteration, can account for 25% of the variance of $\hat{\alpha}$ (after Taylor *et al*, 1978).

Soil type, floral composition, altitude, exposure and geographical isolation are the most obvious of the remaining environmental variables to be accounted for and there are undoubtedly others such as competition (Taylor & Woiwod, 1975) which may be at the root of Hutchinson's question but are not yet adequately defined for this purpose. These all need to be removed before the remaining variance can be assessed effectively in relation to $\log N$ (see Fig. 1.4).

There remains one, so far unconsidered, technicality; $\hat{\alpha}$ must be statistically independent of $\log N$ before its biological dependence can be assessed. Fig. 1.16. shows that for some samples with $N > ca\,1000$, $\hat{\alpha}$ is independent of N. This level of independence, however, interacts with the effect of the commonest species (see Taylor *et al*, 1976) and only when the variance of $\hat{\alpha}$ has been fully accounted for can the limits of tolerance to N be finally defined.

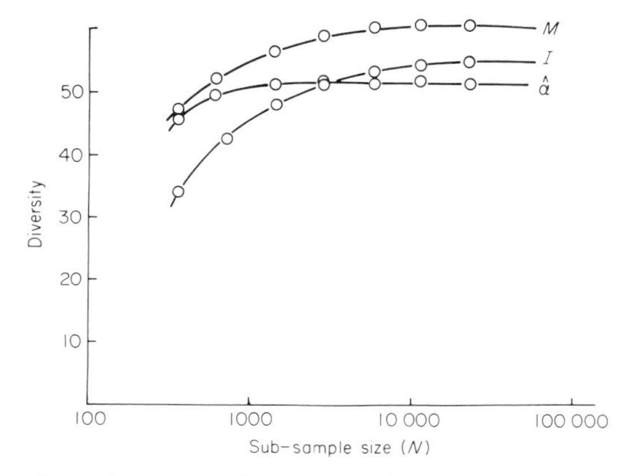

FIG. 1.16. $\hat{\alpha}$ is less dependent on N than are I and M, but even so, samples of less than a thousand individuals are suspect. Data accumulated over 1966–1973 from Stratfield Mortimer and sub-sampled, but see text (after Taylor *et al*, 1976).

Nevertheless, using $\hat{\alpha}$ there is at least a prospect of success in this search for an answer to Hutchinson's question that is not offered by λ, I or M.

Conclusions

In her recent review of the state of the art, Pielou (1975) asked if there really are any discoverable general laws about many-species communities. She separated potential answers into mathematical and statistical, with some discerning comments on both. Mathematical ecologists, she explained, devise dynamic models to simulate rates of population processes based on apparently reasonable postulates, and they hope to deduce from these what happens in the real world; in other words, they adopt a synthetic approach. Statistical ecologists look at the real world, judge which questions will receive unambiguous answers and collect the necessary data, using the analytical approach. Shrewdly, Pielou comments that mathematicians may still construct unreal models, perhaps because nobody yet knows the relevant and valid premises for the synthesis; and statisticians may provide clear concise answers to uninteresting questions, perhaps because a question that is known to have an answer is already half-way to being understood and is that much less interesting.

There are, however, those other students of diversity, biologists of many kinds, general or applied entomologists, ecologists, practical field geneticists, taxonomists, palaeontologists, like most of the authors in this Symposium. All, in some degree, are naturalists as were those former colleagues and Fellows of this Society, Darwin, Wallace and Bates and, between them, the present authors cover the same variety of diversities as the earlier generation. They have for their expertise a wide knowledge and penetrating grasp of what animals, or plants, are able and likely to do, in other words of their intrinsic behaviour, and of the vast range of differences and interactions between individuals; components which are lacking in most dynamic models. Accordingly, their questions are often intuitive, inspired by original curiosity, and rarely uninteresting or unreal. They may, however, turn out to be unanswerable or, even when answers are found, their authors may lack the technical virtuosity to formulate them succinctly.

We, therefore, have three approaches to the study of diversity, mathematical, statistical and biological, all of which are necessary and may be found in this Symposium, but which only occasionally come together in this way. This paper may appear to be statistical by Pielou's criteria, so it is perhaps as well to point out that, although it has been essential to look closely at all the sampling methodology, including the practical techniques not considered here (Taylor & Brown, 1972; Taylor & French, 1974) as well as these statistical ones, the investigation remains essentially exploratory and biological. It was originally motivated by a naturalist's conviction that the dynamics of single species populations hinge on behavioural interactions between uniquely different individuals, mediated by movement (Taylor & Taylor, 1978), rather than on the simplistic birth and death processes then current in population dynamics, and it is consequently from this behaviour that the structure of multi-species populations results.

If the variety of diversities discussed at this Symposium draws attention to other, different views of the process involved in multi-species populations by encouraging naturalists to persist with confidence in their own expertise, it will have been appropriate to have done so at the Royal Entomological Society of London, where it all began a century ago.

Acknowledgements

It is a great pleasure to have this opportunity to thank the staff and volunteers connected with the Rothamsted Insect Survey, without whom this work would be impossible, as well as those colleagues, especially R. A. French, R. A. Kempton, R. A. J. Taylor and I. P. Woiwod, whose collaboration has made it so rewarding.

References

Bates H. W. (1892) *The Naturalist on the River Amazons: A Record of Adventures, Habits of Animals, Sketches of Brazilian and Indian Life, and Aspects of Nature under the Equator, during Eleven Years of Travel* Reprint of the Unabridged Edition, John Murray, London.

Bazzaz F. A. (1975) Plant species diversity in old-field successional ecosystems in southern Illinois. *Ecology* **56**, 485–488.

Bliss C. I. (1963) An analysis of some insect trap records. *Classical and Contagious Discrete Distributions*, Ed. G. P. Patil, pp 385–397. Statistical Publishing Society, Calcutta.

Corbet A. S. (1941) The distribution of butterflies in the Malay Peninsula (Lepid.). *Proc. R. ent. Soc. Lond.* (A) **16**, 101–116.

Darwin Charles (1890) *Journal of Researches into the Natural History and Geology of the Countries visited during the Voyage of H.M.S. Beagle round the World* (New Edition). John Murray, London.

Darwin Francis (1887) *The Life and Letters of Charles Darwin* Vol. 1 (Third Edition). John Murray, London.

Fisher R. A., Corbet A. S. & Williams C. B. (1943) The relation between the number of species and the number of individuals in a random sample of an animal population. *J. anim. Ecol.* **12**, 42–58.

Hurlbert S. H. (1971) The nonconcept of species diversity: a critique and alternative parameters. *Ecology* **52**, 577–586.

Hutchinson G. E. (1962) Homage to Santa Rosalia or, Why are there so many kinds of animals? Presidential Address to the American Society of Naturalists, 1958. *The Enchanted Voyage and Other Studies*. Yale University Press, New Haven.

Kempton R. A. & Taylor L. R. (1974) Log-series and Log-normal parameters as diversity discriminants for the Lepidoptera. *J. anim. Ecol.* **43**, 381–399.

Kempton R. A. and Taylor L. R. (1976) Models and statistics for species diversity. *Nature, Lond.* **262**, 818–820.

Kempton R. A. & Taylor L. R. (1978) The Q-statistic and the diversity of floras. *Nature, Lond.* (in press).

Kempton R. A. & Wedderburn R. W. M. (1978) The measurement of species diversity. *Biometrics* **34**, 25–37.

Lamont B. B., Downes S. & Fox J. E. D. (1977) Importance-value curves and diversity indices applied to a species-rich heath-land in Western Australia. *Nature, Lond.* **265**, 438–441.

May R. M. (1976) Patterns in multi-species communities. In *Theoretical Ecology: Principles and Applications*, Ed. R. M. May. Blackwell Scientific Publications, Oxford.

Pielou E. C. (1975) *Ecological Diversity*. John Wiley, New York.

Preston F. W. (1948) The commonness and rarity of species. *Ecology* **29**, 254–283.

Shannon C. E. & Weaver W. (1949) *The Mathematical Theory of Communication*. University of Illinois Press, Urbana.

Simpson E. H. (1949) Measurement of diversity. *Nature, Lond.* **163**, 688.

Taylor L. R. (1963) Analysis of the effect of temperature on insects in flight. *J. anim. Ecol.* **32**, 99–117.

Taylor L. R. (1974) Monitoring change in the distribution and abundance of insects. *Rep. Rothamsted exp. Stn for 1973, Part 2*, 202–239.

Taylor L. R. & Brown E. S. (1972) Effects of light-trap design and illumination on samples of moths in the Kenya highlands. *Bull. ent. Res.* **62**, 91–112.

Taylor L. R. & French R. A. (1974) Effects of light-trap design and illumination on samples of moths in an English woodland. *Bull. ent. Res.* **63**, 583–594.

Taylor L. R. & Woiwod I. P. (1975) Competition and species abundance. *Nature, Lond.* **257**, 160.

Taylor L. R., French R. A. & Woiwod I. P. (1978) The Rothamsted Insect Survey and the urbanisation of land. In *Perspectives in Urban Entomology* Ed. G. W. Frankie & C. S. Keshler. Schmid-McCormick, Berwyn, Pa.

Taylor L. R., Kempton R. A. & Woiwod I. P. (1976) Diversity statistics and the Log-series model. *J. anim. Ecol.* **45**, 255–271.

Taylor L. R. & Taylor R. A. J. (1977) Aggregation, migration and population mechanics. *Nature, Lond.* **265**, 415–421.

Taylor L. R. & Taylor R. A. J. (1978) The dynamics of spatial behaviour. In *Population Control by Social Behaviour*, Ed. F. J. Ebling & D. M. Stoddart, pp. 181–212. Institute of Biology, London.

Wallace A. R. (1853) *A Narrative of Travels on the Amazon and Rio Negro, with an Account of the Native Tribes and Observations on the Climate, Geology and Natural History of the Amazon Valley.* 1st Edn, Lovell Reeve, London.

Wallace A. R. (1894) *The Malay Archipelago: The Land of the Orang-Utan and the Bird of Paradise: A Narrative of Travel with Studies of Man and Nature.* (New Edition). Macmillan, London.

Wallace A. R. (1910) *The World of Life: A Manifestation of Creative Power, Directive Mind and Ultimate Purpose.* Chapman and Hall, London.

Whittaker R. H. (1972) Evolution and measurement of species diversity. *Taxon* **21**, 213–251.

Williams C. B. (1964) *Patterns in the Balance of Nature and related Problems in Quantitative Ecology.* Academic Press, London.

Yule G. U. (1944) *The Statistical Study of Literary Vocabulary.* Cambridge University Press, London.

2 • The components of diversity

T. R. E. SOUTHWOOD

Imperial College, University of London

It does not require any deep perception to guarantee that the most recurring theme in this Symposium will be the great diversity of insects. If our concepts of evolution and ecology are valid this is synonymous with saying that a very large number of niches are occupied by insects. Three processes will be involved in this: speciation, colonisation and coexistence.

Diversity in insects is expressed both within species and between species: intraspecific diversity, which is perhaps synonymous with polymorphism, was the subject of the Society's first Symposium (Kennedy, 1961) and is extremely well represented in the Insecta (Richards, 1961). The dominance of the Animal Kingdom, in terms of number of species, by this class is striking (Fig. 2.1). I suggest the fact that both intra- and interspecific diversity are so well represented in insects is a direct reflection of certain of their biological features that predisposes them to diversity in response to environmental heterogeneity. This claim is strengthened by a recent study that has demonstrated the substitution of interspecific diversity for intraspecific diversity as the habitat changes.

Davidson (1978) studied Harvester Ants in various deserts in the USA. In the least productive-habitats there are only two species of ant; one harvests individually and the other, *Veromessor pergandei*, in columns. In these deserts *V. pergandei* shows a great variation in size, including that of the mandibles; this enables them to handle seeds of a wide range of sizes. However, as the deserts become more productive, and there are more seeds to harvest, certain more specialised species of ant are able to exist. These specialists then out-compete the generalist *V. pergandei* for particular ranges of seed size. In these habitats *V. pergandei* still exists but it is much less polymorphic in size. As the interspecific diversity of the ant fauna rises, so the intraspecific diversity of the component species falls. This elegant study emphasises the general principle that the 'fractional niches' to which intraspecific diversity is a response, may, if they come to contain more resources, become full niches for interspecific diversity. But note, this is permissive, not obligatory.

INTRASPECIFIC DIVERSITY

Within a species the same individual will have different forms at different stages of its development and different individuals of the same species may have different forms: the first is commonly referred to as metamorphosis and the second as polymorphism. However, as the earlier Symposium showed (Kennedy, 1961), the phenomena have many of the same features and in some instances, as in termites, where the castes represent individuals in

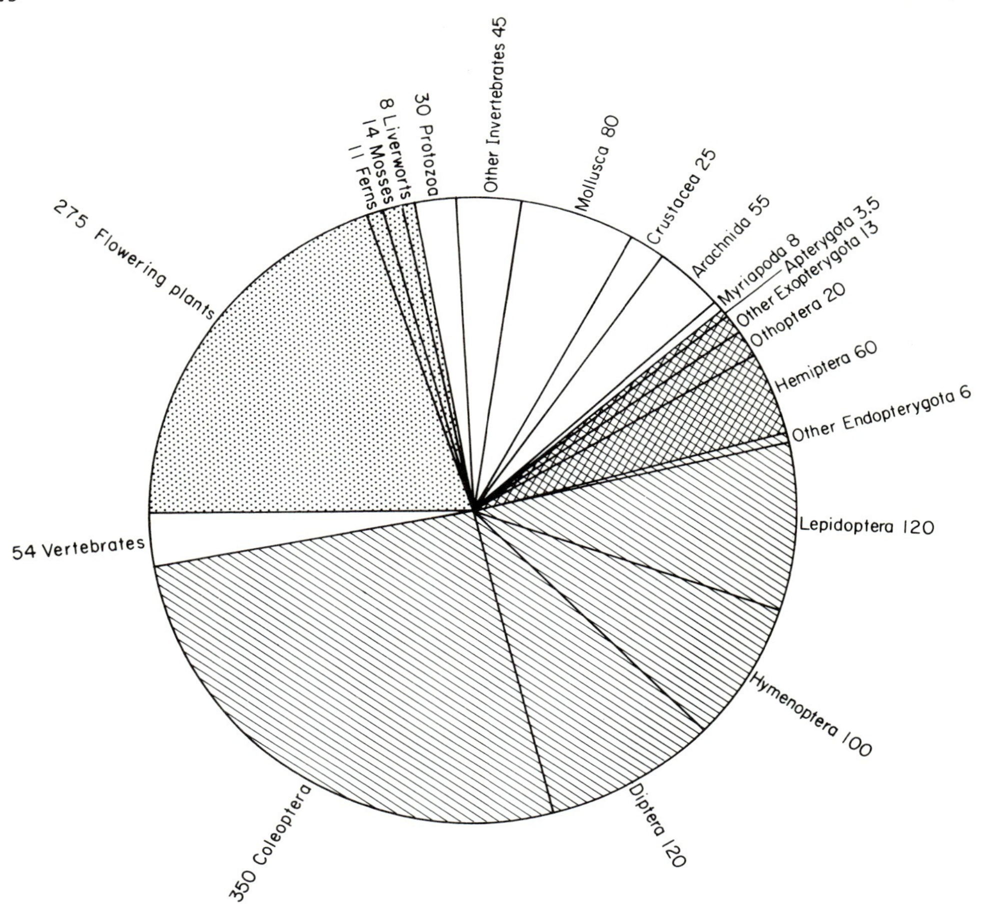

FIG. 2.1. The number of species in different taxa (excluding Fungi, Algae & microbes) (Numbers are thousands of species).

different stages of development, they are indistinguishable (Richards, 1961). For convenience, I will separate the two here and continue to utilise the classical labels.

METAMORPHOSIS – DEVELOPMENTAL POLYMORPHISM

Complex life-cycles are found in several groups of animals; they are often associated with internal parasitism or, in for example Amphibia, the occupancy of aquatic and terrestrial environments. In the Apterygota young stages and adults appear to have similar ecological requirements. In the Exopterygota the larval stages may be confined to aquatic habitats, but otherwise the distinctions between the requirements of the stages are slight. It is the Endopterygotes, with their pupae, that are uniquely well suited to exploit entirely different terrestrial habitats as larvae and adults (Labeyrie, 1977). About 88% of known species of insects are Endopterygotes and Hinton (1948, 1977) has forcefully argued that this success is witness to the evolutionary value of the pupal stage allowing this 'disassociation in form and structure between the feeding and growing stages and distributive and reproductive stages'.

In this statement Hinton (1977) highlighted the importance of the adult stage of insects as the dispersal stage. Habitats, particularly those that may be described as temporary or ephemeral (Southwood, 1962, 1977*b*), vary in their location with time; organisms that exploit them need to be able to move from the old sites to the new: the high mobility and often sophisticated host-finding behaviour of the winged adults of many insects are a measure of the success of their adaptation to this role. The few species (e.g. the winter moth, *Operophtera brumata*) that rely on ballooning on silk threads, like spiders and most mites, have large 'targets' of relatively high durational stability. This dispersive role for a stage in the life history is one of the reasons why complex life-cycles persist in evolutionary time (Slade & Wassersug, 1975; Istock, 1975), when a simple model considering population levels in a constant environment indicates that they would be unstable (Istock, 1967).

However, food may be harvested in both larval and adult stages and one may suppose that this is the primitive condition, as exemplified by the Apterygota and some Exopterygota groups such as Phasmida. In others, even where the habitat is similar, small differences in diet may be detected; in predators, such as Mantids, due to size changes with development, in plant feeders such as Coreids, the increased mobility of the adult makes other food sources available. But complete metamorphosis permits quite different food to be harvested from different habitats by the different stages. This is particularly important when food is available in a particular habitat for only a short time in a season or when adult feeding appears to compensate for inadequacies in the 'nutritional capital' carried forward from larval life.

The utilisation of a short-lived seasonal resource may be illustrated by the apple blossom weevil (*Anthonomus pomorum*). The larvae feed on the stamens and pistils of the apple flower, which are in the correct stage for about three weeks each year; hatching is even more narrowly limited, larvae that hatch too late to prevent the petals opening die (Greenslade, 1945). The adults have three phases of feeding; in mid-summer and early autumn mainly on the leaves of the apple tree and in the early spring on the unopened buds of the apple. In the tropical *Heliconius* butterflies the long adult life, sustained by pollen feeding enables the species to survive through periods when larval food resources are unavailable (Gilbert, 1977). The larval stage itself may sequentially be adapted for a number of different roles. Striking examples are provided by the hypermetamorphosis of Strepsiptera, Meloid beetles, Bombylid flies and other parasites; however, here the diversity of form is associated with first the discovery and then the exploitation of a single food resource. Although not sufficiently distinct structurally to justify the term hypermetamorphosis, the last instar larvae of the Large Blue butterfly (*Maculinea arion*) exploit an entirely different food source (ant larvae) and shelter (ants' nests) than the early instars that feed and live on wild thyme (*Thymus*) flowers. The larvae of many Lepidoptera change form and habitat when they move from feeding on one part of the host plant to another (e.g. *Prays curtisellus* on ash).

The role of adult feeding as an essential supplement to larval feeding is well shown in Diptera, by the bloodmeals required by the females of many families (Downes, 1958), by the nectar and possibly protein essential to some Anthomyiids (Jones, 1970; Coaker & Finch, 1973) and by the protein necessary for blow-flies (Harlow, 1956; Kitching, 1977). In general, in pterygote insects, the emphasis is on food harvesting by the larval stage and in the Ephemeroptera and few other groups it is confined to that stage. There are even fewer species in which the activity is restricted to the adult and these are either blood-feeders (e.g.

Tsetse-flies, Hippoboscids) or semisocial or social species. In parenthesis I will note that adult food gathering and social organisation are part of a suite of characters that these insects share with some vertebrates.

The generality with which food is harvested in larval and adult stages of the Insecta indicates the evolutionary advantages and stability of this strategy. We may conclude that metamorphosis enables species of insects to avail themselves of a sequence of niches, utilising resources (food, shelter) that may be separated in both time and space and might not, on their own, be adequate for a whole generation or season.

POLYMORPHISM OF THE SAME STAGE

That type of polymorphism so well exhibited by aphids (Blackman, 1974) and Cynipid gall wasps, the alternation of generations, is a further example of diversity within a single species allowing the sequential exploitation of seasonal resources. Aphids, with their short, telescoped, generations, additionally have the opportunity to sequentially optimise their adaptation to seasonal conditions by facultative change. The sycamore aphid (*Drepanosiphum platanoides*) has individuals in the spring and autumn with melanic bands on the abdomen, these are virtually absent in aphids developing in the summer and if the temperature is high (16–27°C) increasing numbers of a red form occur. Dixon (1972) has documented these changes and shown how the body temperature of the melanic form is raised relative to the other forms through the more efficient absorption of radiation.

The production of apterous or alate virginoparae in response to conditions on the particular host plant (Lees, 1961) occurs in many Aphididae. This is, of course, a special instance of polymorphism with regard to migratory ability (Johnson, 1969) which is particularly common in insects with habitats that are variable in their favourableness (Southwood, 1977*b*). It may manifest itself in strikingly different morphs, as in the solitary and gregarious phases of locusts, either through structural wing polymorphism that may even be confined to the musculature, or to behavioural types – 'good fliers' and 'poor fliers' (Richards, 1961; Young, 1965*a* & *b*; Acton & Scudder, 1969; Shaw, 1970; Vepsäläinen, 1971; Andersen, 1973; Waloff, 1973; Y. Y. May, 1975; Järvinen & Vepsäläinen, 1976).

Habitat favourableness (Southwood, 1977*b*) is generally thought of with respect to food and shelter, but mate-finding is another aspect (Kitching, 1977). In a fascinating study on fig insects Hamilton (1977) shows how males may be dimorphic in relation to flight and fighting ability. In species with intermediate densities in the fig this dimorphism is well marked. However, in species where the population density in figs is high, the advantage to the male is to stay where he is and he will be flightless and if the degree of relatedness to other males in the fig is low, a fighter. In contrast, when population density in the fig is always low there will be many figs with only females maturing, under these circumstances the males are always winged; it is better for them to go somewhere else (in terms of Southwood's, 1977*b*, favourableness matrix $r_c < r_a$). This work underlines the importance of small discrete habitats (i.e. figs) in varying favourableness in time and space for insect diversity.

Defence is another requirement that has led to polymorphism. Now is not the time to go into the various types of mimicry and crypsis and their associated theories; a constructive review has been undertaken by Vane-Wright (1976). The concept is, however, well established that the number of mimics must be limited in relation to the number of models and thus this polymorphism, like metamorphosis, 'allows' the species to put a number of small (evolutionary) opportunities together to make one larger 'package'. One of the most highly polymorphic insects is the female of the swallowtail butterfly, *Papilio dardanus*, that has been so carefully investigated by Clarke and Sheppard (1959, 1960, 1962*a* & *b*, 1963); 26 different forms, 10 or 11 of them mimetic, have been recognised (Sheppard, 1961). Polymorphism in respect of cryptic colouration is, for example, shown by the adults and larvae of the now well-known peppered moth, *Biston betularia* (Kettlewell, 1973).

Considerable diversity is often found in the castes of social or semisocial insects permitting a series of specialist functions. Colonies of social insects are not only evolutionary entities, but in many ways they are ecological entities. It has already been noted that in adult food gathering they share a trait with vertebrates; because of the size of their colonies, social insects may like vertebrates mould and modify their habitats (Janzen, 1967, 1977). Thus they 'open' to the insects an additional range of niches for larger animals, here they may compete with both herbivorous (Brown & Davidson, 1977) and insectivorous vertebrates.

In considering the diversity within a species, I have attempted to show the various ways that members of the same species, either different individuals at the same time or the same individual at different times of its life, may fill different ecological niches. We may perhaps envisage the occupancy of these niches as a series of stepping stones, forward and sideways, in time and space, that enable a generation to survive the temporal and spatial heterogeneity of their environment and pass their genes on to their successors. In other words intraspecific diversity enables the combination of resources scattered in time and space to provide a composite niche for a species.

INTERSPECIFIC DIVERSITY

Although there is still some doubt as to the extent of gene flow throughout the population of a species (Ehrlich & Raven, 1969; Ehrlich *et al.*, 1975), there is general agreement that the attenuation of gene flow to a low level, when any 'foreign' genes will quickly be removed by selection, is an essential feature of speciation (Bush, 1975*b*). Heterogeneity would be expected to contribute to this attenuation (isolation) and it may seem surprising that controversy surrounded the roles of temporal isolation and sympatric speciation (Mayr, 1963; Alexander, 1963; Blackith, 1974; White, 1974). In a recent comprehensive review Bush (1975*b*) recognises four types of speciation:

1. *Allopatric, type 1a*, speciation by subdivision, due to geographical isolation.
2. *Allopatric, type 1b*, speciation by founder effect.
3. *Parapatric* or *stasipatric* speciation by selection acting on individuals that penetrate new habitats within the range of a viscous (non-mobile) species, e.g. morabine grasshoppers (Key, 1974; White, 1974).
4. *Sympatric* speciation by an isolating mechanism within the species range.

However, in this section I will seek to identify those features of the biology of the Insecta that have contributed, against environmental heterogeneity, to:

a. The establishment of barriers to gene flow ('isolation') in space and time i.e. to speciation.
b. The 'mixing' of species formed in other geographical locations.
c. The continued coexistence of the diversity of species resulting from [a] and [b] (whilst recognising, of course, the phenomenon of extinction).

SPACE AND SIZE

The patchiness of the habitat of an organism must be scaled to two spatial characters of the organism: its trivial range, the area over which it gathers its food, and its migratory range (Southwood, 1977*b*). The trivial range is very approximately related to the size of the organism and hence insects being of small size have smaller trivial ranges than, say, mammals. Thus the same natural environment is more patchy for an insect than a mammal and so insects can evolve adaptations to particular patches.

The relationship of the trivial range of an animal to its body size is shown in Fig. 2.2 for some ambulatory grassland-scrubland animals found at Silwood Park. The trivial range has been expressed as the linear diameter of an assumed circular range. The ranges of the rodents

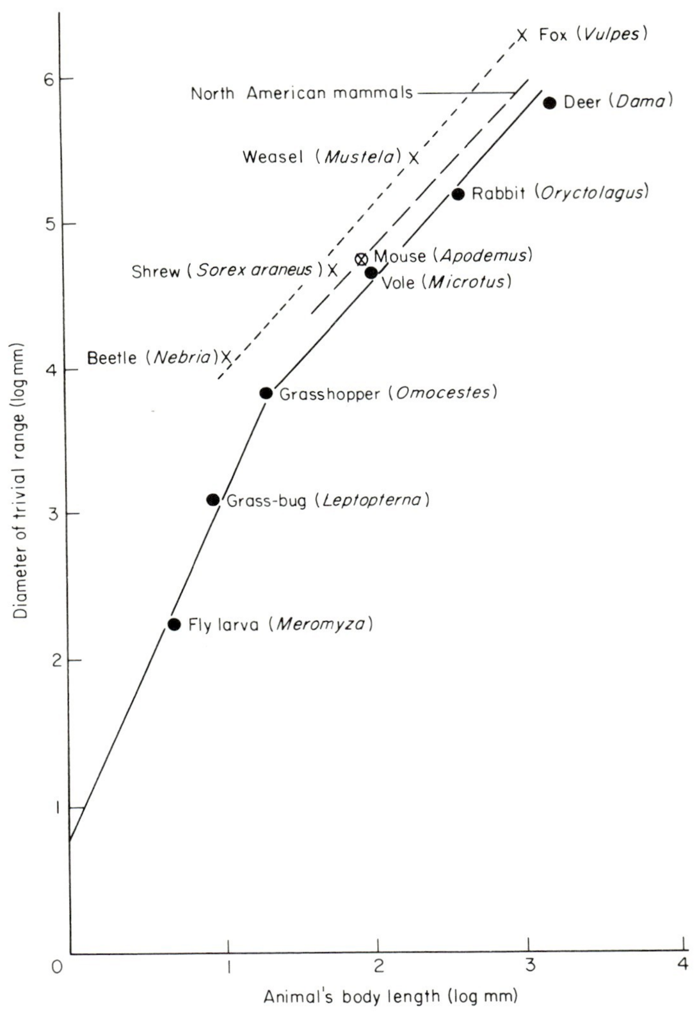

FIG. 2.2. The relationship of trivial range (linear diameter) to body length for ambulatory herbivores (solid line) and predators (dotted line) at Silwood Park. The corresponding relationship for North American mammals [after Hutchinson & MacArthur (1959)] is shown dashed.

are based on the detailed studies at Silwood by Brown (1956), for the weasel by King (1975), for other predatory mammals on data given in Southern (1964), for the ground beetle *Nebria brevicollis* on Greenslade's (1961) measurement made at Silwood, and for the other animals on estimates made by persons who have worked on them. Also on the figure is the line from Hutchinson and MacArthur's (1959) comparison for North American mammals (converted to the present units). Caution must be exercised because of the nature of the estimates and the lines have been fitted by eye, but the following comments seem justified.

1. When herbivores are considered separately from predators two distinct relationships are derived suggesting that for the body length range 15–1500 mm, the linear diameter of the trivial range of an ambulatory herbivore, in the type of habitat there is at Silwood, is about 400 times the body length and for a predator three to three and a half times greater. That the data point for *Apodemus sylvaticus*, which has a mixed diet of plant material and insects, and the 'Hutchinson & MacArthur line', which was based on both predatory and herbivorous mammals, are intermediate, strengthens this conclusion.
2. The line for herbivores, where animals under 15 mm in length are included, appears to turn in this region and gives an intercept predicting, for a 1 mm herbivore (say a spider mite) a trivial range of about 80 mm (a grassblade). The non-linearity is probably due not to the taxonomic shift from mammals to insects, but to the increasing addition of the surfaces of the vegetation to the areas traversed and to the restriction due to the vegetative cover, on horizontal movement of the small organisms, i.e. impedance due to structural complexity of the vegetation. The smaller the organism the more trivial movement there is in the vertical plane, rather than the horizontal plane and at the size of the stem-boring fly larva the range has become restricted by the vegetation so that it is linear and vertical rather than circular and horizontal.

Body length is only one indicator of structural size in relation to movement, although it is the simplest. The best measurement might be the area spanned by the animal's legs. The Opiliones are a group where the exceptional prolongation of the legs has clearly given them a much greater trivial range than body size alone would suggest. Todd (1949) showed that *Leiobunum rotundum* moves down the trunks of oak trees at dusk and hunts over vegetation, ascending again at dawn. The distance covered is clearly considerably greater than the 17 cm predicted from *L. rotundum's* body length of 4 mm.

A relationship between foraging range and size has been remarked upon in desert harvester ants (Davidson, 1977); the ranges (Hölldobler, 1977) are relatively greater than those for similar sized herbivores in temperate grasslands and this undoubtedly reflects both the structural simplicity of the habitat and evolved anatomical features.

It may also be noted from Fig. 2.2 that there is a general tendency for the variety of food items taken to increase with increase in trivial range. Over 15 species of Dipterous stem borer are found commonly at Silwood Park; with two exceptions they specialise on two or three species of grass (Southwood & Jepson, 1962). *Leptopterna dolabrata* feeds on both the growing shoots and the flowering parts of several species of grass; although showing preferences, grasshoppers feed on most species of grass and occasionally other plants (Richards & Waloff, 1954; Williams, 1954). *Microtus* 'prefers' grass shoots but will feed on other plants, roots and bark, whilst the rabbit feeds on a wide range of herbaceous vegetation at all seasons and the deer includes browsing on shrubs.

Mattson (1977) has analysed the implications of size on host plant range and type within the Lepidoptera. He found a weak trend for larger species to utilise larger plants; the exceptions involved special strategies (few eggs per plant) or special conditions (plants adjacent in large stands). This study supports the general principle that it is 'harder' for the larger species to be a food specialist than it is for the smaller.

The small size of insects has facilitated their adaptation to many special patches in the natural environment. Because of the structure provided by the vegetation many of these habitats present surface areas that are not available to the larger organisms (Martin, 1969). This is illustrated by the data in Table 2.1. Even in the winter months for any insect of up to 3 mm the tree offers an area about four times that of the soil surface overhung by the canopy; in the summer months it is in the region of 50 times and this excludes the volume of habitat offered by the rotting wood, various bracket fungi, the fruits of the beech and the epiphytes, the leaf tissues and the heart wood. This great area, volume and variety of micro-habitats sprouts from a mere 6 m^2 of the soil surface. In woodland types, e.g. oak, with a herb layer, the total surface area will be greater, but in a field less. This further increases the potential numbers of small animals to large ones over and above the effect of their smaller trivial ranges (Fig. 2.3). The general form of the two relationships can be seen in Fig. 2.3, but it will be noted that the increase in number of ranges for small animals is due more to their smaller ranges than to the increased surfaces available from vegetational complexity. It may be concluded that the effects of the qualitative differences due to

TABLE 2.1. Habitat areas associated with an old beech (*Fagus sylvaticus*) tree. (Basic data gathered as a student field exercise by A. Mill, L. Butters, J. Downing, A. Leclezio, S. Roe and T. R. E. Southwood)

	m^2	
Ground area covered by canopy		200
Living parts of Beech tree		
Trunk (surface area)	52	
Branches and twigs (surface area)	230	
Leaves (both surfaces)	10 200	
		10 482
Dead parts of Beech tree		
Branches (surface area)	80	
Branches (surface area under bark where present)	60	
Epiphytes		140
Ivy (*Hedera helix*) leaves (both surfaces)	92	
Lonicera + *Rubus* leaves (both surfaces)	2	
Fern (*Polypodium*)	0.5	
Mosses, foliose and fructose lichens (especially *Iosthecium, Parmelia, Ramalina* and *Usnea*)	283	
		380.5
Total *surface area presented by 'tree' in summer		11 000
Total* surface area presented by tree in winter (excluding leaves and deciduous epiphytes, but plus area bud surface)		824

*This excludes surface increased due to bracket fungi, crustose lichens, bark fissures, the stems of epiphytes and fruits (beech nuts, blackberries, etc).

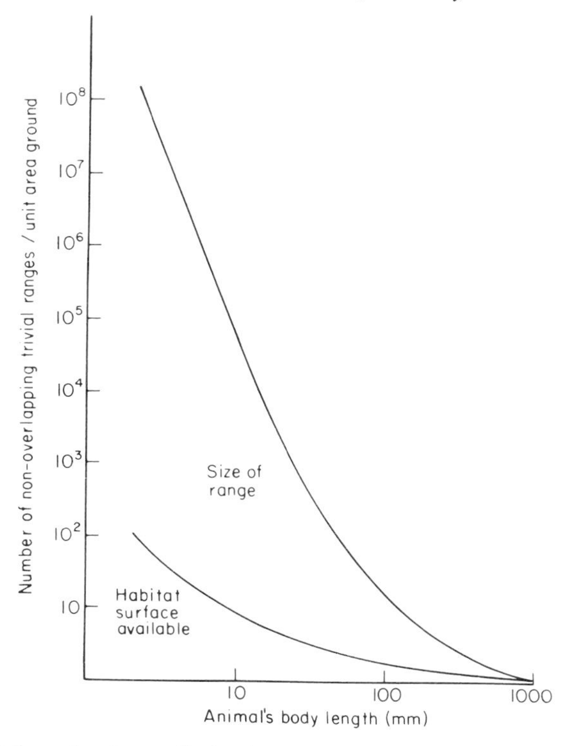

FIG. 2.3. The total number (upper line) of non-overlapping trivial ranges available to animals of different body lengths in a unit area ($4 \times 10^5 m^2$) of ground surface in a beech woodland in the summer. The lower line separates the contributions to this total of the surface available and the size of the trivial range. (Note, litter surfaces excluded.)

vegetational structure outweigh its quantitative impact (see also Lawton, this symposium). The qualitative differences, the spatial heterogeneity, will be further exaggerated by the tendency, already noted, for the smaller animals to be more specialised in their resource utilisation. Thus we can envisage that the relationship of the number of niche types in an area to an animals' size would be of the same general form as the curves in Fig. 2.3.

It is not easy to envisage how such spatial heterogeneity will alone provide the degree of isolation necessary to attenuate gene flow for sympatric speciation, although larger scale heterogeneity, with a rather non-mobile organism, could lead to parapatric speciation as in morabine grasshoppers (Key, 1974).

The essential role of spatial heterogeneity, on its own, is probably more to maintain the coexistence of different potentially mutually exclusive competitors when they are brought together after evolving in geographical or temporal isolation. May (this symposium, Chapter 12) elaborates on the dynamics of the process, but the greater the possibility that each species will have an adequate competitive advantage in a particular place (refuge), or at one time, the greater the potential for stable coexistence.

There are many examples of closely allied species coexisting with contemporary overlapping ranges. What are the causes of this mixing? Past climatic changes, in the Palaearctic region, especially those associated with the Ice Ages, are often considered to have isolated populations, that then formed separate species; these have remained distinct when further changes (usually amelioration) allowed them to mix again (but see Coope,

Chapter 11). The oak mirids *Phylus melanocephalus* and *P. palliceps* probably represent such a case (Southwood, 1957); there are further putative examples in the Coleoptera (Lindroth, 1945–9) and Lepidoptera (Beirne, 1952). In other instances speciation or incipient speciation arose as species spread north after the retreating ice sheet e.g. in *Erebia epipsodea* (Ehrlich, 1955). Adams (1977) has suggested a similar mechanism for the Pronophiline butterflies of Columbia; here the warm spells isolated the populations by driving them up the mountains and only in cold spells could interchange occur across the lowlands leading to the present day distribution of several often altitudinally zoned species on each mountain.

Other groups of animals have been equally exposed to the faunal mixing and separation that arises from these climatic changes and continental drift, and some of these animals are of comparable size to insects. The small size of insects has also contributed to another cause of the mixing of faunal elements: their carriage round the world by man (Elton, 1958; Waloff, 1966), but again in this they are not unique.

Probably the greatest cause of this mixing, this invasion of new areas, by insects, is their remarkable power of dispersal stemming from their possession of wings: a unique characteristic in animals of their size. Between Coad's (1931) delightfully titled paper 'Insects captured by airplane are found at surprising heights' and the sophisticated studies reported at this Society's recent Symposium (Rainey, 1976) clear evidence of the frequency and extent of insect aerial movement has been established (Johnson, 1969), although the extent of trans-oceanic movement under the present climatic regime is very limited (Johnson & Bowden, 1973). The insects that are found moving in this way are not simply random samples of the insect fauna, but samples that are biassed towards those species that are denizens of temporary habitats (Southwood, 1962; Howden, 1977). Leston (1957*a*) pointed out that this behavioural trait may also be recognised zoogeographically; among related families those with many species on oceanic islands were those that in general had a high level of dispersal. He termed this 'spread potential' and calculated its value for various families of Heteroptera against the 'standard family' (Miridae); Becker (1975) makes the important point that after invasion, which depends on flight and mobility, survival will depend on finding a suitable food, and this is more likely in a carnivore than herbivore. Insects as a whole, and certain groups in particular have therefore been most successful in colonising isolated islands. After invasion there may be considerable speciation; by founder effects on adjacent islands, and sympatrically through adaptative radiation, with isolating mechanisms of the types discussed below. The Hawaiian islands provide an outstanding example of this; Zimmerman (1970) has estimated that there were about 6500 endemic species descended from 250 immigrants (which only represented about one third of the insect orders).

The high mobility of insects and its particular aerial form makes another contribution to insect diversity besides colonisation and mixing on a geographical scale discussed above. After dispersal flights insects will often land on plants that are not their hosts (or habitats); such insects will normally take off again. However, eventually there will be no further flight reserves, the threshold for feeding, or at least drinking, will be low and a few individuals may manage to feed. Even rarer will be the individuals that remain on the new host (or in the new habitat) and reproduce there, but such individuals will extend the host range of the species and may through further spatial and temporal (see below) isolation form new species. Richardson (1974) has strong evidence to suggest that this was the first step in the evolution of one of a pair of essentially sympatric *Drosophila* species in Hawaii.

Thus we may visualise the Insecta as continually probing the environment, 'throwing together' new combinations of insects and habitat. This is the mechanism that underlies the relationship between the abundance of a tree and the number of insect species associated with it (Southwood, 1960*b*, 1961*a* & *b*; Strong, 1974*a*), a relationship that seems to extend to other types of plant (Lawton & Schroder, 1977; Southwood, 1977*a*). When man introduces a new species and plants it densely the 'normal complement', the 'species equilibrium', seems to be reached in around 500–1000 insect generations (Strong, 1974*b*; Southwood, 1977*a*).

In conclusion then, on the spatial scale the small trivial range and the large migratory range of the average insect are both components that have made significant contributions to the diversity of insect faunas. The former stems from their small size, the latter additionally reflects the evolution of wings (Wigglesworth, 1963).

TIME AND PHENOLOGY

Most habitats are also heterogeneous in time: this ranges from the regular and predictable seasonal changes of, say, a temperate forest to the irreversible successional changes that are most marked in ephemeral habitats like a dung pat (Southwood, 1977*b*). The corresponding relationships between organisms and habitats hold for time as for space, that is the greater the longevity of the organism the less it perceives temporal fluctuations in the habitat (May, 1974*b*, 1976*b*, Southwood, 1976). Longevity is related to size (Bonner, 1965) and so as insects are in general relatively small they are comparatively short lived, with short generation times (some insects of timber are exceptional, Smith, 1962; Shaw, 1961) and fast rates of evolution. One could therefore develop a series of arguments in relation to time paralleling those for space outlined above. There are, however, two additional features that deserve special emphasis.

Firstly, metamorphosis, as described previously, allows a species to occupy a habitat for only part of its generation time and to spend the rest somewhere else doing something entirely different or perhaps 'nothing at all' i.e. in diapause. Many other invertebrate groups, and a few vertebrates, have physiological mechanisms that allow them to pass through unfavourable periods in a quiescent state, but taken with metamorphosis, diapause greatly increases the insects' ability to use a series of 'stepping stones' through the environment. From the aspect of interspecific diversity the importance of this is the enormous scope it provides to reduce interspecific competition, to reduce the proportion of the niche that overlaps with another species. When closely allied species occurring in the same habitat are studied it is often found that their phenologies differ slightly.

Therefore viewing time as the resource, the concepts initiated by MacArthur & Levins (1967) on the utilisation of a resource by competing species may be applied to determine the extent of niche overlap. The latter may be expressed by the ratio d/w, where d = distance between mean resource values of the two species and w = the standard deviation of the utilisation of the resource; May (1973, 1976*a*) suggests that for coexistence $d/w \geqslant 1$. Precise measurements of the standard deviation of the mean time of occurrence are not normally available in phenological data, but the range and the period of maximum abundance are usually indicated: w will be taken as one-third of the range, the mid-point of the period of maximum abundance is taken as the mean. This assumption of a Gaussian emergence pattern is approximately justified and the emphasis on the pre-peak time as representing $3w$ eliminates the influence of the long post-reproductive life of some insects.

There are five species of mirid that occur together on broom (*Sarothamnus scoparius*) at Silwood Park, the periods of occurrence of adults and larvae are given by Waloff & Southwood (1960). The utilisation of 'time' by the adults and the extent of overlap,

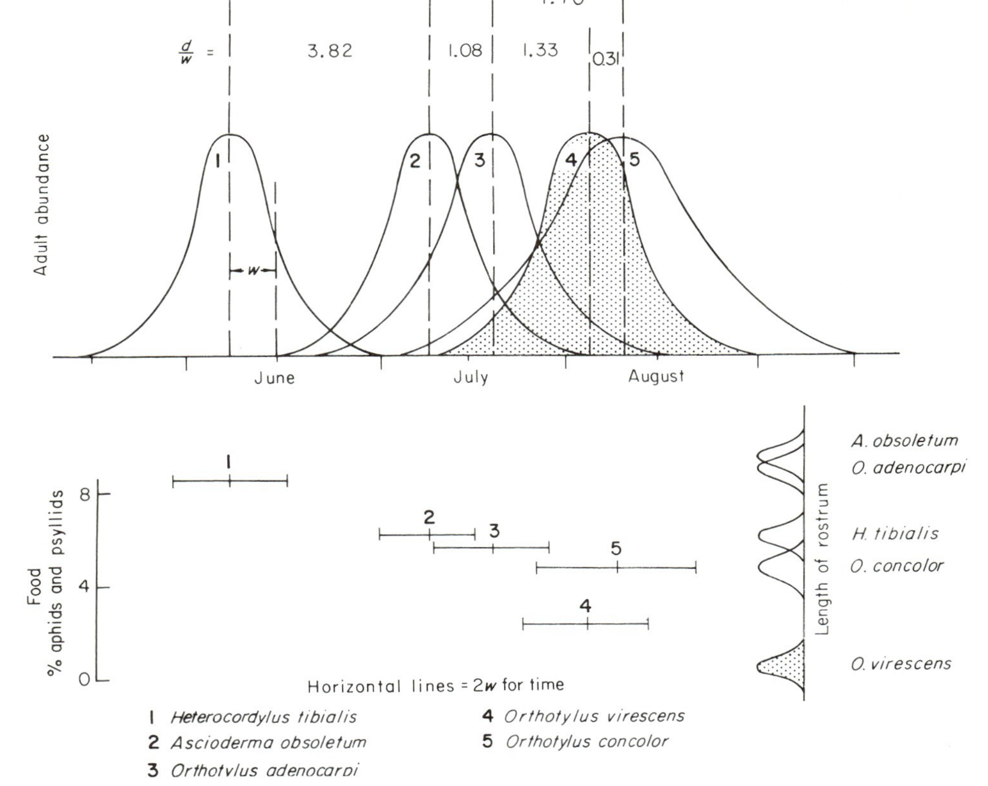

FIG. 2.4. Niche relationships between five species of mirid on broom (*Sarothamnus scoparius*) at Silwood Park in terms of phenology and feeding habits [data from Waloff & Southwood (1960) and Dempster (1964)].

calculated as indicated above, is shown in Fig. 2.4. On the basis of May's hypothesis ($d/w \geqslant 1$ for stable coexistence) one would predict that *Orthotylus virescens* and *O. concolor* must, since they coexist, be separated by a resource other than time. The lower part of Fig. 2.4 shows this to be true; *O. virescens* has a much shorter rostrum, it feeds on the mesophyll tissues of the leaf and not into the vascular bundles as the other four species do; it also takes animal food less frequently (Dempster, 1964). The niche overlap between *Asciodema obsoletum* and *O. adenocarpi* is in the region of one and their feeding habits do not show any significant separation: however, the former species also lives on gorse (*Ulex europaeus*), whereas the latter species has the single host plant, broom. However, these species interact more than this analysis suggests, because they will also compete, in the normal sense, as larvae and as partial predators; earlier (i.e. larger) individuals will prey on smaller ones (i.e. early species on those later). Waloff's (1966) observations on the unusual abundance of *O. concolor* in California, in the absence of the other species, is further evidence of the reality and extent of this interaction.

Another aspect of time that may be partitioned is the diel cycle, although this is not unique to insects. Within one family this is well shown in ground beetles, Carabidae, whose diel cycles and seasonal patterns at Silwood Park have been studied by P. J. M. Greenslade

(1961, 1963, 1965) and elsewhere by Evans (1962), Grum (1976) and others. A diagrammatic representation of time and space (as reflected in the beetle's size) resource utilisation in this group of 28 common species is given in Fig. 2.5. A full analysis of these data, in which over fifty pairs of species interactions are significant (i.e. $d/w \geqslant 3$), is beyond the scope of this review, but the following points can be made:

1. Seasonal time is tightly packed from March to November. (These are only times, calculated as for the mirids, of peak activity. Many of the species are present as aestivating or hibernating adults or, in small numbers, as post-reproductive throughout the year.) Species packing in the larger species is 'tighter' in summer than in early spring or late autumn which presumably reflects the greater food resources at mid-season.
2. Diel time partitioning is dependent on size. The larger species except *Pterostichus madidus* are nocturnal, whilst diurnal activity predominates among the smallest species. This may be a reflection of predator pressure; the larger species being vulnerable to vertebrate predators in daylight (there are, of course, insectivorous vertebrates active at night, but less than in the day). The smaller species may avoid being preyed upon by the larger ones if they are diurnal (and their small size makes them less conspicuous to vertebrates). Against this trend are several instances where, on other resource utilisation coefficients, species are in 'intolerable' competition, but niche separation also has some contribution from the diel rhythm. The relatively large *P. madidus* is partly diurnal and this seems to be an important niche character

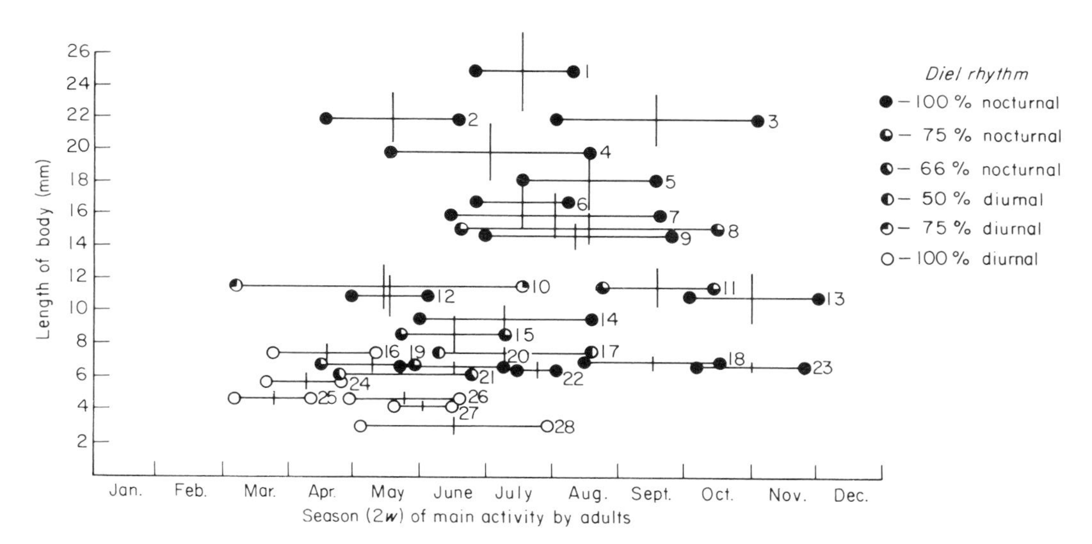

FIG. 2.5. Species packing in ground beetles (Carabidae) in terms of season of maximum adult activity, adult size range and diel periodicity. Data for woodland-grassland areas at Silwood Park from P. J. M. Greenslade (1961, 1963, 1965). The numerals identify species as follows:

1, *Carabus violaceus*; 2, *Carabus nemoralis*; 3, *Carabus problematicus*; 4, *Abax parallelepipedus*; 5, *Pterostichus niger*; 6, *Cychrus rostratus*; 7. *Pterostichus melanarius*; 8, *Pterostichus madidus*; 9, *Harpalus rufipes*; 10, *Pterostichus caerulescens*; 11, *Calathus fuscipes*; 12, *Nebria brevicollis* spring; 13, *Nebria brevicollis* autumn; 14, *Calathus piceus*; 15, *Harpalus affinis*; 16, *Loricera pilicornis*; 17, *Amara lunicollis*; 18, *Calathus melanocephalus*; 19, *Agonum dorsale*; 20, *Stomis pumicatus*; 21, *Amara communis*; 22, *Synuchus nivalis*; 23, *Leistus ferrugineus*; 24, *Notiophilus rufipes*; 25, *Notiophilus substriatus*; 26, *Notiophilus biguttatus*; 27, *Asaphidion flavipes*; 28, *Bembidion lampros*.

separating it from *P. melanarius* and *Harpalus rufipes.* Likewise, small *Agonum dorsale* is three-quarters nocturnal and this seems to provide an important separation from *Loricera pilicornis* (wholly diurnal) and some from *Amara communis* (one third diurnal). The resource utilisation coefficients (d/w) for diel time partitioning are greater than one in 10 of the 35 species pair interactions calculated for beetles in the size range 5–11.5 mm, and are zero in only eight of these. In contrast, zero values were found for 19 of the 22 coefficients calculated for species pairs whose lengths were above or below this range.

3. Spatial partitioning, expressed by the size of the beetle which influences both trivial range and prey size, also seems tight from 25 mm down to about 5 mm, below which other types of general predator become relatively more common. There is, however, a very obvious 'gap' between about 12–14 mm. It is noteworthy that just above and just below the 'gap' are the two carabid species that are most abundant at Silwood, *P. madidus* and *Nebria brevicollis.* Thus there will be more individuals of these species in the size ranges beyond one standard deviation of the mean than of some of the rarer species altogether. If the effect was entirely of the normal competitive type one might expect the gap to be around both species, rather than between and below them. However, like the Broom mirids, these beetles may actually prey on their smaller competitors and these common species do seem to leave a 'wake' of empty 'niche space'. Species packing is particularly tight on the edges of this gap. Perhaps as a reflection of this the species at the 'bottom' of larger group, *Harpalus rufipes*, exceptionally for a carabid, is partly herbivorous as an adult (Lindroth, 1945–9). The 'gap' seems to be general, there are only 10 carabids on the British list with mean lengths in the range 12–14 mm, and these are mostly rare species; yet this size must be close to the median for the families' range.

4. Spatial partitioning in terms of habitat specialisation (e.g. arable land, beechwood, litter) also occurs (Greenslade, 1964). This is complete ($d/w = 3$) in only eight of the calculated 57 species pair interactions. Taken with diel time partitioning, it seems most significant at the 'top' of the lower group, i.e. between *N. brevicollis* (spring) and *P. caerulescens* and between *N. brevicollis* (autumn) and *Calthus fuscipes*.

Similar studies on species packing have been made in leaf-hoppers on sycamore (McClure & Price, 1976), bumble-bees (Heinrich, 1976) and dung-beetles (Hanski & Koskela, 1977). A dramatic example of temporal separation is provided by the periodic *Magicicada*, brilliantly analysed by Alexander and Moore (1962) and Dybas and Lloyd (1974); there are six species, three with 17-year life-cycles and three with 13-year life cycles; each 17-year species has a 13-year sibling from which it cannot be separated structurally. The 13-year species tend to be more southern, though there is a narrow zone of overlap; however, the two broods of adults synchronise in this zone only once every 221 years. The 17-year species, where they overlap are separated microspatially by having different habitats in the same woodland.

Besides maintaining diversity by the reduction of competition, temporal heterogeneity also makes a major contribution to insect diversity through sympatric speciation. Bush (1975*a* & *b*) shows that this type of speciation is most frequent in animals that live on or in their food supply, parasites – broadly defined to include many phytophagous insects. It has already been pointed out that the vagility of insects, facilitated by the evolution of wings, has led to them being frequently exposed to non-hosts, within their original host's range, and individuals that have the appropriate allele to overcome the resistance of the new host may arrive on it and leave progeny (Southwood, 1961*b*). Whether such 'pioneers' merely contribute to the polymorphism of the species or to sympatric speciation depends, as

Maynard-Smith (1966) has shown theoretically, on whether mating is biassed within 'pioneer colony' (assortative mating) or whether they mate randomly within the rest of the population. The different nutritional and other conditions of the new host could lead to slight changes in the development times so that the 'pioneers' were either slightly ahead or behind the rest of the population. As the mating period of so many insects is so short, and with discrete generations so synchronised, such temporal shifts could effectively 'isolate' pioneers. Studies on arboreal leafhoppers (Le Quesne, 1963, 1965; Ahmad & Abror, 1977) sawflies (Knerer & Atwood, 1973; Knerer & Marchant, 1973) and Trypetid fruit flies (Bush, 1974, 1975*a*) point towards the importance of this allochronic isolation. Price and Willson (1976) showed host plant induced changes in development time in a cerambycid beetle.

The effectiveness of allochronic isolation due to developmental differences will be enhanced if:

1. There is no diapause or other mechanism for ensuring synchrony of *adult* emergence (Corbet, 1954).
2. Pre-reproductive movement is minimal which will be particularly frequent in: (a) insects where the normal mode of locomotion for trivial movement is walking (e.g. Coleoptera and Hemiptera in temperate regions), rather than flight (e.g. Lepidoptera); or if it is flight, then courtship and mating occur around the host plant (e.g. Lycaenidae) (Gilbert, 1978); and (b) insects where the level of migratory flight is low; this is the general condition in arboreal insects (Southwood, 1960*a*, 1962).

If this hypothesis is correct arboreal species that fulfil these conditions should show more speciation and narrower host ranges than those that do not. Futuyma (1976) showed a wide host range of arboreal Lepidoptera in the northern temperate regions, but these do not fulfil either condition (1) or (2). However, (1) and (2) both apply to the Heteroptera and Table 2.2 shows analysis of host plant range for the major subfamilies of British Miridae. The array supports the conclusion that host range is related to host plant type when conditions (1), (2a) and (2b) all apply. The spatial isolation associated with the arboreal habitat arises, I suggest, both from its high durational stability (and the therefore low migratory level of tree denizens) and from the structural characters of trees (notably their large three dimensional size), and complex architecture (see Lawton, this Symposium, Chapter 7).

Evidence for narrow host plant ranges in various arboreal Homoptera is given by Eastop (1973) and Hodkinson (1974). Leston (1975*b*) pointed out the remarkable 'explosive speciation' in the arboreal species of *Lygocoris* (*Neolygus*). There are large numbers of

TABLE 2.2. Host plant range in relation to plant type in British Miridae [excluding small specialist groups: Bryocorinae, Deraeocorinae, Dicyphinae and Hallodapini] (based on data in Southwood & Leston, 1959)

Host plant range of mirid species	Type of host plant		
	Trees	Shrubs	Herbs
Within one genus	55	8	28*
Within one family	10	3	33
Two or more families polyphagous	10	3	24

*A third of these are attached to either *Galium* or *Urtica*.

species in North America and others in Europe; the form of the male claspers allows one to group the species and Leston concluded that host plant associations had been separately evolved in the two continents, i.e. the North American species on *Tilia* is much closer to other North American species than to the European species on *Tilia*. Another case of explosive speciation is provided by the North American mirid genus *Lopidea* which has over 100 species (Knight & Schaffner, 1975); it could well repay study by evolutionary biologists.

There is also some evidence that phytophagous insects with herbaceous, and often ephemeral, host plants may show remarkable phenotypic variation without speciation, e.g. Mound's (1963) studies on the whitefly, *Bemisia tabaci*, Southwood and Blackith's (1960) on a mirid on *Artemesia*, Mittler and Sutherland's (1969) on aphids on various vegetables, and Price and Willson's (1976) on cerambycids on *Asclepias*. This seems to emphasise the importance of the level of migratory activity, characteristic of host plant type, in maintaining or attenuating gene flow following host plant change. I conclude that allochronic speciation is most likely to occur in species living on trees, whether they are herbivores, predators or parasitoids on other insects, that also fulfil conditions (1) and (2a) above. This may be a factor, additional to spatial complexity, influencing the larger number of herbivorous species on trees compared with other plants (Lawton and Schroder, 1977; Southwood, 1977*a*).

The combination of temporal isolation with climatic changes has been proposed by Alexander (1968) as a method of sympatric speciation in crickets; some species diapause as eggs and some as larvae and there are variations in voltinism. He postulates that as species moved in response to post-glacial climatic changes, these different strategies would provide effective isolation allowing more species packing.

Conclusion

The diversity of insects stems from them having been extraordinarily successful in carving out niches from their environment. The niche can be defined in terms of all the required resources; in space (shelter, including defence, food and a mate) and time. Insects seem to have been adept at utilising 'fractional niches' either in time (at different life history stages) or concurrently (polymorphism). This ability to sub-divide and recombine the environment, like a child's modular building set, gaining one resource from here and another from there, is a major contributor to insect diversity and depends heavily on the evolution of metamorphosis, especially complete metamorphosis. These evolutionary pressures lead to intraspecific diversity, but, as Davidson's (1978) work exemplifies, when the fractional niches become large enough interspecific diversity may replace it.

High interspecific diversity depends on high levels of speciation and of colonisation and mixing, but these in themselves are not enough; there must be features of insects' dynamics that cause them to have relatively high species equilibrium levels (MacArthur & Wilson, 1967) to reduce competitive exclusion. May (this Symposium, Chapter 12) explores faunal dynamics and will draw the final conclusion. At this point we may note that:

1. Metamorphosis, which may include long periods of diapause, provides temporal separation which is an important component in both sympatric speciation and niche separation ('or species packing').
2. The possession of wings increases the rate of colonisation (spread potential) on a geographical scale, both to islands, with subsequent endemism (after allopatric speciation), and on land masses following periods of climatic or man made change. The low level of

migratory flight associated with habitats such as trees of high durational stability is adequate to expose individuals to new habitats, but taken with developmental asynchrony is often inadequate to maintain gene flow against the new selective pressures. Thirdly, flight contributes to diversity by allowing a species to 'link-up' small pieces of similar, but spatially discontinuous resource to form the habitat for which, under evolutionary pressures, it specialises. Again, noting that a level of migratory flight adequate for colonisation may not be adequate to maintain full gene flow, insect flight contributes to the formation of isolated colonies that through founder effects or drift or differential selection are a facet of intraspecific diversity (Gilbert and Singer ,1975; Hallka this Symposium, Chapter 3).

3. Because of their small size insects (range 0.2–160 mm) have small trivial ranges (Fig. 2.3) and therefore have the evolutionary opportunity to specialise for a peculiar local combination of resources. Within the same surface area 3 mm long animals can have a million trivial ranges, 100 mm long animals but ten (Fig. 2.3); the former will appear more heterogeneous than the latter and offer more scope for speciation. Small animals can also utilise far more of the additional surface area provided by the complexity of vegetational cover (Table 2.1; Fig. 2.3), so that if it is an area of woodland the 3 mm long animals will have an extra eleven million ranges (Fig. 2.3).

4. The adaptation to living on and feeding on vegetation, necessary to utilise this space, is an important component of this diversity; relatively few groups of insects have become adapted to this way of life (Southwood, 1973).

5. Small animals have short generation times and hence faster rates of adaptation; if a new resource appears a small animal is likely to evolve and become adapted to it before a large one.

6. By the evolution of social life some insects are able to have some of the ecological attributes of larger animals, including more control of their environment, than one would predict from their size (Janzen, 1977).

Within these features there are, I believe, the keys to the diversity of the Insects. In considering how many species of insects there are (Fig. 2.1) it should not be forgotten that the greater part of this diversity is due to the Endopterygote orders and if we include the plant feeding Exopterygota, particularly the Hemiptera and the Acridoidea, the remainder have little, in terms of diversity, to mark them out from the Arachnida.

To return to my original question, what are the qualitative components of their biology that enabled the Pterygote insects to monopolise the majority of the vast number of niches and 'fractional niches' offered by the development of vegetation; why have they rather than the Vertebrates, Arachnida, Diplopoda, Mollusca, Nematoda or Apterygota, achieved this? The answer, I suggest, is: 'size, metamorphosis and wings'.

Acknowledgements

I am grateful to Drs D. W. Davidson and W. D. Hamilton for permission to quote from their unpublished papers and to many colleagues for useful discussions and help, especially G. R. Conway, M. P. Hassell, R. M. May, S. McNeill, P. M. Reader and N. Waloff.

References

Acton A. B. & Scudder G. G. E. (1969) The ultrastructure of the flight muscle polymorphism in *Cenocorixa bifida* (Hung.) (Heteroptera, Corixidae). *Z. Morph. Okol. Tiere* **65**, 327–335.

Adams M. (1977) Trapped in a Columbian Sierra. *Geographical Magazine* **19**, 250–254.

Ahmad I. & Abror I. (1977) A study of eggs and larvae of two closely related sympatric species of *Oxrhachis* Germar (Homoptera: Membracidae). *Pakistan J. Sci. Ind. Res.* **19**, 233–238.

Alexander R. D. (1963) Animal species, evolution and geographic isolation. *Syst. Zool.* **12**, 202–204.

Alexander R. D. (1968) Life cycle origins, speciation and elated phenomena in crickets. *Q. Rev. Biol.* **43**, 1–41.

Alexander R. D. & Moore T. E. (1962) The evolutionary relationships of 17-year and 13-year Cicadas, and three new species (Homoptera, Cicadidae, Magicicada). *Misc. Publs. Univ. Michigan* **121**, 5–57.

Andersen N. M. (1973) Seasonal polymorphism and developmental changes in organs of flight and reproduction in bivoltine pond skaters (Hem. Gerridae). *Ent. Scand.* **4**, 1–20.

Becker P. (1975) Island colonisation by carnivorous and herbivorous Coleoptera. *J. anim. Ecol.* **44**, 893–906.

Beirne B. P. (1952) *The origin and history of the British fauna.* 164 pp., London.

Blackith R. E. (1974) Strategies and Tactics in Evolution. *Res. Biologiq. Contemp.* 427–435.

Blackman R. (1974) *Aphids*, 175 pp. Ginn, London.

Bonner J. T. (1965) *Size and cycle: an essay on the structure of biology.* 219 pp., Princeton Univ. Press.

Brown L. E. (1956) Movements of some British small mammals. *J. anim. Ecol.* **25**, 54–71.

Brown J. H. & Davidson D. W. (1977) Competition between seed-eating rodents and ants in desert ecosystems. *Science N.Y.* **196**, 880–882.

Bush G. L. (1974) The mechanism of sympatric host race formation in the true frit flies (*Tephritidae*). In *Genetic Mechanisms of Speciation in Insects*, Ed. M. J. D. White, p. 3–23. Sydney.

Bush G. L. (1975*a*) Sympatric speciation in phytophagous parasitic insects. In *Evolutionary Strategies of Parasitic Insects,* Ed. P. W. Price, pp. 187–206. Plenum Press, London.

Bush G. L. (1975*b*) Modes of animal speciation. *A. Rev. Ecol. Syst.* **6**; 339–364.

Clarke C. A. & Sheppard P. M. (1959) The genetics of *Papilio dardanus* Brown. *Genetics* **44**, 1347–1358.

Clarke C. A. & Sheppard P. M. (1960) The genetics of *Papilio dardanus* Brown. *Genetics* **45**, 439–457, 683–698.

Clarke C. A. & Sheppard P. M. (1962*a*) Disruptive selection and its effect on a metrical character in the butterfly, *Papilio dardanus. Evolution* **16**, 214–226.

Clarke C. A. & Sheppard P. M. (1962*b*) The genetics of *Papilio dardanus* Brown. *Genetics* **47**, 909–920.

Clarke C. A. & Sheppard P. M. (1963) Interactions between major genes and polygenes in the determination of the mimetic patterns of *Papilio dardanus. Evolution* **17**, 404–413.

Coad B. R. (1931) Insects captured by airplane are found at surprising heights. *Yb. of Agric. U.S. Dep. Agric.* (**1931**): 320–3.

Coaker T. H. & Finch S. (1973) The association of the cabbage root fly with its food and host plants. *Symp. R. ent. Soc. Lond.* **6**, 119–128.

Corbet P. S. (1954) Seasonal regulation in British Dragonflies. *Nature Lond.* **174**, 655.

Davidson D. W. (1977) Species diversity and community organization in desert seed-eating ants. *Ecology* **58**, 711–724.

Davidson D. W. (1978) Worker size variation in a social insect as a function of the competitive environment. *Am. Nat.* (In press).

Dempster J. P. (1964) The feeding habits of the Miridae (Heteroptera living on broom (*Sarothamnus iscoparius* (L.) Wimm.) *Entomologia exp. appl.* 7, 149–154.

Dixon A. F. G. (1972) Control and significance of the seasonal development of colour forms in the Sycamore aphid *Drepanosiphum platanoides* (Schr.). *J. anim. Ecol.* **41**, 689–697.

Downes J. A. (1958) The feeding habits of biting flies and their significance in classification. *A. Rev. Ent.* **3**, 249–266.

Dybas H. S. & Lloyd M. (1974) The habitats of 17-year periodical Cicadas (Homoptera: Cicadidae: *Magicicada* spp.). *Ecol. Monogr.* **44**, 279–324

Eastop V. F. (1973) Deducting from the present day host plants of aphids and related insects. *Symp. R. ent. Soc. Lond.* **6**, 157–178.

Ehrlich P. R. (1955) The distribution and subspeciation of *Erebia epipsodea* Butler *(Lepidoptera: Satyridae). Univ. Kansas Sci. Bull.* **37**, 175–194.

Ehrlich P. R. & Raven P. H. (1969) Differentiation of Populations. *Science N.Y.* **165**, 1228–32.

Ehrlich P. R., White R. R., Singer M. C., McKechnie S. W. & Gilbert L. E. (1975) Checkerspot Butterflies: a historical perspective. *Science N.Y.* **188**, 221–8.

Elton C. S. (1958) *The Ecology of Invasions by Plants and Animals.* London.

Evans M. G. G. (1962) The surface activity of beetles in a northern English wood. *Trans. Soc. Brit. Ent.* **18**, 247–262.

Futuyma D. J. (1976) Food plant specialization and environmental predictability in Lepidoptera. *Am. Nat.* **110**, 285–292.

Gilbert L. E. (1977) The role of insect-plant coevolution in the organization of ecosystems. In V. Labeyrie (Ed) *Comp. Insect Milieu Trop. Coll. Int. C.N.R.S.*, **265**, 429–442.

Gilbert L. E. (1978) Development of theory in the analysis of insect/plant interactions. In *Analysis of Ecological Systems*, Eds. D. J. Horne, R. Mitchell & G. R. Stairs, Ohio State Univ. Press (In press).

Gilbert L. E. & Singer M. (1975) Butterfly Ecology. *A. Rev. Ecol. & Syst.* **6**, 365–397.

Greenslade R. M. (1945) Observations on the life cycle of the Apple Blossom Weavil (*Anthonomus pomorum* (L.) Curt.) *A Rep. East Malling Res. Sta.* 83–92.

Greenslade P. J. M. (1961) *Studies in the ecology of Carabidae* (Coleoptera). Ph.D. thesis, Univ. of London.

Greenslade P. J. M. (1963) Daily rhythms of locomotor activity in some Carabidae (Coleoptera). *Entomologia exp. appl.* **6**, 171–180.

Greenslade P. J. M. (1964) The habitats of some Carabidae (Coleoptera *Entomologists mon. Mag.* **94**, 129–132.

Greenslade P. J. M. (1965) On the ecology of some British Carabid beetles, with special reference to life histories. *Trans. Soc. Brit. Ent.* **16**, 149–179.

Grum L. (1976) An attempt to characterize matter transfer by carabid communities inhabiting forests. *Ekol. Polska* **24**, 265–375.

Hamilton W. D. (1978) Wingless and fighting males in fig wasps and other insects. In *Reproductive Competition and Selection in Insects*, Eds. M. S. Blum and N. A. Blum. Academic Press, New York.

Hanski I. & Koskela H. (1977) Niche relations among dung-inhabiting beetles. *Oecologia* **28**, 203–31.

Harlow P. M. (1956) A study of the ovial development and its relations to adult nutrition in the blowfly, *Protophromia terrae-novae* (R–D. *J. exp. Biol.* **33**, 777–797.

Heinrich B. (1976) Resource partioning among some ensocial insects. Bumblebees. *Ecology* **57**, 874–889.

Hinton H. E. (1948) On the function, origin and classification of pupae. *Proc. S. Lond. ent. nat. Hist Soc.* **1947–48**: 111–154.

Hinton H. E. (1977) Enabling mechanisms *Proceedings XVth Int. Cong. Ent., Washington,* **71–83.**

Hodkinson I. D. (1974) The biology of the Psylloidea (Homoptera): a review. *Bull. ent. Res.* **64**, 325–339.

Hölldobler B. (1977) Recruitment behaviour, home range orientation and territoriality in harvester ants, *Pogonomyrmex. Behav. Ecol. Sociobiol.* **1**, 3–44.

Howden H. F. (1977) Beetles, Beach drift, and island biogeography. *Biotropica* **9**, 53–57.

Hutchinson G. E. & MacArthur R. H. (1959) A theoretical ecological model of size distributions among species of animals. *Am. Nat.* **93**, 117–125.

Istock C. A. (1967) The evolution of complex life cycle phenomena: an ecological perspective. *Evolution* **21**, 592–605.

Istock C. A. (1975) More on complex life cycles. *Evolution* **29**, 572–574.

Janzen D. H. (1967) Fire, vegetation structure and the ant X acacia interactions in Central America *Ecology* **48**, 26–35.

Janzen, D. H. (1977) Why are there so many species of insects? *Proceedings XVth Int. Cong. Ent. Washington*, 84–96.

Järvinen O. & Vepsäläinen K. (1976) Wing dimorphism as an adaptive strategy in water-striders. (*Gerris*). *Hereditas* **84**, 61–68.

Johnson C. G. (1969) *Migration and Dispersal of Insects by Flight*. Methuen, London.

Johnson C. G. & Bowden J. (1973) Problems related to the trans-oceanic transport of insects, especially between the Amazon and Congo areas. In *Tropical Forest Ecosystems in Africa and South America: A Comparative Review*, Eds. B. J. Magyers, E. S. Ayensu & W. D. Duckworth. pp. 207–222. Washington.

Jones M. G. (1970) Observations on feeding and egg development of the wheat bulb fly *Leptohylenyia coaretata* (Fall.) *Bull. ent. Res.* **60**, 199–207.

Kennedy J. S. (1961) Continuous polymorphism in locusts. *Symp. R. ent. Soc. Lond.* **1**, 80–90.

Kettlewell H. B. D. (1973) *The evolution of Melanism.* Clarendon Press, Oxford.

Key K. H. L. (1974) Speciation in the Australian Morabine Grasshoppers – taxonomy and ecology. *Genetic Mechanisms of Speciation in Insects* Ed. M. J. D. White, pp. 43–56. Sydney.

King C. (1975) The home range of the Weasel (*Mustela* nivalis) in an English woodland. *J. anim. Ecol.* **44**, 639–668.

Kitching R. L. (1977) Time, resources and population dynamics in insects. *Aust. J. Ecol.* **2**, 31–42.

Knerer G. & Atwood C. E. (1973) Diprionid sawflies: polymorphism and speciation. *Science N.Y.* **179**, 1090–1099.

Knerer G. & Marchant R. (1973) Diapause induction in the sawfly *Neodiprion rugifrons* Middleton (Hymenoptera: Diprionidae). *Can. J. Zool.* **51**, 105–108.

Knight H. H. & Schaffner J. C. (1975) Additional species of *Lopidea* Uhler from Mexico and Guatamala (Hemiptera: Miridae) *Iowa St. J. Res.* **49**, 413–422.

Labeyrie V. (1977) Sensorial environment and coevolution of insects (French paper) *Comp. Insect. Milieu Trop. Coll. Int. C.N.R.S.* **265**, 15–30.

Lawton J. H. & Schroder D. (1977) Effects of plant type, size of geographical range and taxonomic isolation of number of insect species associated with British plants. *Nature* **265**, 137–140.

Lees A. D. (1961) Clonal polymorphism in aphids. *Symp. R. ent. Soc. Lond.* **1**, 68–79.

Le Quesne W. J. (1963) Some observations on *Macropsis marginata* (Herrich-Schaeffer) & *M. albae* Wagner (Hem., Cicadellidae). *Entomologists mon. Mag.* **99**, 128.

Le Quesne W. J. (1965) The establishment of the relative status of sympatric forms, with special reference to cases among the Hemiptera. *Zool. Beitr. (N.F.)* **11**, 117–128.

Leston D. (1957*a*) Spread potential and the colonisation of islands. *Syst. Zool.* **6**, 41–46.

Leston D. (1957*b*) The British *Lygocoris* Reuter (Hem: Miridae) including a new species. *Entomologist* **90**, 128–135.

Lindroth C. H. (1945–49) Die fennoskandischen Carabidae. Eine tiergeographische studie. *Medd. Göteborgs mus. Zool. Avd.* **109**, 110, 122.

MacArthur R. H. & Levins R. (1967) The limiting similarity, convergence and divergence of coexisting species. *Am. Nat.* **101**, 377–85.

MacArthur R. H. & Wilson E. O. (1967) *The Theory of Island Biogeography*. Mongraphs in Population Biology (1), 203 pp. Princeton Univ. Press.

McClure M. S. & Price P. W. (1976) Ecotope characteristics of coexisting *Erythroneura* leafhoppers (Homoptera: Cicadellidae) on Sycamore. *Ecology* **57**, 928–40.

Martin F. J. (1969) Searching success of predators in artificial leaflitter. *Am. Midl. Nat.* **81**, 218–227.

Mattson W. J. (1977) Size and abundance of forest Lepidoptera in relation to host plant resources. In Labeyrie V. (ed) *Comp. Insect Milieu Trop. Coll. Int. C.N.R.S.* **265**, 429–442.

May R. M. (1973) *Stability and complexity in model ecosystems*, Princeton, U.S.A.

May R. M. (1974*a*) Biological populations with non-overlapping generations, stable points, stable cycles and chaos. *Science, N.Y.* **186**, 645–7.

May R. M. (1974*b*) *Ecosystem patterns in randomly fluctuating environments.* In *Progress in Theoretical Biology* (Ed) R. Rosen & F. Snell, pp. 50. Academic Press, New York.

May R. M. (1976*a*) Models for two interacting populations. In *Theoretical Ecology*, Ed. R. M. May, W. B. Saunders Co., Philadelphia.

May R. M. (1976*b*) Mathematical aspects of the dynamics of animal populations. In *Studies in Mathematical Biology,* Ed. S. A. Levin. American Mathematical Society.

May Y. Y. (1975) Study of two forms of the adult *Stenocranus minutus. Trans. R. ent. Soc. Lond.* **127**, 241–254.

Maynard-Smith J. (1966) Sympatric speciation. *Am. Nat.* **100**, 37–65.

Mayr, E. (1963) Animal Species and Evolution. *Harvard Univ. Press.*

Mittler T. E. & Sutherland O. R. W. (1969) Dietary influences on aphid polymorphism. *Entomologia exp. appl.* **12**, 703–713.

Mound L. A. (1963) Host-correlated variation in *Bemisia tabaci* (Gennadius) (Homoptera: Aleyrodidae) *Proc. R. Ent. Soc. Lond.* (A) **38**, 171–180.

Price P. W. & Willson M. F. (1976) Some consequences for a parasitic herbivore, the Milkweed Longhorn Beetle, *Tetraopes tetrophthalmus* of a host-plant shift from *Asclepias syriaca* to *A. verticillata. Oecologia (Berl.)* **25**, 331–340.

Rainey R. C. (ed.) (1976) Flight behaviour and features of the atmospheric environment. *Symp. R. ent. Soc. Lond.* **7**, 75–112.

Richards O. W. (1961) An introduction to the study of polymorphism in insects. *Symp. R. ent. Soc. Lond.* **1**, 1–10.

Richards O. W. & Waloff N. (1954) Studies on the biology of population dynamics of British grasshoppers. *Anti Locust Bull.* **17**, 1–184.

Richardson R. H. (1974) Effects of dispersal, habitat selection and competition on a speciation pattern of Drosophila endemic to Hawaii. In *Genetic Mechanisms of Speciation in Insects*, Ed. M. J. D. White pp. 140–164. Sydney.

Shaw M. J. P. (1970) Effects of population density on alienicolae of *Aphis fabae* Scop. *Ann. Appl. Biol.* **65**, 191–212.

Shaw M.W. (1961) The golden Buprestid *Buprestis* aurulenta L. (Col, Buprestidae) in Britain. *Entomologist's mon. Mag.* **97**, 97–98.

Sheppard, P. M. (1961) Recent genetical work on polymorphic mimetic *Papilios. Symp. R. ent. Soc. Lond.* **1**, 20–29.

Slade N. A. & Wassersug R. J. (1975) On the evolution of complex life cycles. *Evolution* **29**, 568–571.

Smith S. N. (1962) Prolonged larval development in *Buprestis aurulenta L. (Coleoptera: Buprestidae). A review with new cases.* Can. Ent. **94**, 586–593.

Southern H. N. (ed) (1964) *Handbook of British Mammals.* Blackwell Scientific Publications, Oxford.

Southwood T. R. E. (1957) The zoogeography of the British Hemiptera-Heteroptera. *Proc. S. Lond. ent. nat. Hist. Soc.* **1957**, 111–136.

Southwood T. R. E. (1960*a*) The flight activity of Heteroptera. *Trans. R. ent. Soc. Lond.* **112**, 173–220.

Southwood T. R. E. (1960*b*) The abundance of the Hawaiian trees and the number of their associated insect species. *Proc. Hawaii ent. Soc.* **17** 299–303.

Southwood T. R. E. (1961*a*) The number of species of insect associated with various trees. *J. anim. Ecol.* **30**, 1–8.

Southwood T. R. E. (1961*b*) The evolution of the insect-host tree relationship – a new approach. *Proc. XIth Int. Cong. Ent.* **1**, 651–4.

Southwood T. R. E. (1962) Migration of terrestrial arthropods in relation to habitat. *Biol. Rev.* **37**: 171–214.

Southwood T. R. E. (1973) The insect/plant relationship – an evolutionary perspective. *Symp. R. ent. Soc. Lond.* **6**, 3–30.

Southwood T. R. E. (1976) Bionomic strategies and population parameters. In *Theoretical Ecology* Ed. R. M. May, pp. 26–48. Blackwell Scientific Publications, Oxford.

Southwood T. R. E. (1977*a*) The stability of the trophic milieu, its influence on the evolution of behaviour and of responsiveness to trophic signals. In *Comp. Insect Milieu Trop. Coll. Int. C.N.R.S.* Ed. V. Labeyrie **265**, 471–493.

Southwood T. R. E. (1977*b*) Habitat, the templet for ecological strategies. Presidential Address to British Ecological Society *J. anim. Ecol.* **46**, 337–365.

Southwood T. R. E. & Blackith R. E. (1960) Inter- and intra-specific variation in the mirid genus, *Plagiognathus. Proc. R. ent. Soc. Lond.* (C) **25**, 9–10.

Southwood T. R. E. & Jepson W. F. (1962) The productivity of grasslands in England for *Oxinella frit* (L.) (Chloropidae and other stem-boring Diptera. *Bull. ent. Res.* **53**, 395–407.

Southwood T. R. E. & Leston D. (1959) *Land and Water Bugs of the British Isles.* 436 pp. Warne, London.

Strong D. R. (1974*a*) Nonasymptomic species richness models and the insects of British trees. *Proc. natn. Acad. Sci. USA* **71**, 2766–69.

Strong D. R. (1974*b*) Rapid asymptotic species accumulation in phytophagous insect communities: the pests of cacao. *Science, N.Y.* **185**, 1064–6.

Todd V. (1949) The habits and ecology of the British Harvestmen (Arachnida, Opilionis) with special reference to those of the Oxford district. *J. anim. Ecol.* **18**, 209–229.

Vane-Wright R. I. (1976) A unified classification of mimetic resemblances. *Biol. J. Linn. Soc.* 8, 25–56.

Vepsäläinen K. (1971) The role of gradually changing day length in determination of wing length, alary dimorphism and diapause in *Gerris odontogaster* (Zett.) population (Gerridae, Heteroptera) in South Finland. *Ann. Acad. Sci. Finn. A* IV. *Biologia* **183**, 1–25.

Waloff N. (1966) Scotch broom (*Sarothamnus scoparius* (L.) Wimmer) and its insect fauna introduced into the Pacific Northwest of America. *J. appl. Ecol.* **3**, 293–311.

Waloff N. (1973) Dispersal by flight of Leafhoppers (Auchenorrhyncha Homoptera) *J. appl. Ecol.* 10, 705–730.

Waloff N. & Southwood T. R. E. (1960) The immature stages of Mirids (Het.) occurring on broom (*Sarothamnus scoparius* (L.) Wimmer) with some remarks on their biology. *Proc. R. ent. Soc. Lond.* (A) **35**, 39–46.

White M. J. D. (ed.) (1974) *Genetic Mechanisms of Speciation in Insects* Sydney.

Wigglesworth V. B. (1963) The origin of flight in insects. *Proc. R. ent. Soc. Lond.* (C) **28**, 23–4.

Williams L. H. (1954) The feeding habits and food preferences of *Acrididae* and the factors which determine them. *Trans. R. ent. Soc. Lond.* **105**: 423–454.

Young E. C. (1965*a*) The incidence of flight polymorphism in British Corixidae and description of the morphs. *J. Zool.* **146**, 567–576.

Young, E. C. (1965*b*) Flight muscle polymorphism in British Corixidae, ecological observations. *J. anim. Ecol.* **34**, 353–396.

Zimmerman E. C. (1970) Adaptive radiation in Hawaii with special reference to insects. *Biotropica* **2**, 32–38.

3 • Influence of spatial and host-plant isolation on polymorphism in *Philaenus spumarius*

OLLI HALKKA

Department of Genetics, University of Helsinki, Finland

All the spittle-bug species of northern Europe belong to the family Aphrophoridae. As nymphs, as well as adults, they live mainly on herbs and grasses; however, some species favour bushes such as alders and willows. The spittle masses ('cuckoo spit', 'frog spit') produced by the nymphs at the feeding sites are often very conspicuous. The habit of forming spittle clumps renders the nymphal stage amenable to studies on feeding ecology within and between species. Not only is it easy to locate each nymph on its food plant, but nymphs belonging to different species are rather easy to identify.

Eight species of the family Aphrophoridae are known from Finland (Linnavuori, 1969). The following four are the subjects of investigation in the present study: *Philaenus spumarius, Neophilaenus lineatus, Lepyronia coleoptrata* and *Aphrophora alni*.

Many of the northern Aphrophoridae show colour polymorphism as adults. Some species are also, curiously enough, polymorphic with regard to colour in the nymphal stages (e.g., *L. coleoptrata*). The colour polymorphism of adults is most pronounced in *Philaenus spumarius*, the most common species of the family (Fig. 3.1).

The colour morphs of *P. spumarius* are determined by a series of genes located at the same site (= allelic to each other) on one of the 12 chromosomes of the haploid set. In the

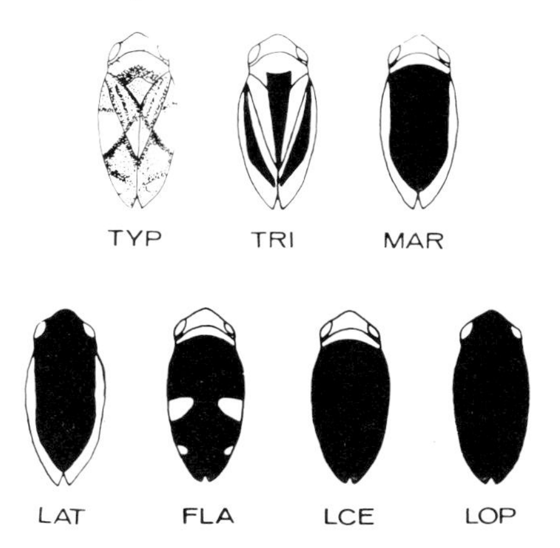

FIG. 3.1. The colour morphs of *Philaenus spumarius*. TYP = *typicus*, TRI = *trilineatus*, MAR = *marginellus*, LAT = *lateralis*, FLA = *flavicollis*, LCE = *leucocephalus*, LOP = *leucophthalmus*.

populations of continental Finland, seven alleles constituting a system of complex dominance and co-dominance relationships (Halkka *et al.* 1973, 1975*a*) govern the polymorphism. One of these alleles (called p^t) is very frequent (0.9), all the others (called p^T, p^M, p^L, p^F, p^C, p^O) sharing the remaining 0.1. This means that a great majority of individuals carrying the rare alleles are heterozygotes, mostly with p^t in one and the rare allele in the other of the two homologous chromosomes.

The colour morphs (Fig. 3.1) and the allele combinations thus generally correspond to each other as follows: *typicus* ($p^t p^t$), *trilineatus* ($p^T p^t$), *marginellus* ($p^M p^t$), *lateralis* ($p^L p^t$), *flavicollis* ($p^F p^t$), *leucocephalus* ($p^C p^t$), and *leucophthalmus* ($p^O p^t$). (A more detailed description of the genotype–phenotype relationships is included in a short review of the genetics of *Philaenus* colour polymorphism (Halkka *et al*, 1975*a*).

In Finland, the full series of colour morphs is expressed only in the females. In the southern part of the country, the males are practically dimorphic, with about 96% *typicus* and 4% *trilineatus* in most of the populations. One possible explanation for this difference between the females and the males is to be sought in the fact that the females must enjoy protection from enemies during the rather long (4–6 weeks) preoviposition period. The first males emerge earlier than the first females. The males are mainly *typicus*, and thus this colour phenotype constitutes the searching image adopted by predators during the emergence and preoviposition periods. About 80% of the females belong to the *typicus* morph, and this morph is probably subject to more intense predation than the six other morphs sharing the remaining 20%. This would mean that apostatic selection (most intense predation of the most common morph, see Clarke, 1969) is one of the mechanisms maintaining colour variability in *Philaenus spumarius*.

There is considerable evidence, admittedly mainly indirect, indicating that visual selection is not the only mode of natural selection responsible for maintenance of the colour morphs. In Finland, and also in Sweden, the morphs have dissimilar northern limits of distribution, *trilineatus* not reaching further than about 67 °N. Certain other morphs go slightly farther north, and *lateralis* and *typicus* reach about 68 °N (Halkka *et al*, 1974*a*). The morphs thus appear to differ in their reactions to the very exacting conditions of Finnish and Swedish Lapland.

Morphs which are unlike in distribution may also be expected to differ in their ecological requirements when living in the same habitat. For many reasons, it is rather difficult to reveal such dissimilarities, and even more difficult to trace their causes. In the present study, it is attempted to show that *Philaenus* colour morphs do diverge in their ecological requirements. To provide a background for a comparative ecological study of the morphs, an hypothesis is given as to coexistence of species of the family Aphrophoridae. A possible means of studying *Philaenus* colour morph ecology with the help of the other spittle-bug species is also described.

Material and methods

The total number of *Philaenus spumarius* individuals investigated for studies on spatial and host-plant isolation is about 120 000. In addition to these, about 6000 individuals belonging to the species *Lepyronia coleoptrata, Aphrophora alni* and *Neophilaenus lineatus* are included in studies dealing with the significance of host-plants for the ecology and ecological genetics of *P. spumarius*. The studies were carried out mainly during the years 1969–77, but some pilot investigations were made in 1967–68.

All the methods used in this study are based on the capture – release principle. Spittle-bugs are captured, analysed as to their sex and (in *Philaenus*) colour polymorphism and released to contribute to the production of the next generation. This, of course, means minimal possible interference with the structure of the population. The principle has been practised in 16 isolated *P. spumarius* populations for nine generations, and in 14 additional populations for eight or less generations.

For capture at the nymphal stage, the 'minicage method' and for capture at the adult stage, the 'sweep net method' have been used.

In the minicage method (Halkka *et al*, 1970), the nymphs in the spittle masses are isolated in small (5 cm) plastic cages attached to the plants. After emergence of the nymphs into adults, the cages are opened, and the individuals inspected and released. The process is repeated in the following generations, and in small populations it is possible to capture more than 90% of the total number of individuals in each generation.

The sweep net method is less accurate, but has been found to produce good results if practised in exactly the same way in successive generations. The spittle-bugs are collected with a sweep net and taken into an aspirator. From the aspirator, small groups of spittle-bugs are forced through a funnel into a long narrow transparent plastic tube in which they are inspected for sex and colour polymorphism. From the tube, the insects are blown back onto the food plants.

Study area and habitats

The study area (about 59°50′N, 23°23′E) comprises parts of the archipelago and the adjoining mainland of the Finnish side of the Gulf of Finland. The islands investigated belong to the Tvärminne Zoological station of the University of Helsinki.

The habitats available to the spittle-bugs in the islands are very small in area and restricted in biotic diversity. This is because all the islands consist of hard rock which is mostly bare or covered by a very thin layer of algae and lichens. Sites favourable for maintenance of meadows with tall herbs are rather restricted in area. Such herbs are favoured by *P. spumarius* as food plants. In most of the island meadows, Aphrophoridae other than *Philaenus* are sparse or absent.

Brännskär, (16E and 16W)

In the largest meadow investigated, on the wooded island Brännskär, there are about 300 m^2 of vegetation suitable for *Philaenus.* About 50 m west of this meadow, on the other side of a small bay, there is a smaller meadow about 80 m^2 in area. These two meadows, Brännskär-E (East) and Brännskär-W (West) are quite dissimilar ecologically. Meadow Brännskär-E is rich in *Filipendula ulmaria* and *Galium palustre*, meadow Brännskär-W in *Lysimachia vulgaris.*

Rovholmen (18)

The island Rovholmen, small but wooded, contains about 250 m^2 of meadowlike vegetation capable of supporting a spittle-bug population. This meadow, like those of Brännskär, gently slopes to the shore.

Stora Västra Långgrundet (22W and 22E)

The meadows on the western and eastern sides of the very small island Stora Västra Långgrundet (=SVL) are also shore meadows. Here they are denoted as SVL-W and SVL-E. The whole island is only about 70 m long and 40 m broad, and both meadows are only about 150 m^2 in area. The meadows are quite dissimilar in the composition of the vegetation: SVL-W is rich in *Myrica gale*, but this is absent from SVL-E which is rich in *Chrysanthemum vulgare.*

Mellanspiken (8A and 8B)

The treeless island Mellanspiken is about 200 m long and 50 m broad and contains about 400 m^2 of meadowlike vegetation scattered in small plots. These 'minimeadows' are small spots of grass and herb associations in depressions and crevices of the prevalent granite surface. One of the minimeadows, M-A, has twice been devastated; by voles (*Microtus agrestis*) in 1969, and by erosion through flooding and waves in 1975. The meadow M-B was very badly damaged by voles in 1969, but did not suffer appreciably from the 1975 flood.

Both the 'vole disaster' and 'flood disaster' caused profound changes in the plant species composition of meadows. In meadow M-A, *Filipendula ulmaria* was greatly reduced leaving *Lythrum salicaria* and graminaceous plants prevalent after both events. Between the two disasters, *Filipendula* temporarily regained part of its former area. In meadow M-B the voles demolished practically all vegetation from the upper part of this tiny (2.5 m x 7 m) meadow. The changes which took place in the plant species composition were similar to those in meadow M-A, with temporary reduction of *Filipendula ulmaria* as one of the main features of the succession.

Svanvik (37A)

The mainland site of investigation, Svanvik (= S), is an approximately 300 m long section of a seashore meadow which is about 1600 m long and 50–100 m wide. The meadow, with the exception of some small-sized elevations, is inundated several times each year. It comprises a large number of plant associations generally dominated by monocotyledons close to the shore and mostly by dicotyledons close to the meadow–forest borderline. The four Aphrophoridae, *Philaenus spumarius*, *Neophilaenus lineatus*, *Lepyronia coleoptrata* and *Aphrophora alni* live together in those parts of the meadow in which mixed stands of dicotyledons and monocotyledons grow.

RESULTS

CONSTANCY OF ALLELE FREQUENCIES IN A STABLE HABITAT

Among the island populations investigated, there are several in which the gene frequencies for colour have remained practically constant for the nine generations between 1969 and 1977. Being univoltine, *Philaenus* reacts in every generation to the peculiar climatic features of a whole year. The period 1969–77 was long enough to include years very dissimilar in both temperature and humidity. As the most critical period in the life of *Philaenus* appears to be the nymphal stage, the weather conditions of the nymphal period, June and early July, may well be decisive to the success of the populations. In the islands of the Tvärminne

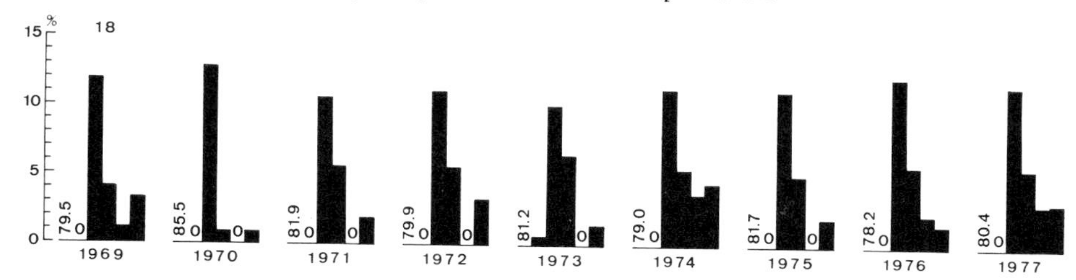

FIG. 3.2. Colour morph frequency histograms (for females) of the Rovholmen population. The numbers and bars, from left to right, denote percentages of the morphs *typicus*, *trilineatus*, *marginellus*, *lateralis*, *leucocephalus* and *leucophthalmus*.

archipelago this part of the summer was moderately humid in 1969–72, dry or very dry in 1973–76 and quite humid in 1977. On many islets in which the soil layer of the minimeadows upon the rock surface is very thin (10–20 cm), in the dry periods of 1973 and 1975 many of the food plants of *Philaenus* died or at least suffered withering of the green parts. This, of course, caused high nymphal mortality. Consequently, the size of many of the adult *Philaenus* populations fluctuated widely in some islands during the study period.

In the Rovholmen population (18), the observed extremes of numbers of individuals were 203 (1975) and 2035 (1977), a fluctuation about tenfold. In spite of this variability in size, the frequencies of the colour alleles remained almost unchanged throughout the study (Fig 3.2). This stability has been maintained with an astonishing constancy beginning with 1971; the frequency of *typicus* varying between 78–82%, that of *marginellus* 10–12% and that of *lateralis* 5–7%. This very narrow fluctuation must contain some sampling error owing to the inaccuracy of taking samples with a sweep net on Rovholmen.

The plant composition of Rovholmen meadow has not changed appreciably since 1969. Along the shoreline there is always enough water in the soil to prevent withering of the food plants of *Philaenus*. The salinity of sea water in Tvärminne is only about 0.6%, and there is a relatively good supply of water in Rovholmen trickling downwards from the upper parts of the slope. The food plants of *Philaenus* in the upper part of the meadow include both drought-sensitive species, such as *Lysimachia vulgaris*, and drought tolerant ones like *Achillea millefolium*. The stands of both those prevalent food plants have remained constant throughout the study period.

Constancy of colour gene frequencies thus parallels apparent constancy of the habitat. The Rovholmen population is the best example of stability of gene frequencies among the island populations investigated in Tvärminne archipelago. But Rovholmen is paralleled by many other populations living in habitats which have retained their original coverages of the most important food-plants. In these additional instances of allele frequency stability, the fluctuation is only slightly greater than in Rovholmen.

CONSTANCY OF ALLELE FREQUENCIES IN SEMI-ISOLATED SUBPOPULATIONS

The *Philaenus* population of Stora Västra Långgrundet (SVL, 22) is divided into two subpopulations living on the west and east sides of the island. The subpopulations 22W and 22E inhabit meadows SVL-W and SVL-E separated by a low rocky ridge about 5 m broad. Both meadows are sharply bordered by the glacier-smoothed rock on three sides and by the sea on the fourth.

The meadows are ecologically quite dissimilar. The W meadow is quite rich in *Lysimachia vulgaris* (about 70%) and *Myrica gale* (30%) and has a small stand of *Filipendula ulmaria*

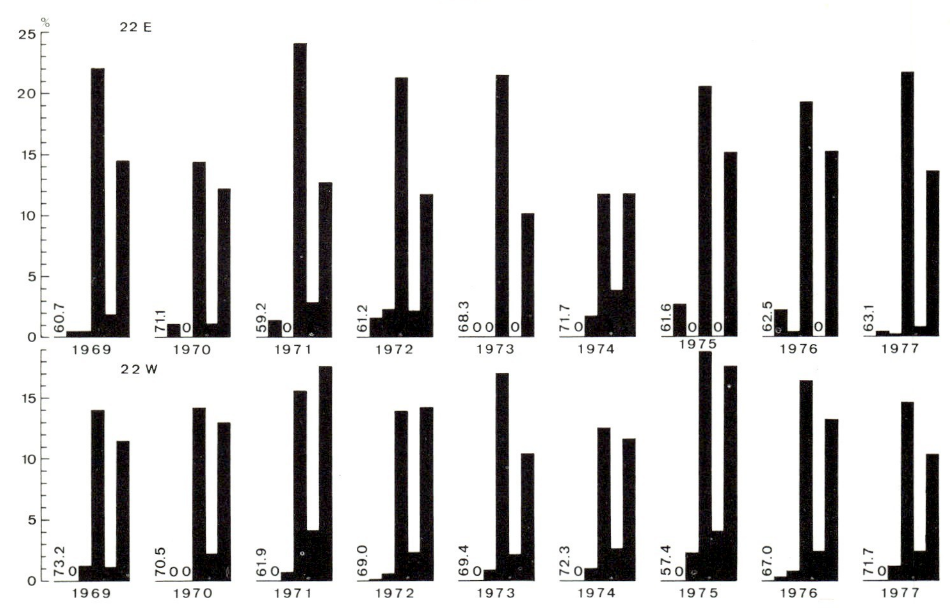

FIG. 3.3. Colour morph frequency histograms (for females) of the Stora Västra Långgrundet (SVL) population 22E and 22W (see the legend to Fig. 3.2). Extremes of sample size (females): 60 (1974) and 485 (1977) in 22E, 147 (1971) and 836 (1972) in 22W.

(5%). The E meadow is poor in these plants but rich in *Crysanthemum vulgare* (60%) and *Rubus idaeus* (30%). The seaside parts of the meadows are less dissimilar than the upper parts.

Among the 43 populations investigated, SVL displays a unique pattern of morph frequencies, both subpopulations being rich in *lateralis* and *leucophthalmus*. But, although similar in overall frequency distribution, the subpopulations 22W (in SVL-W) and 22E (in SVL-E) maintain significant differences in the exact adjustment. During 1969–77, the phenotype frequencies in 22W and 22E were divergent at the $P < 0.001$ level ($\chi^2 = 27.5$, d.f. 3; over 1969–77 neither 22W nor 22E was internally heterogeneous). The subpopulations were sampled by the sweep net method.

The two subpopulations differ in the frequency of *lateralis*, which is higher in 22E (Fig. 3.3). Thus it seems that gene flow over the narrow ridge between the SVL-W and SVL-E meadows is not intensive enough to neutralise completely the selection pressures tending to maintain dissimilar allele frequencies in the subpopulations 22W and 22E. These selection pressures probably are contingent upon the ecological dissimilarity of the habitats, in particular upon the existing dissimilarities in coverages of such important food plants as *Myrica* and *Chrysanthemum*.

BREAKDOWN OF CONSTANCY OF ALLELE FREQUENCIES

The Mellanspiken minimeadows M-A and M-B lie only about 12 m from each other. They are separated by a flat rocky plateau with tiny stands of herbs and grasses in depressions and crevices. The two minimeadows are ecologically quite dissimilar, as seen from their plant communities. The *Philaenus* subpopulations 8A (in M-A) and 8B (in M-B) were investigated in both meadows by the minicage method. This means that the food plants of the nymphs

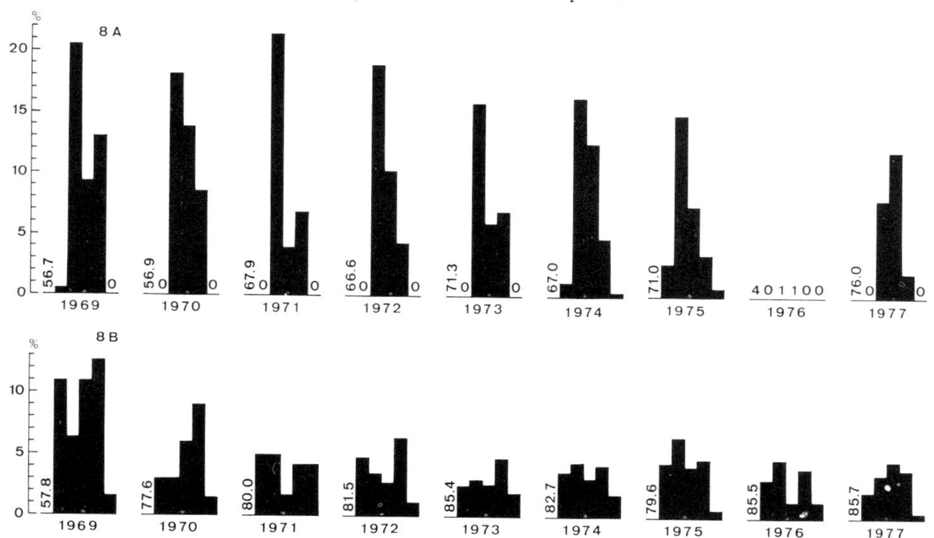

FIG. 3.4. Colour morph frequency histograms for females of the Mellanspiken populations 8A and 8B. (see the legend to Fig. 3.2). Extremes of sample size (females): 6 (1976) and 618 (1974) in 8A, 64 (1969) and 642 (1973) in 8B.

are on record for all the nine generations (years). Considerable changes in the coverage of food plants took place after the vole disaster (1969) in both M-A and M-B and after the flood disaster in M-A (see section 'Study area').

After the vole disaster, the colour morph frequencies of population 8B changed considerably in two generations (Fig. 3.4). Since 1971, 8B has remained rather stable in size and in gene frequencies. The meadow M-B inhabited by 8B is bordered on one side by a small but permanent pond and is thus relatively insensitive to droughts. The 8B population tends to move towards the wet end of the meadow in dry summers, and in the opposite direction in humid summers. Changes in food plants in 8B after the year 1971 reflect these movements, rather than food plant coverages. The *Philaenus* population obviously has adjusted its allele frequencies according to the average conditions of the minimeadow.

The vole disaster was less severe in meadow M-A and changed the coverage of the food-plants of *Philaenus* less than in M-B. But M-A is very drought sensitive and suffered badly from the dry early summer of 1973. This drought, and the 1975 flood, completely changed the ecology of the meadow. These changes in the environment probably are responsible for the instability of allele frequencies in population 8A (Fig. 3.4). Except for the last years, the alleles for *typicus* and *marginellus* seem to have retained their frequencies much better than the alleles for *lateralis* and *flavicollis*.

Throughout the years, the populations 8A and 8B have remained very dissimilar in allele frequencies. The differences are evident from a comparison of allele frequency histogram sets for the two populations (Fig. 3.4). In nine years, the alleles p^T (for *trilineatus*) and p^O (for *leucophthalmus*), stably represented in 8B, have not gained permanent foothold in 8A. The 12 m of flat rock surface separating 8A and 8B are enough to keep the gene pools distinctly dissimilar. In this instance, contrary to 22W and 22E, adjustment of allele frequencies according to habitat characteristics completely over-rides gene flow, and the frequencies do not bear the slightest resemblance to each other. It seems that the niche complexes offered by meadows M-A and M-B favour dissimilar colour morph frequencies.

CONSISTENCY OF ALLELE FREQUENCIES: THE ABSENCE OR PRESENCE OF LEPYRONIA

The *Philaenus* population inhabiting the meadow Brännskär-E retained rather stable allele frequencies in 1969–74 (Fig. 3.5). The population, 16E, lives in a shore meadow which in many respects resembles ecologically the mainland site of investigation (Svanvik). In the Tvärminne area, Svanvik is the only meadow in which dense populations of *Philaenus* and *Lepyronia* coexist. Obviously, due to a weak migratory capacity, *Lepyronia* does not occur in Brännskär or in any other of the islands investigated. (A small immigrant population of 26 individuals was found in Mellanspiken in 1974, but disappeared after the 1975 flood.)

To study the possible effect of a coexisting Aphrophorid on *Philaenus* allele frequencies, 1203 individuals of adult *Lepyronia* were transferred to the Brännskär-E meadow in 1975 and a further 2109 in 1976. These introduced founders established a permanent population in the meadow. In the early summer of 1976, and especially 1977, the *Lepyronia* nymph population exceeded *Philaenus* in density in those parts of the meadow closest to the shoreline. In 1977, the meadow as a whole yielded *Lepyronia* and *Philaenus* in the ratio 0.3 : 1 by sweep net sampling.

It might be expected that competition by *Lepyronia* would have affected primarily that part of the *Philaenus* nymph population which lives close to the shoreline. If the various colour morphs of *Philaenus* have dissimilar niche requirements, then some of these niches may overlap with the *Lepyronia* niche to a greater extent than others. This would perhaps mean a decrease in the frequency of the morphs in question as a consequence of introduction of *Lepyronia* to the habitat.

The frequencies of the main colour morphs in *Philaenus* population 16E (Fig. 3.5) show a slight decrease in morph *typicus* in the 1976 and 1977 generations as compared to the earlier years. A chi-square analysis for the difference between the pooled sample from 1969–75 and that from 1976–77 yielded, with 4 d.f. (morphs *typicus, trilineatus, marginellus, lateralis, leucocephalus* and *leucophthalmus* included), a $\chi^2 = 16.9, P < 0.01$.

But in spite of the statistically significant difference in colour morph frequencies before and after *Lepyronia* introduction, it is by no means clear that the introduction has resulted in a shift in *Philaenus* allele balance. The change in gene frequencies is, perhaps, temporary, and may have resulted from some cause other than the *Lepyronia* introduction. To give a

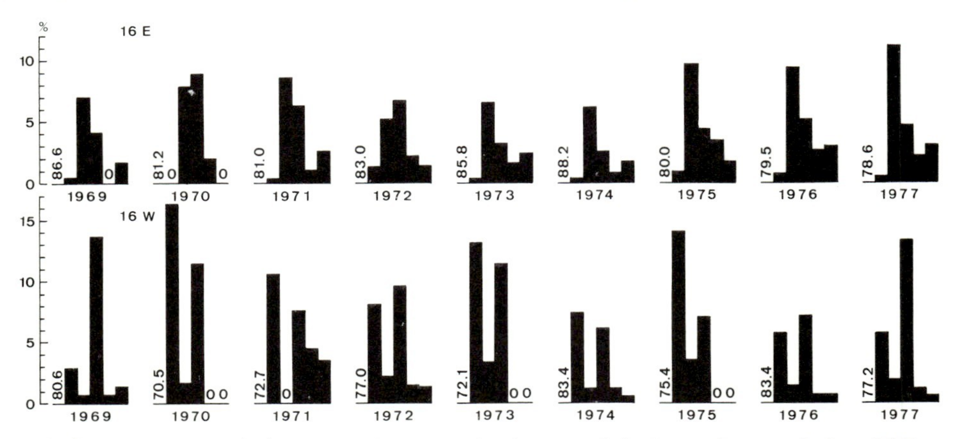

FIG. 3.5. Colour morph frequency histograms for females of the Brännskär populations 16 E and 16W (see the legend to Fig. 3.2). Extremes of sample size (females): 101 (1970) and 590 (1976) in 16E, 61 (1970, 1973) and 163 (1974) in 16W.

basis for evaluating possible effects of the introduction, some facts about niche overlap and coexistence of *Philaenus* and *Lepyronia* are provided in the next section.

COEXISTENCE OF PHILAENUS, LEPYRONIA AND THE OTHER APHROPHORIDS IN SVANVIK

Coexistence of four species of the Aphrophoridae: *Philaenus spumarius*, *Neophilaenus lineatus*, *Lepyronia coleoptrata* and *Aphrophora alni* was investigated in the Svanvik meadow. Distribution of the nymphs on the food plant species available in Svanvik was studied by opening spittle masses and recording the nymphs by species. Table 3.1 shows the number of nymphs of the four species on 22 food plant species in 1976. *N. lineatus* was found on monocotyledons only, *A. alni* exclusively on dicotyledons, and *P. spumarius* and *L. coleoptrata* on both. *Lepyronia* was, however, much more frequent on monocotyledons than *Philaenus*. With regard to dicotyledons, *Lepyronia* was more frequent than *Philaenus* on short herbs such as *Potentilla anserina* and *Galium palustre*. The favourite food plant of *Philaenus* in Svanvik was *Filipendula ulmaria*.

These frequencies in the distribution of *Philaenus* and *Lepyronia* nymphs on the food plants may be called the 'taxonomic dimension of coexistence'. As shown by Table 3.1, niche overlap appears considerable along this dimension.

But there are other dimensions of coexistence which reduce the extent of niche overlap from that apparent from Table 3.1. The vertical distribution of the nymphs of the different Aphrophoridae on the same food plants was investigated in 2 m x 2 m quadrats. *Filipendula ulmaria* was the prevalent food plant species of the spittle-bugs in quadrat number 1 in 1976,

TABLE 3.1. Distribution of nymphs on four food plants (Taxonomic dimension of coexistence)

	Philaenus spumarius	*Lepyronia coleoptrata*	*Aphrophora alni*	*Neophilaenus lineatus*
Triglochin maritimum	3	15	–	–
Scirpus tabernaemontani	8	88	–	5
Scirpus uniglumis	–	15	–	11
Carex nigra	2	63	–	16
Agrostis stolonifera	–	27	–	152
Molinia coerulea	–	23	–	8
Phragmites communis	3	55	–	5
Festuca ovina	–	–	–	16
Festuca rubra	1	8	–	181
Myrica gale	25	77	1	–
Thalictrum flavum	–	14	3	–
Potentilla anserina	22	166	1	–
Potentilla palustris	35	70	1	–
Filipendula ulmaria	259	109	49	–
Trifolium pratense	3	20	–	–
Lythrum salicaria	13	38	2	2
Angelica silvestris	10	27	13	–
Lysimachia vulgaris	38	71	2	–
Lysimachia thyrsiflora	28	19	–	–
Lycopus europaeus	4	10	–	–
Galium palustre	77	188	–	–
Chrysanthemum vulgare	8	16	–	1

Table 3.2 Vertical distribution of nymphs on *Filipendula ulmaria*

	Number of Nymphs		
Height (cm)	*P. spumarius*	*L.coleoptrata*	*A.alni*
55			
	1	–	–
50			
	–	–	–
45			
	7	–	–
40			
	20	–	–
35			
	16	–	–
30			
	27	–	–
25			
	26	2	–
20			
	11	5	1
15			
	8	4	1
10			
	3	4	–
5			
	1	1	11
Soil level			
Totals	120	16	13

$F = 69.02, P < 0.001$.

and Table 3.2 shows the vertical distribution of nymphs of three species on this plant. Clearly, *Philaenus* lives higher up on the plant than *Lepyronia*, which lives higher up than *Aphrophora*. This result was corroborated by the distribution of nymphs in a large number of other quadrats, both on *Filipendula* as well as on other food plant species. The distribution observed may be called the 'vertical dimension of coexistence'.

There are other dimensions of coexistence along which *Philaenus* and *Lepyronia* appear to differ in distribution, the time dimension for instance. In the present context, only the significance of the taxonomic and vertical dimensions will be evaluated. It seems that if there is any competition between *Philaenus* and *Lepyronia* for feeding sites, such competition is most intense on short-stalked plants like *Galium palustre*.

Both *Philaenus* and *Lepyronia* favour *Galium palustre* as a food-plant, and the number of feeding sites on this tender plant is limited. *Galium palustre* is the dominant herb in the wet parts of the Brännskär E meadow. It is quite possible that *Lepyronia* now excludes *Philaenus* from some of its feeding sites on *Galium* in the Brännskär E meadow.

Discussion

In lush meadows, nymph populations of *Philaenus* may reach densities amounting to more than 100 nymphs per square metre. In normal years this is very far from saturating the food plants with nymphs, but during droughts it may mean oversaturation. If there is any intra- or interspecific competition, it may often cause some of the nymphs to seek feeding sites other than those originally chosen by them. Dividing the *Philaenus* population into genotypes with dissimilar modes of resource utilisation could perhaps reduce competition and enhance use of the resources.

The minicage method has been used in Tvärminne in studies on colour morph food-plant combinations. In the years 1969–77, altogether about 50 000 *Philaenus* nymphs were isolated on their food-plants in minicages. From these minicages, about 34 000 adults emerged and were classed by sex and colour morph. It would seem that such a large number of observations should suffice for establishing food-plant preferences of the colour morphs. But this huge material consists of numerous relatively small samples. Some of the colour morphs occur at a very low frequency, which means that a very small absolute number of a given morph is found yearly in a minimeadow on a given plant species. Thus, to be able to perform a statistical test on colour morph food-plant combinations, it is unavoidable to pool samples from different years, sometimes also from different meadows. This procedure may, however, lead to erroneous conclusions (Cochran, 1954; Halkka & Mikkola, 1977).

However significant χ^2 values may be obtained for many of the colour morph food-plant combinations in a 2 x 2 setting during a single generation of a population inhabiting an isolated meadow. In each 2 x 2 setting, the number of colour morph 'x' on food-plant 'y' and on all other food plants is compared with the number of all other colour morphs on y and on all other food plants. It is legitimate to sum chi-squares for successive generations and to calculate a composite chi-square value (Cochran, 1954).

The composite χ^2 values are often statistically significant when separate values are summed for three or four successive generations, but generally not for longer periods. It seems that the environments provided by the food-plants do not remain constant in time (Halkka & Mikkola, 1977). The plant/colour-morph combinations appear to depend on the weather of the study year and on the microclimatically varying qualities of the sites in which the populations live.

It thus seems impossible to prove statistically that a given colour morph favours or disfavours a certain food plant species. The disproportionate occurrence of some of the colour morphs on certain food-plants, as indicated by Table 3.3, may result from factors independent of possible differences between morphs in food-plant utilisation. In some of the meadows, owing to the founder principle, some of the morphs and some of the food plants may have attained exceptionally high frequencies independently of each other (see Halkka & Mikkola, 1977).

From Table 3.3 it appears that *trilineatus* (tri) is disproportionately common on *Lythrum salicaria, marginellus* (mar) and *lateralis* (lat) on *Filipendula ulmaria, leucocephalus* (lce) on *Solidago virgaurea* and *leucophthalmus* (lop) on *Lysimachia vulgaris* and *Chrysanthemum vulgare.* The frequencies indicated in the table, have been obtained by the minicage method. They are supported, in some instances, by simultaneous high frequency of a colour morph and high coverage of a food-plant in meadows investigated by the sweep net method. To give an example, *leucophthalmus* is unusually frequent in the three islands (Porskobben, Stora Västra Långgrundet and Klovaskär 2; see Halkka *et al*, 1970) in which there are dense stands of *Chrysanthemum vulgare*. In this instance, at least, it seems possible that simultaneous high frequencies of a food plant and a colour morph are not due to the founder principle. The high frequencies depend either on ecological conditions suitable for both the plant and the spittle-bug or on high fitness of *leucophthalmus* on *Chrysanthemum*.

The total niche of *Philaenus spumarius* can be divided into subniches, labelled according to the names of the food-plants. The subniches so defined can be characterised in terms of

Table 3.3. Niche width and niche overlap of phenotypes of female *Philaenus* as shown by percentages bred from nymphs feeding on different plant species. The phenotype *flavicollis* (*fla*, p^F or p^C) is included in the totals. On account of the high frequency of *typ* (p^t/p^t), most non-*typ* phenotypes are heterozygous for p^t, and hence the table shows the distribution of genotypes (alleles).

	Filipendula ulmaria	*Lysimachia vulgaris*	*Solidago virgaurea*	*Lythrum salicaria*	*Chrysanthemum vulgare*	Other food-plants	Number of plant species utilised	Total number of females	Frequency of morph (%)
typ (p^t)	23.8	21.3	13.7	13.2	4.2	23.8	44	10235	80.5
tri (p^T)	9.4	26.4	12.7	25.0	1.4	25.0	22	212	1.7
mar (p^M)	27.5	17.6	12.3	13.7	3.6	25.1	30	839	6.6
lat (p^L)	33.9	9.8	15.5	17.4	2.2	21.1	27	407	3.2
lce (p^C)	28.2	16.8	19.7	6.0	4.3	25.1	23	351	2.8
lop (p^O)	12.4	28.4	20.0	7.0	6.5	25.6	22	429	3.4
Totals	3037	2632	1764	1708	525	2123		12710	
Food-plant utilisation %	23.9	21.0	13.9	13.4	4.1	24.0			100.1

the ecology of the plant species. It has been found that survival rates of nymphs isolated in minicages at the third to fifth instar vary greatly according to the plant species (Halkka & Mikkola 1977; Raatikainen *et al*, 1977). As subniches, the food plant species vary in carrying capacity from year to year. The minimal, or bad-year (dry summer) carrying capacities, are much more variable than the maximal, from about 8% on *Solidago virgaurea* to about 63% on *Potentilla palustris*. The survival percentages in favourable years vary between 90–100% on the food-plants named in Table 3.3.

The nymphs of *Philaenus* in their frothy 'cuckoo-spit' masses change position only about once a week (Halkka *et al*, 1967), and then seldom crawl more than 0.6 m from the previous sucking site. The adults move over a much wider range, as they jump vigorously from food-plant to food-plant. This mode of distribution of individuals at different stages of the life cycle corresponds nicely to the basis of Levene's (1953) well-known model of selection in a spatially heterogeneous environment. The dissimilar motilities of the nymphs and the adults render ecological specialisation possible at the nymphal stage, without leading to divergence of species into reproductively isolated ecological races (cf. Soans *et al*, 1974).

The divergence of *Philaenus*, as it seems, into colour morphs with dissimilar ecological requirements, may mean that the species as a whole has no very exacting niche requirements. *Philaenus* is able to maintain constant populations on very small islands in the absence of the other Aphrophoridae. It may seem from Table 3.1 that *Lepyronia* has greater niche width than *Philaenus*, but the great drought tolerance of *Philaenus* increases its niche to cover habitats in which no other Aphrophoridae may persist. Fluctuation in humidity sets a limit to niche overlap (c.f. May & MacArthur 1972).

In short, the existence of ecologically dissimilar morphs means that there is diversity *within Philaenus*. This diversity is of maximum significance during periods of minimum fitness for the whole population, for example in droughts. As survival rates vary greatly on different food-plant species the individual colour morphs may be expected to show their maximum fitness in combinations of subniches (i.e. in communities of food plants) particularly suitable to each morph.

Macro- and microclimatic conditions, together with the structure of the plant communities may often dictate both the frequencies of the various *Philaenus* colour morphs and the frequencies of the coexisting Aphrophoridae.

Gene frequencies in two spatially isolated populations seem to deviate as a product of interpopulation distance and degree of dissimilarity of the meadows inhabited. Subpopulations 22W and 22E, separated by 5 m of rock surface and living in the large ecologically dissimilar meadows SVL-W and SVL-E, show statistically significant difference ($\chi^2 = 27.5$, d.f. 3, $P < 0.001$). But taking population 22 as a whole, it shows a unique pattern of morph frequencies, both subpopulations being unusually rich in *lateralis* and *leucophthalmus* (Fig. 3.3). This similarity must be due to migration between 22W and 22E. Migration counteracts selection brought about by dissimilar food-plant compositions in meadows SVL-W and SVL-E.

The populations 8A and 8B, on the island Mellanspiken, isolated by 12 m of rock surface, show profoundly dissimilar allele frequencies ($\chi^2 = 338.1$, d.f. 3, $P > 0.001$). In this instance, the effect of migration is negligible and ecological dissimilarity of the meadows alone determines the allele and morph frequencies.

During evolution, utilisation of resources has been divided between *Philaenus* colour morphs and between the various aphrophorid species in a way which seems to minimise competition (c.f. Pianka, 1974). It is not unusual to encounter, within a single spittle mass, nymphs of two or three different species. Provided there is no shortage of xylem, coexistence and true niche overlap in this instance mean a net profit to each species. Diversity, as expressed by the presence of many colour morphs in the same population, and by the coexistence of two or more aphrophorid species, is maintained by external conditions rather than by intra- or interspecific competition.

Summary

Populations of *Philaenus spumarius* were investigated in the absence and presence of other species belonging to the same family (Aphrophoridae; Homoptera). The populations of this species are polymorphic with regard to colour. The polymorphism is governed by a gene locus with a different number of alleles in populations living in different geographical regions.

Philaenus populations were studied in southern Finland, in the Tvärminne archipelago and the adjacent mainland. Populations living on small islands tend to maintain allele frequencies which are island-specific. The allele frequencies may remain practically unchanged for at least nine years. The very constancy of the frequencies indicates that they are maintained by island-specific selection pressures.

In many of the small rocky islands, the meadows suitable for *Philaenus* lie scattered over the island in small plots called 'minimeadows'. The islands differ profoundly in the composition of the plant communities within the minimeadows. The island-specific allele frequencies of *Philaenus* appear to be contingent on the specific characteristics of the minimeadow(s) inhabited. Each colour morph seems to favour a certain combination of food-plants. The diversity existing *within the species* in the form of ecologically dissimilar colour morphs may allow *Philaenus* to persist in a greater number of subniches than is possible for other, related aphrophorid species.

During evolution, the rate of gene flow between spittle-bugs inhabiting adjacent plant stands has been adjusted to be restricted enough not to interfere with multiniche polymorphism stabilisation. A distance of five or 12 metres between adjacent but isolated minimeadows suffices for maintenance of dissimilar allele frequencies in two *Philaenus* populations. The prerequisite for this is, of course, that the meadows are ecologically dissimilar.

In some of the meadows *Philaenus* coexists with one or more of the following species: *Lepyronia coleoptrata, Aphrophora alni* and *Neophilaenus lineatus*. It seems that evolution has imposed restrictions on interspecific niche overlap with regard to food plant utilisation and with regard to the height above soil level of nymphal feeding sites. Diversity in mixed populations of aphrophorids is perhaps regulated mainly by (micro)climatic conditions and only secondarily by rules of coexistence.

Through an experimental introduction of large numbers of *Lepyronia* individuals, it was attempted to change colour morph frequencies in a *Philaenus* population.

Acknowledgements

Grants for this study have been received from the National Research Council for Sciences (Academy of Finland). This is report Number 573 from the Tvärminne Zoological Station.

References

Clarke B. (1969) The evidence for apostatic selection. *Heredity, Lond.* **24**, 347–352.

Cochran W. G. (1954) Some methods for strengthening the common χ^2 tests. *Biometrics* **10**, 417–451.

Halkka O., Raatikainen M., Vasarainen A. & Heinonen L. (1967) Ecology and ecological genetics of *Philaenus spumarius* (L.) (Homoptera). *Ann. Zool. Fenn.* **4**, 1–18.

Halkka O., Raatikainen M, Halkka L. & Hovinen R. (1970) The founder principle, genetic drift and selection in isolated populations of *Philaenus spumarius* (L.) (Homoptera). *Ann. Zool. Fenn.* **7**, 221–238.

Halkka O., Halkka L., Raatikainen M. & Hovinen R. (1973) The genetic basis of balanced polymorphism in *Philaenus* (Homoptera). *Hereditas* 74, 69–80.

Halkka O., Raatikainen M. & Halkka L. (1974*a*) Radial and peripheral clines in northern polymorphic populations of *Philaenus spumarius*. *Hereditas* 78, 85–96.

Halkka O., Raatikainen M. & Halkka L. (1974*b*) The founder principle, founder selection, and evolutionary divergence and convergence in natural populations of *Philaenus*. *Hereditas* 78, 73–84.

Halkka O., Halkka L., Hovinen R., Raatikainen M. & Vasarainen A. (1975*a*) Genetics of *Philaenus* colour polymorphism: the 28 genotypes. *Hereditas* 79, 308–310.

Halkka O., Raatikainen M., Halkka L. & Hovinen R. (1975*b*) The genetic composition of *Philaenus spumarius* populations in island habitats variably affected by voles. *Evolution Lancaster, Pa.* **29**, 700–706.

Halkka O., Raatikainen M. & Halkka L. (1976) Conditions requisite for stability of polymorphic balance in *Philaenus spumarius* (L.) (Homoptera). *Genetica* **46**, 67–76.

Halkka O. & Mikkola E. (1977) The selection regime of *Philaenus spumarius* (L.) (Homoptera). Ove Frydenberg Memorial Symposium Volume.

Halkka O., Raatikainen M., Halkka L. & Raatikainen T. (1977) Coexistence of four species of spittle-producing Homoptera. *Ann. Zool. Fenn.* **14**, 228–231.

Levene H. (1953) Genetic equilibrium when more than one ecological niche is available. *Am. Nat.* 87, 331–333.

Linnavuori R. (1969) Hemiptera IV. *Animalia Fennica* **13**, 1–312.

May R. M. & MacArthur, R. H. (1972) Niche overlap as a function of environmental variability. *Proc. Natn. Acad. Sci. U.S.A.* **69**, 1109–1113.

Pianka E. R. (1974) Niche overlap and diffuse competition. *Proc. Natn. Acad. Sci. U.S.A.* **71**, 2141–2145.

Raatikainen M., Halkka O., Vasarainen A. & Halkka L. (1977) Abundance of *Philaenus spumarius* in relation to types of plant community in the Tvärminne archipelago, southern Finland. *Biol. Res. Rep. Univ. Jyväskylä* **3**, 3–38.

Soans A. B., Pimentel D. & Soans J. S. (1974) Evolution of reproductive isolation in allopatric and sympatric populations. *Am. Nat.* **108**, 117–124.

4 • Ecological and behavioural origins of diversity in butterflies

R. I. VANE-WRIGHT

Department of Entomology, British Museum (Natural History), London SW7 5BD

'Any mode of thought which attempts to attribute to man or any other organism any form of unchanging essence, or any character that is conceived as *being* rather than *becoming*, flies in the face of our whole understanding of biology.'
C. H. Waddington, in *The Humanist Frame* (ed. J. S. Huxley), 1961.

A major factor in considering faunal diversity is merely S_T, the total number of species present. This measure, coupled with the area available to the given fauna, is sufficient to raise some interesting questions. For example, why is S_T for the butterflies of Peru (approximately ½ million square miles) about 4000, whereas for Western Europe S_T is less than one tenth of this figure (fewer than 400 species in over one million square miles)?

The ecologist usually has need of a more refined measure of diversity than that provided by S_T alone, and endeavours to incorporate at least factors for individual species abundance, or equitability. The taxonomist and the biogeographer usually demand refinement in another direction – some estimate or reflection of the number of higher taxa represented in a given fauna.

Here I will be concerned largely with the origins and coexistence *of species within a single fauna*, that is the origins and dynamics of species which coexist in sympatry. To study this, it is necessary to establish working hypotheses in two realms – the realm of time (phylogenetic hypotheses) – and the realm of multidimensional ecospace (hypotheses of community ecology).

The focus to which attention is directed concerns the processes by which speciation events are either generated, maintained, or eliminated within faunas. This demands that we be interested in the operation of genetic isolating mechanisms, and of competition. In a group such as butterflies, both of these are spectacularly *behavioural* processes. First, however, I would like to look briefly at the idea of accounting for butterfly diversity through coevolution with flowering plants.

Butterflies and plants

There are approximately 200 000 species of Lepidoptera, and a similar number of flowering plants. The dependence of almost all Lepidoptera on plants for larval development suggests this too neat correspondence may not be entirely fortuitous.

Milkweeds (Danainae – ca 150 spp.) – Apocynaceae/Asclepiadaceae/Moracceae
Glass wings (Ithomiinae – ca 400 spp.) – mainly Solanaceae
Fritillaries & Heliconians (part Nymphalinae – ca 500 spp.)
– Passifloraceae/Flacourtiaceae/Violaceae/Turneraceae
Satyrs (Satyrinae – ca 2 000 spp.) – mainly monocotyledons, esp. Gramineae & Palmae
Whites (part Pierinae – ca 400 spp.) – Cruciferae/Capparidaceae
Jezebels (part Pierinae – ca 300 spp.) – Loranthaceae
Swallowtails (Papilioninae ca 650 spp.) – mainly Aristolochaceae; Rutaceae; Umbelliferae

TABLE 4.1. Examples of some of the larger groups of papilionoid butterflies which have either radiated on or co-evolved with certain groups of flowering plants (from Ehrlich & Raven, 1965). See text.

The 20 000 butterflies (Hesperioidea and Papilionoidea) comprise about 10% of known Lepidoptera. Individual species show various levels of host-specificity, from monophagy, through oligophagy, to limited polyphagy. Ehrlich and Raven (1965), in their classic paper on the coevolution of butterflies and plants, suggested that mutual interactions of plants and phytophagous insects might be responsible for generating much of the species diversity of both groups. Within butterflies they noted several well-marked subgroups which appear to have undergone adaptive radiation in relationship with either phylogenetically or chemically related groups of plants (Table 4.1).

There is little doubt that co-evolution with plants has been very important in the evolution of butterflies. But how does it relate to the general maintenance of diversity? If specific host/herbivore coevolution were all-dominant, at the extreme we might expect to see a situation where each major group of butterflies was restricted to one set of plants, with no overlap, and that within faunas a 'one species, one host-plant/group' relationship existed. We might further expect that the species richness of a flora, with respect to those plant families successfully colonised by butterflies, would give a direct prediction of associated butterfly species richness. Although these are naive, extreme predictions, their failure to fit the facts suggests that relationships other than host/herbivore coevolution must be important in maintaining and generating butterfly faunal diversity. Four aspects of this problem, which lead me to reject a simplistic co-evolution thesis, are discussed below.

SIZE OF PLANT FAMILIES AS HOST TARGETS

Our knowledge of butterfly host plants is very incomplete. At most, we have records for about one quarter of the species. Ehrlich and Raven (1965) record papilionoid butterflies from 141 families of dicotyledons, monocotyledons and gymnosperms, and discuss their absence or virtual absence from a number of plant groups of quite large size. Willis (1973) lists some 520 families of spermatophytes, varying in size from one included species (66 families) to 17 000 species (Orchidaceae; most botanists would now accept the Compositae as the largest family). From these data, Fig. 4.1 plots the number of plant families for 15 family size classes, the number of plant families within each size class utilised by papilionids, and the total number of plant species contained by each family size class.

Although the data so presented are very crude, it can be seen that a wide spectrum of the diversity offered by plants has been used by butterflies. The vast majority of plant families for which there are no butterfly records are small in terms of included species. I think the

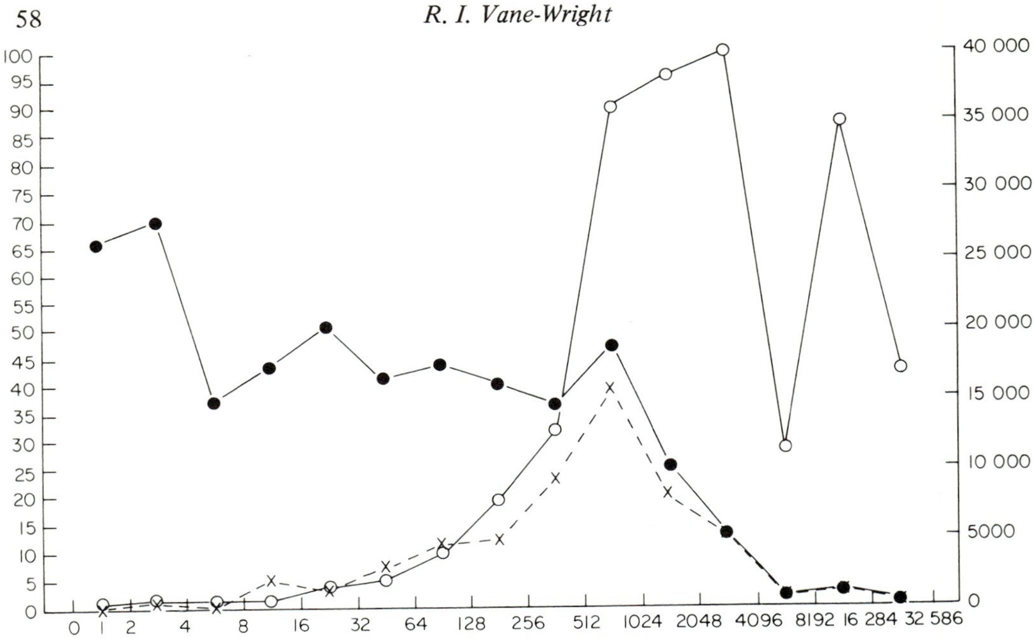

FIG. 4.1. Size classes of spermatophyte families (abscissa: 1 sp., 2–3 spp., 4–7 spp., 8–15 spp., etc.) plotted against (a) the number of families in each size class (solid circles, left ordinate), (b) the number of families within each size class recorded to include butterfly host plants (crosses, left ordinate), and (c) the total number of plant species represented by each family size class (open circles, right ordinate). Data from Ehrlich & Raven (1965) and Willis (1973). See also text.

graph shows that the likelihood of recording butterflies on a particular family of plants can be predicted, to a first approximation, merely by the size of that family in terms of included species. This does not, of course, deny coevolution (it could well be explained by it!), but coevolution must surely be with the whole spectrum of phytophages. My point is that it equally suggests a probabilistic element – the size of a plant family in some sense indicates its 'target area' for attack by butterfly larvae. This corresponds to the suggestion of Benson *et al* (1976, p. 669) that observed coevolutionary 'patterns' might just reflect initial plant radiation followed only later by herbivore occupation.

NON-EXCLUSIVE HOST-PLANT FAMILIES

Within the Papilionoidea we presently recognise only four *monophyletic* family divisions (Fig. 4.2; Kristensen, 1976). Out of the total of 141 families of known host plants, the Papilionidae are recorded from 37, Pieridae from 30, Lycaenidae from 85, and Nymphalidae from 94 (data from Ehrlich and Raven, 1965). The distribution of the butterfly records within this assemblage is not random – providing the data base of the coevolutionary hypothesis.

If we accept that the existence of host plant families exclusive to various subgroups of butterflies may reflect coevolution, what can be said of the non-exclusive host-plant groups? We can score the number of plant families shared by members of any pair of butterfly families, for all six possible papilionoid family pairings. If the occurrence of shared hosts were at random, we could predict the number of shared families from the product

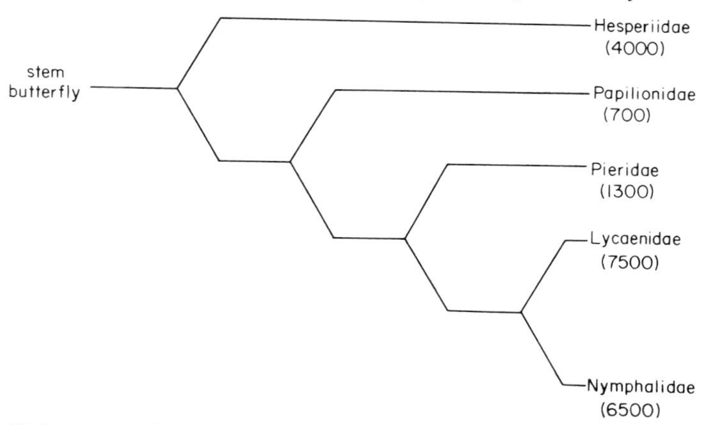

FIG. 4.2. Cladogenetic relationships of butterfly families (after Kristensen, 1976). Estimates of the total number of species within each family are given in parentheses.

TABLE 4.2. Number out of the total of 141 host plant families utilised by the Papilionoidea, as recorded for each of the four constituent families (diagonal). The intersections give the number of shared host families for each pair of butterfly families, based on random shift model (figures in parentheses). Figures not in parentheses give observed number of shared host families for each pairing (data from Ehrlich & Raven, 1965). See text.

	Pa	Pi	N	L
Papilionidae (700)	37/141			
Pieridae (1300)	10 (8)	30/141		
Nymphalidae (7500)	27 (25)	19 (20)	94/141	
Lycaenidae (6500)	19 (22)	19 (18)	52 (57)	85/141

percentage of families utilised by both members of a pair, for all six pairings. Table 4.2 shows the application of this calculation (figures in parentheses), together with the corresponding numbers of shared host families for each pair as recorded in Ehrlich and Raven. The predicted and observed figures match very closely. Although such a treatment can and should be criticised as gross, it is still suggestive that notions of extreme coevolutionary dependence between butterflies and their larval host plant species are not tenable. It is possible that butterflies have made repeated, occasionally successful, attempts to colonise other food plant 'niches'. Even within good examples of extreme coevolution, there is evidence of re-radiation (Benson *et al*, 1976, p. 674) – facilitated through advances in the insect behaviour.

FLORAL AND FAUNAL RICHNESS

Despite the comments above, specialisation on different host plants is a major component in subdividing butterfly ecospace. We may broadly expect that the more species of plants within a system, or the more families present known to be used by butterflies ('butterfly' families), the more species of butterflies we may expect to find in that system. Tropical South America and South-east Asia are both areas of exceptional plant and butterfly diversity. But if we look at islands, the picture is less clear, due mainly to classic MacArthur-Wilson island biogeographic effects. A few examples may be considered.

New Zealand (259 000 km^2, over 1600 spermatophyte species in 100 families, including 71 'butterfly' families) supports only 12 resident butterflies (Miller, 1971). This might be explained by recent glaciation coupled with extreme remoteness. But Hudson (1928, p. 372) notes over 1200 resident moth species. This then compares less unfavourably with *Great Britain* (233 100 km^2, 2500 spermatophytes in 110 families, including 68 'butterfly' families), also recently glaciated, but much closer to a faunal source, and having 57 resident butterflies (Dennis, 1977) and 2300 moths.

Jamaica (11 400 km^2, 3000 spermatophytes in 170 families, including 110 'butterfly' families) has 114 butterfly species (Riley, 1975). Its small size and distance from source areas might be sufficient to explain this rather low S_T for a tropical island. But the source areas (North, Central and South America, plus the rest of the West Indies) are, taken together, extremely rich. Trinidad, less than half the area, probably has over 600 species (Barcant, 1970).

Sri Lanka (64 750 km^2, 2800 spermatophytes in 150 families, including 103 'butterfly' families) has 242 butterfly species (Woodhouse, 1950). The island is close to a major faunal source, but southern India is relatively poor for a tropical area.

The examples given above, and many others, suggest that although a suitable and diverse flora is probably necessary to maintain a diverse butterfly fauna, it does not, in the case of islands, generate one. Island butterfly faunas, although harbouring some striking endemics (many of which are probably relicts), are strongly affected by the size of the island and its height, its distance to the nearest faunal source, and the richness of that source. It is not apparent that island butterfly faunas are directly responsive to the absolute size, quality or diversity of their floras. Even Australia, with an old and diverse native flora covering a large area, has a poor butterfly fauna (excluding the recent New Guinea derived element).

FAUNAL DIVERSITY WITHOUT FLORAL DIVERSITY

There are probably many cases where, within a fauna, several species of related butterflies coexist on a single species of larval food plant. For example, five argynnids native to the British Isles, *Argynnis aglaja, A. adippe, A. paphia, Boloria selene* and *B. euphrosyne*, are restricted locally to seven *Viola* species (*riviniana, odorata, canina, hirta, lutea, tricolor,* and *palustris juressi*). All five fritillaries can feed on the Common Violet, *V. riviniana*, and there are a number of areas where all five species of butterflies have been recorded, but only *riviniana* amongst the violets. Assuming that all five butterflies breed in such areas, do they compete for the tiny host plants, and if not, how do they avoid doing so? There are many questions we could pose about the dynamics of this situation, but it has yet to be investigated. The two *Boloria* have rather similar total distributions, but all the *Argynnis* show individual differences. Interestingly, the distribution of *V. riviniana* is considerably greater than that of all five butterflies (data from the Biological Records Centre, Monks Wood, UK).

The presence of food plant is a necessary, but not sufficient condition for the breeding presence of a butterfly. On the other hand, several butterflies may coexist where there is only one species of larval food plant. Thus food plant species diversity may be neither sufficient nor necessary for maintaining faunal diversity. If so, we may question if it is a prerequisite for generating butterfly diversity. But before proceeding with further ecological discussion, I would like to turn to the phylogenetic realm of argument (Hennig, 1966).

PLATE 1. 1–2, *Mycalesis drusillodes* (1, ♂; 2, ♀). 3, *Mycalesis durga*, ♂. 4, *Mycalesis bazochii*, ♀. 5–7, *Papilio dardanus* (5, ♂; 6, ♂-like ♀; 7, mimetic ♀). 8–10, *Papilio phorcas* (8, ♂; 9, ♂-like ♀; 10 *constantinus*-like ♀). 11–12, *Nessaea hewitsoni* from Iquitos (11, ♀; 12, ♂). 13–14, *Nessaea ancaea* (13,♂ from Iquitos; 14, ♂ from French Guiana).

5
8
9
6
10
7
3
4
14
1
2
13
12
11

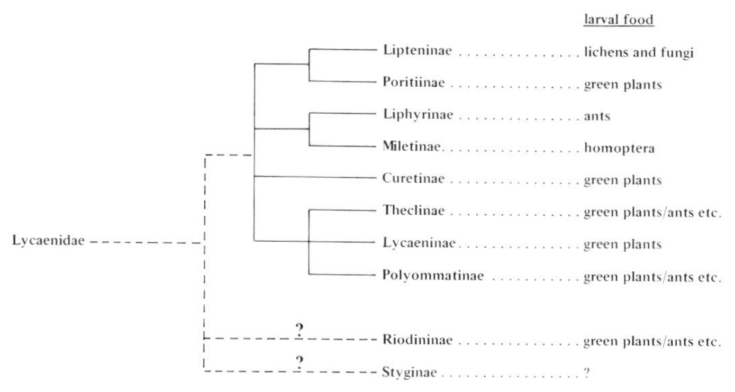

FIG. 4.3. Classification of Lycaenidae (solid line–after Eliot, 1973), to which must be added the Riodininae and Styginae to render the family monophyletic (Kristensen, 1976). Column on right indicates known larval food range of each subfamily.

Phylogeny, lycaenids and ants

The Lycaenidae, representing nearly 40% of all known butterfly species, have recently been the subject of an excellent, though not strictly phylogenetic reclassification by John Eliot (1973). He recognises eight subfamilies, grouped as in the upper part of Fig. 4.3. To this assemblage must be added the Riodininae and Styginae (Ehrlich, 1958). As indicated by Kristensen (1976), to exclude the Riodininae, on present evidence, renders the Lycaenidae paraphyletic (i.e., it would not include all the living descendants of its stem species).

According to group and species, some lycaenid larvae feed partly or wholly on ant larvae or pupae, or on various Homoptera tended by ants (Eliot, 1973). Lycaenid adults may be found feeding on homopteran honeydew, and sometimes even the honeydew produced by lycaenid larvae tended by ants (Gilbert, 1976). At present, five of the 10 recognised subfamilies are known to include species regularly involved with ants or ant symbionts (Fig. 4.3).

In the past, many workers have imagined that lycaenid/ant associations were few, and that each was independently evolved (e.g., Cottrell, 1965). But it is now apparent that hundreds of species are involved. This, coupled with the underlying phylogeny partially reflected by Fig. 4.3, leads me to conclude, with Eliot (1976, *in litt.*), that 'the primitive ancestors of Lycaenidae were vegetable feeders . . . the carnivorous habit is a secondary specialisation arising as a result of symbiosis with ants . . . [and] this symbiosis was a very early development in the evolution' of the Lycaenidae.

From our present understanding of butterfly phylogeny, I am thus prompted to suggest that the development of the Lycaenidae, the largest family of butterflies, may have been intimately connected with their larval association with myrmicine ants, initially as symbionts. This single step may have triggered the diversification of the whole lycaenid group. It would be interesting to know to what extent the food plants favoured by various groups of lycaenids for the initial larval stages have been affected, or even 'chosen', by their availability at ant nesting sites, rather than as the result of more direct plant/insect coevolution.

Speciation and phylogeny

Imagine the primal butterfly. What processes are involved in building a butterfly fauna from this one species? Clearly, speciation, and development of the ability to coexist. If our

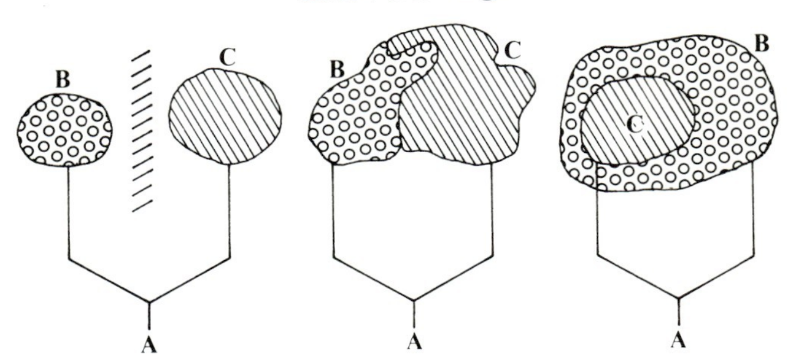

FIG. 4.4. Graphic representation of allopatric, parapatric and sympatric modes of speciation (left to right).

growing fauna is to have any permanence, our species must be able to exist sympatrically more or less indefinitely. Thus we might reasonably expect that specialisation will either occur as a result of competitive displacement, or will be a prerequisite to avoid it.

Theoretically, there are three modes of speciation open to our primal butterfly: allopatric, parapatric and sympatric (Murray, 1972; Bush, 1975). Of these, only sympatric speciation leads directly to faunal enrichment (Fig. 4.4). The dynamics of such a process must relate to disruptive selection, and genic problems of specialisation.

The development of a fauna through allopatric speciation involves subsequent dispersal, coupled with the secondary development of isolating mechanisms. Presumably propagules from allopatric speciation events, when arriving in another fauna, are often out-competed by closely related species already present. Parapatric speciation, like allopatric speciation, is at its inception faunistically trivial. But, unlike allopatric speciation, there is a dynamical evolution of isolating mechanisms from the start, and a suggestion that, despite the specialisation that is proceeding in the two parts of the parental species range, this is not directly related to any ability to coexist. Later coexistence of parapatrically evolved species would presumably relate to disruptive effects at the boundary zone, with mutual ecological displacement. Parapatric speciation is thus rather peculiar: initially the siblings are unable to coexist, but because of the opportunities offered by the maintenance of a dynamic boundary between the component semi-species, it would seem likely that parapatric speciation could rapidly give rise to subsequent faunal enrichment.

At present I do not think we know enough to accept or reject any of these models as more, or less, important. For a long period in the recent past it has been dogmatically asserted that allopatric speciation is the major, if not the only speciation process in animals. But in terms of faunation, allopatric speciation *per se* is trivial speciation: further steps have to be invoked. And there is a paradox. If we invoke the potential for random distribution, so beloved by the centres-of-origin school of biogeographers, why then are the undoubted biogeographic patterns of allopatric deviation still so persistent even on continental land masses? Even if opportunities for allopatric speciation abound, how much of this leads to eco-shifted forms which can coexist? Sympatric speciation might be a much rarer event, but if and when it occurs all the conditions have been met for a direct increase in local species richness.

It seems reasonable to me to suppose that all three types of speciation have, or may have played a part. But however the process gets going, at the moment of faunal enrichment, when two closely related species come to share the environment formerly enjoyed by one, two processes must be operating or starting to operate: genetic isolation and ecological displacement (this assumes that no two butterfly species with exactly the same ecological requirements can coexist, essentially an untested assumption). The problem then becomes one of determining what genetic isolating mechanisms, and what ecological shifts, accompany these steps.

THE STUDY OF SISTER-SPECIES PAIRS AS A TOOL IN EVOLUTIONARY BIOLOGY

Benson *et al* (1976, p. 659) state: 'We believe that searching for patterns among insect/plant relationships at finer taxonomic levels (populations, races, species within a tribe, rather than tribes and subfamilies or families within orders) might help further in the understanding of the relation between evolutionary historic events and ongoing dynamic processes of coevolution'. I agree with this sentiment, but in some ways in a more extreme form. To understand faunal growth, I believe it is imperative to recognise sister-species pairs (Fig. 4.5), and examine their patrial, isolating and ecological characteristics.

A sister-species pair is defined as a pair of species arising from a common ancestor which has given rise to no other living species. The two components may be considered the direct products of a single speciation event, so long as no subsequent extinctions have occurred (Fig. 4.6). Even if extinction has taken place, this will often be because the further speciation events were trivial, not giving rise to species which could coexist. All other types of species pairs necessarily involve at least two speciation events (Fig. 4.7), and are not, I believe, likely to give as much insight as the study of sister-species pairs (except perhaps

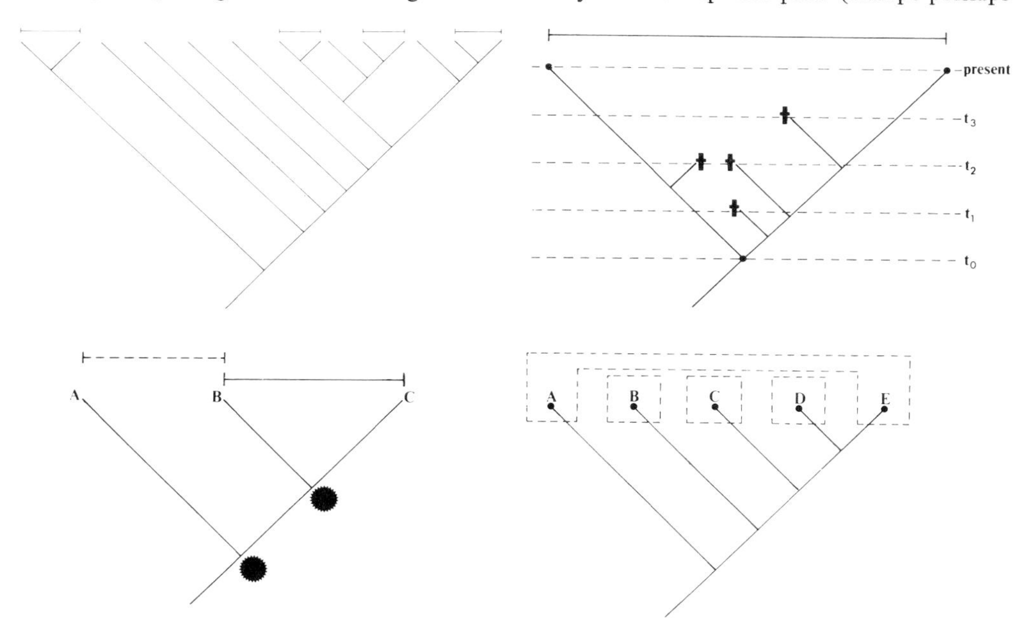

FIG. 4.5. (*top left*) Topology of hypothetical phylogeny showing four sister-species pairs (linked by bars).

FIG. 4.6. (*top right*) Hypothetical phylogeny of an extant sister-species pair, showing former speciations and four extinctions (at t_1, t_2 and t_3), and the original dichotomy of the apir at t_0. The possibility of such extinctions must always be borne in mind when making use of sister-species pairs for investigating speciation processes.

FIG. 4.7. (*bottom left*) Despite the possibilities of unknown extinctions (Fig. 4.6), it is better to use a sister-species pair (B/C) for investigating speciation processes, rather than any other type of pair (e.g. A/B), because other pairs necessarily involve at least two speciation events.

FIG. 4.8. (*bottom right*) Representation of hypothetical phylogeny where A, B, C and D are all allopatrically or parapatrically distributed, but A and E are sympatric. Pairs such as A/E may be of similar value to sister-species pairs in investigating faunal enrichment processes.

where only two members of a monophyletic assemblage are sympatric – Fig. 4.8). In particular, where sister-species are found to coexist sympatrically, we are likely to be closest to seeing the minimal conditions for *significant* speciation – speciation capable of enriching a fauna, and the isolating mechanisms and ecological shifts involved.

The ideal would be to know, without doubt, all extant sister-species pairs throughout the butterflies. A reasonable hope might be to recognise, with some confidence and tested by independent means, all sister-species pairs throughout a number of contrasting subgroups – say one tribe from each of the five families. At present I have not even approached the latter for one subgroup. Here I will have to be content to discuss some of those few sister-species pairs that have come to my attention during 10 years' work with butterfly phylogenetics. This is not a random sample, but I do not think it is so distorted as to invalidate all the ideas to be drawn from it.

Mycalesis drusillodes/M. bazochii/M. durga

The dual mimic species *Mycalesis drusillodes* (plate 1, Figs. 1, 2) is probably the sister-species of either *M. durga* (see Vane-Wright, 1971) or *M. bazochii* (overlooked at the time of the 1971 paper), or is the sister-group of that pair of species (plate 1, Figs. 3, 4). All three are broadly sympatric in West Irian, *M. bazochii* and *M. drusillodes* having been collected at the same locality on a number of occasions. *M. bazochii* and *M. durga* are similar, weakly dimorphic, typically cryptic-looking satyrids. Their underside wing patterns include conspicuous marginal ocelli, which I presume to function defensively as deflection marks (Blest, 1957; Vane-Wright, 1976). *M. drusillodes* is mimetic, the females belonging to the *Taenaris* ring, the males to the *Tellervo* ring (Vane-Wright, 1974). All three species have a similar and complex array of presumed male scent organs. Nothing is known of their ecology. It is likely that all three are Gramineae feeders. *M. drusillodes* has dramatically shifted its pattern defence mechanism from crypsis/deflection to dual mimicry, and it is probable that the adults have markedly different behaviour to the other two species. There is some morphological evidence of a change in the presumed olfactory isolating mechanism of *drusillodes* (Vane-Wright, 1971, p. 108).

Papilio dardanus/Papilio phorcas

I believe there is very good evidence that these two species form a sister-species pair (Vane-Wright, in preparation). *P. dardanus* (plate 1, Figs. 5–7) is the celebrated polymorphic, sex-limited mimic species of Africa, with monomorphic, non-mimetic populations in Somalia, the Comoros and Madagascar (Sheppard, 1961). *P. phorcas* is also a sex-limited polymorph, but it is not mimetic (plate 1, Figs. 8–10). It is restricted to the African continent, not extending into Ethiopia. Generally it is sympatric with *dardanus*, and they may be seen flying together in many places. The known foodplants of *P. phorcas* are three species of *Teclea* (Rutaceae), all of which are also utilised by *P. dardanus*. The latter is also recorded from a wider range of Rutaceae, including *Citrus*, and some non-Rutaceous plants.

Contrary to general belief, the pattern of male *dardanus* must be regarded as highly specialised (Vane-Wright, in preparation), as originally pointed out by Bernardi (1963). The male colour patterns of both *dardanus* (plate 1, Fig. 5) and *phorcas* (plate 1, Fig. 8) are extremely divergent, both from related Papilios and from each other. It is relevant that the sister group of *dardanus/phorcas* is a single species, *Papilio constantinus*, which is very similar in pattern to the female-limited form of *phorcas* (plate 1, Fig. 10). The males of *constantinus* have a conspicuous scent organ, absent from *dardanus* and *phorcas.*

3
2
1
6
5
11
12
4
13
10
8
14
9
7

P. constantinus is sympatric with *dardanus* and *phorcas* over a part of its purely East African range, but it generally frequents drier areas.

The *phorcas/dardanus* pair suggest a speciation event involving a shift of defence pattern from the cryptic/disruptive type of *constantinus*, to a mimetic type in *dardanus*. This has apparently necessitated, for isolation purposes, a striking divergence in the male colour pattern of both species, giving rise in turn to the phenomenon of partial female-limited polymorphism in both (Vane-Wright, 1975; and in press *a*). The wider distribution of *dardanus*, which envelops the range of *phorcas* and extends into zones where the mimicry is not of selective advantage, is probably related to its greater range of host plants, including species cultivated by man. The two species occasionally hybridise in nature, and hybrids have been produced in the laboratory by hand pairing (Clarke & Sheppard, 1975).

Nessaea ancaea/N. hewitsoni

The South American nymphalid genus *Nessaea* comprises five species, the brightly coloured orange and blue males of which are all separable by a glance at their upperside wing patterns. The female upperside patterns of all five species are very similar (plate 1, Fig. 11), as are all individuals on the underside regardless of species or sex. Only two pairings of species involve sympatry, *ancaea* and *batesii* on the lower Amazon/Guyana shield, and *ancaea* and *hewitsoni* on the upper Amazon. The latter are apparently a sister-species pair (Vane-Wright, in press *b*). The signal pattern difference between the males of *ancaea* and *hewitsoni* is almost startling (plate 1, Figs. 12, 13) yet simply achieved through the complete suppression of orange coupled with the addition of extra blue in *hewitsoni*.

Sadly, nothing is known of the biology. Only one rather peculiar observation can be made. The range of *ancaea* coincides broadly with that of *hewitsoni*, but it also extends much further to the east. Where it is allopatric to *hewitsoni*, *ancaea* is a much smaller species (plate 1, Fig. 14), but where the two species are sympatric they are identical in size (Vane-Wright, in press *b*). This is the reverse of character displacement; I interpret this as consistent with a sympatric origin for the two species, the spectacular difference in male patterns being part of the isolating mechanism necessary to achieve this. Can we speculate that the two species arose as a result of adaptation to different host plants?

Hypolimnas misippus/H. bolina

H. misippus is a nymphalid distributed throughout the old world tropics and extending marginally into the neotropics. The females are variable or polymorphic mimics of *Danaus chrysippus*, and are quite different to the monomorphic male (plate 2, Figs. 1–3). It seems probable that *misippus* forms a sister-species pair with *H. bolina*, a species widely sympatric with *misippus* in the orient, but only extending westwards as far as Madagascar. Whereas *misippus* is monotypic throughout its huge range, *bolina* is highly polytypic, although the replacement patterns are disturbed by frequent migrations. The females are highly variable or polymorphic, and mostly either *Euploea*-like in appearance, or male-like (Plate 2, Figs. 4–6). The male patterns of the two species are quite similar, and divergent from other *Hypolimnas*. The host-plant relationships are fairly well documented for both species (Vane-Wright *et al*, 1977). *H. bolina* is known from eleven flowering plant families,

PLATE 2. 1–3, *Hypolimnas misippus* (1, ♂; 2–3, mimetic ♀). *Hypolimnas bolina* (4, ♂; 5, ♂-like ♀; 6, mimetic ♀). 7, *Amauris tartarea*, ♂. 8, *A. tartarea damoclides*, ♂. 9, *Amauris niavius*, ♂. *Amauris echeria*, ♂. 11–12, *Danaus chrysippus*, two morphs. 13–14, *Danaus gilippus*, two subspecies.

and, like most other *Hypolimnas* and related nymphalids, it is most prevalent on Urticaceae (the most likely 'coevolved' group). *H. misippus*, known from seven plant families, is conspicuously absent from the Urticaceae.

If *bolina* and *misippus* are sister-species, it seems possible they coexist through non-competition for Urticaceae, *misippus* apparently forgoing this ancestral speciality. As both are also remarkable colonisers, sympatric over a large area, and plausibly the most 'r-selected' species in the genus (Smith, 1976; Vane-Wright *et al*, 1977), this represents a most interesting situation. There is no obvious clue as to the nature of the genetic isolating mechanism between the two (no hybridisation has yet been attempted): it is possible that this may be purely 'behavioural', as the courtship of these species is complex. The female pattern adaptations of the two species are different, *bolina* females generally either being male *Euploea*-like, whereas *misippus* females are always more or less *Danaus*-like.

Amauris niavius/A. tartarea

The genus *Amauris* consists of about 16 species confined to the Afrotropical region. In common with most danaids, *Amauris* are aposematic, and form a major group of models for mimetic African butterflies. Two common species, *A. niavius* and *A. tartarea*, almost certainly form a sister-species pair (Ackery and Vane-Wright, in prep.). Both are widely distributed through central Africa, with *niavius* extending further into Ethiopia and South Africa. The wing pattern of *niavius* differs from other *Amauris* in the extension of the white fore wing discal area through to the hind margin. *A. tartarea* (Plate 2, Fig. 7) differs from other *Amauris* in the great variability of its pattern, and was regarded by Talbot (1940) as polymorphic. Up to four rather weakly separated morphs may coexist. *A. niavius* is divided into three subspecies, and *A. tartarea* into two. Of the latter, some *A. tartarea damoclides* from Tanzania (Plate 2, Fig. 8) show a fore wing pattern transitional to the *niavius* type, which might be an indication of introgression. With haploid chromosome numbers in the range 33–40, both species may have a history of chromosome fragmentation, in contrast to the more normal fusion pattern displayed by some other *Amauris* species, which have $n = 29$ or 30 (see Bernardi & de Lesse, 1964; de Lesse, 1966; and Robinson, 1971).

A. niavius is recorded as feeding on various asclepiads, normal food plants for danaids. The food plants of *tartarea* are unknown, except for the very surprising record recently published of *tartarea* feeding on Cruciferae (Ouattara *et al*, 1977). Sex pheromones are known to be of great significance in danaids, and it is notable that a particular pheromone component has so far only been found in *tartarea* and *niavius*, whereas another component widespread in danaids also occurs in *niavius* but not in *tartarea* (Meinwald *et al*, 1974; Schneider, 1975).

These two species are thus broadly sympatric, both show pattern deviations despite belonging essentially to a Müllerian assemblage, and they show apparent differences in their sex pheromone chemistry. They may also have a history of chromosome fragmentation and rearrangement. If it is true that *tartarea* has abandoned the Asclepiadaceae to feed on Cruciferae, this might be the root cause of the speciation, but this clearly requires confirmation.

Danaus chrysippus/D. gilippus

These two well-known species cover practically the entire world range of open habitats suitable for danaids, but they are completely allopatric. *D. chrysippus* occurs from Africa, southern Europe and Russia, eastwards across the whole of South-east Asia to the eastern Pacific. *D. gilippus* occurs from the southern USA south through Central America to Uruguay. There is little doubt that the two form a sister-species pair (Ackery and Vane-Wright, in prep.). Both are recorded from a large number of host plants, with rather little overlap in species – this presumably reflects differences in host distribution rather than specialisation *per se*. *D. chrysippus* shows pattern polymorphism in parts of Africa and the orient (Plate 2, Figs. 11, 12), although in many areas it is monomorphic. *D. gilippus*, a generally darker looking species, is perhaps more polytypic than *chrysippus* (Plate 2, Figs. 13, 14). There seems to be no chemical difference between the sex pheromones of the two species, nor is there any significant difference in the known ultrastructure of their scent organs (M. Boppré, *personal communication*). Seibt *et al* (1972) note that both species have identical courtship behaviour.

This pair of species appears to constitute a classic case of allopatric speciation – or does it? It is hoped that hybridisation experiments will be undertaken to study this. Controlled competition studies, if feasible, would also be of great interest. If released together, would the two 'species' merge, or would they be able to divide up the ecospace and avoid hybridisation, and so make a potentially significant contribution to faunal diversity?

A conflict theory of butterfly speciation

Gilbert and Singer (1975) recognise six broad components allowing ecological segregation amongst butterflies: (1) specialisation on host-plant species, (2) partitioning of host-plant parts, (3) phenology and voltinism, (4) habitat partitioning, (5) adult resource partitioning, and (6) alternatives in parasite and predator escape. In (6) they also include mutualism with ants, which I would make a separate category. Gilbert and Singer (1975, p. 376) also state that 'The evolution of multispecies communities has resulted from the fact that no one genotype has high enough fitness over all the different ecological circumstances to prevent more specialised genotypes (and species) from evolving in or invading the community.'

The genotypes of species are not, however, monomorphic, they are highly polymorphic. Genetic polymorphism can go a long way towards giving a species ecological plasticity. Gilbert and Singer's statement becomes true, I think, when we take into account behaviour. The genetic and epigenetic difficulties of an efficient species, the various members of which are, say, adapted to different larval host plants, different adult food sources, or different predator escape devices, become apparent when it is remembered that each alternative requires a different and appropriate pattern of behaviour. To use different host plants may well require a different search and oviposition behaviour for each. Can you imagine the behavioural difficulties involved if adult *Heliconius* were to endeavour to become polymorphic for both pollen and fallen fruit feeding? If a species is polymorphic for three different mimetic morphs, the models for which fly in three different forest strata (Papageorgis, 1974), then ideally each morph must have the appropriate behaviour. And even when this is possible, how do such shifts affect mate locating behaviour?

This last point is particularly important, because all the situations we are discussing involve some form of disruptive selection, coupled with number of frequency-dependent effects due to competition, giving the situation so elegantly discussed by Murray (1972): the problem of the 'unwanted' heterozygote and its solution through either the evolution of dominance, or speciation. If behavioural difficulties place a limit on the number of polymorphisms that a species can cope with, then this leaves speciation as the only solution.

A very important question is whether or not a shift in just one niche component is sufficient to allow two species to coexist. Many ecologists would probably accept a shift in host plant species to be sufficient, but might be doubtful about a shift in say adult mimicry association. In the past it has often been said that mimicry, although of importance to the survival of the individual, is not important for the species, because the main density regulating factors affect the early stages (e.g. Sheppard, 1961). However, Levins' (1975) ideas derived from loop analysis indicate, if I understand him correctly, that predator avoidance strategies may be more important for the continued existence of a herbivore species than increased efficiency in food utilisation.

But in relation to speciation, all these ideas are too static.

I imagine successful species will continually add, given the opportunity, new ways of exploiting their environment, resulting in a tendency to ever increasing polymorphism. Evolution favours generalists. However, for most species there must come a point where they can no longer hold all this together. The conflicting demands, especially on the production of the appropriate behaviour, become too great (man, with his fantastic capacity for non-instinctive behaviour, coupled with the institution of education, avoids this problem—Huxley, 1941).

If this is so the question then becomes, is the addition of the ability to utilise just one more niche subcomponent sufficient to *precipitate* speciation? I imagine a sort of evolutionary straw, which breaks the camel's back, forcing a species to divide into two. The two halves, differing in their representation of the total gene pool, will both tend to be more limited than the parental species; hence the relative phenomenon of specialisation. If one or both halves can increase their range of abilities anew, there will come a point of further division, and so on.

I believe this is well shown by the case of the three *Papilio* species, *dardanus, phorcas* and *constantinus. P. constantinus* evidently uses scent as an isolating signal, but *dardanus* and *phorcas* are apparently dependent on specific male colour patterns. The original *constantinus* pattern, still used by many *phorcas* females, is essentially a cryptic or disruptive defence pattern, requiring appropriate behaviour. *P. dardanus* females are fundamentally mimetic, and presumably require a different behaviour. I suggest that the *P. dardanus/P. phorcas* speciation event was precipitated through the addition of mimicry to the range of patterns of the stem species, causing a conflict of behavioural adjustment, solved by speciation.

The sister-species pairs discussed above suggest that a defensive pattern function shift may often be involved in butterfly speciation. However, I think it likely that the attempt by a species to enter any additional niche subcomponent may be sufficient to precipitate a speciation event through the behavioural conflict engendered (Huxley, 1941).

AN EVOLUTIONARY SCENARIO FOR THE ORIGIN OF BUTTERFLY DIVERSITY

One consequence of the idea of conflict speciation is that, as the total number of species builds up in a fauna, it offers an ever shifting, ever increasing array of niches for potential exploitation. Mimicry is an especially good example of this in butterflies. Despite the convergent effects of Müllerianism, there are more, not fewer mimicry rings within species-rich communities. Thus I would suggest that species diversity begets species diversity. This in turn raises the tempo of evolutionary events, including the coevolution involved in the exploitation of relatively immune hosts, for example. (Presumably an eventual, even greater rise in the extinction rate keeps the absolute level of species diversity within some upper bound: Rosenzweig, 1975).

With respect, for example, to the particular niche of mimicry, we may expect to see the rise of a continual succession of new models as a result of evolution onto new, chemically protected, hosts. In the train of this we may expect the speciation of a number of relatively unprotected species, caught up by chance with the new aposematic associations. As predators and parasites also coevolve to deal with the new aposemes (today's poison is tomorrow's meat), so their star wanes, and their attendant satellites, the mimics, wane with them. So we can picture a continuous succession, involving the rise and fall of groups specialised in various ways, each in its turn bringing about a certain amount of subsidiary speciation. At any one moment we will see our central, highly diverse faunas, with their amazing range of common and rare butterfly species, their generalists and their specialists. In this way I visualize coevolution, ecological separation, opportunity, isolation and behaviour all related in producing the faunal diversity of butterflies, which itself will be an ever shifting pattern dependent partly on its own complexity.

Acknowledgements

Many colleagues and other friends have helped in a variety of ways with the preparation of this paper, including P. R. Ackery, S. Collins, C. B. Cottrell, B. D'Abrera, Vera Dick, John Eliot, John Heath, C. J. Humphries, Theya Molleson, L. A. Mound, M. C. Singer, R. L. Smiles, and A. E. Stubbs. F. Greenaway produced the colour photographs. To all of them I am very grateful.

References

Ackery P. R. & Vane-Wright R. I. The biology and phylogeny of danaid butterflies. B.M.(N.H.). London. In prep.

Barcant M. (1970) *Butterflies of Trinidad and Tobago.* Collins, London.

Benson W. W., Brown, K. S. & Gilbert, L. E. (1976) Coevolution of plants and herbivores: passion flower butterflies. *Evolution, Lancaster, Pa.* 29, 659–680.

Bernardi G. (1963) Quelques aspects zoogéographiques du mimétisme chez les Lépidoptères. *Proc. 16th Int. Congr. Zool.* 4, 161–166.

Bernardi G. & de Lesse H. (1964) Formules chromosomiques de quelques lépidoptères du Gabon. *Biologia gabon.* 1, 65–71.

Blest A. D. (1975) The function of eyespot patterns in the Lepidoptera. *Behaviour* **11**, 209–256.

Bush G. L. (1975) Modes of animal speciation. *A. Rev. Ecol. Syst.* **6**, 339–364.

Clarke C. A. & Sheppard P. M. (1975) [Exhibit report] *Proc. R. ent Soc. Lond.* (C) **39**, 39–40.

Cottrell C. B. (1965) A study of the *methymna*-group of the genus *Lepidochrysops* Hedicke (Lepidoptera: Lycaenidae). *Mem. ent. Soc. sth. Afr.* (9), 110 pp., 3 pls.

Dennis R. L. H. (1977) *The British butterflies, their origin and establishment* Classey, Faringdon, England.

de Lesse, H. (1966) Formules chromosomiques de quelques lépidoptéres rhopalocéres d'Afrique centrale. *Annls Soc. ent Fr.* (N.S.) **2**, 349–353.

Ehrlich P. R. (1958) The comparative morphology, phylogeny and higher classification of the butterflies (Lepidoptera: Papilionoidea). *Kans. Univ. Sci. Bull.* **39**, 305–370.

Ehrlich P. R. & Raven P. H. (1965) Butterflies and plants: a study in coevolution. *Evolution, Lancaster, Pa.* **18**, 586–608.

Eliot J. N. (1973) The higher classifcation of the Lycaenidae (Lepidoptera): a tentative arrangement. *Bull. Br. Mus. nat. Hist.* (Ent.) **28**, 373–506, 6 pls.

Gilbert L. E. (1976) Adult resources in butterflies: African lycaenid *Megalopalpus* feeds on larval nectary. *Biotropica* 8, 282–283.

Gilbert L. E. & Singer M. C. (1975) Butterfly ecology. *A. Rev. Ecol. Syst.* **6**, 365–397.

Hennig W. (1966) *Phylogenetic Systematics* University of Illinois Press, Urbana.

Hudson G. V. (1928) *The butterflies and moths of New Zealand.* Fergusson & Osborn, Wellington.

Huxley J. [S.] (1941) *The uniqueness of man.* Chatto & Windus, London.

Kristensen N. P. (1976) Remarks on the family–level phylogeny of butterflies (Insecta, Lepidoptera, Rhopalocera). *Z. Zool. Syst. EvolForsch.* **14**, 25–33.

Levins R. (1975) Evolution in communities near equilibrium, In *Ecology and evolution of communities*, Eds. M. L. Cody, and J. M. Diamond, pp. 16–50. Belknap Press, Cambridge (Mass.) & London.

Meinwald J., Boriack C. J., Schneider D., Boppré M., Wood W. F. & Eisner T. (1974) Volatile ketones in the hairpencil secretion of danaid butterflies (*Amauris* and *Danaus*). *Experimentia* **30**, 721–722.

Miller D. (1971) *Common insects in New Zealand.* A. H. & A. W. Reed, Wellington, Auckland, Sydney & Melbourne.

Murray J. J. (1972) *Genetic diversity and natural selection.* Oliver & Boyd, Edinburgh.

Ouattara S., Jolivet P. & Parys E. Van. (1977) *Seconde liste des insectes et des plantes-hotes en Haute–Volta et dans les regions limitrophes. Projet der renforcement de la protection de plantes en Haute–Volta.* Bobo–Dioulasso.

Papageorgis C. (1974) *The adaptive significance of wing coloration of mimetic tropical butterflies.* Ph D thesis, Princeton University, Princeton, N.J, U.S.A. 117 pp. [not seen].

Riley N. D. (1975) *A field guide to the butterflies of the West Indies*, 224 pp., 24 pls, end papers. Collins, London.

Robinson R. (1971) *Lepidoptera genetics*, ix + 687 pp. Pergamon, Oxford.

Rosenzweig M. L. (1975) On continental steady states of species diversity. *In* Cody, M. L. & Diamond, J. M. *Ecology and evolution of communities*, pp. 121–140. Belknap Press, Cambridge (Mass.) & London.

Schneider D. (1975) Pheromone communication in moths and butterflies. In *Sensory physiology and behaviour*, Ed. R. Galun, P. Hillman, I. Parnas, & R. Werman, pp. 173–193. Plenum, New York.

Seibt U., Schneider D. & Eisner T. (1972) Duftpinsel, Flügeltaschen und Balz des Tagfalters *Danaus chrysippus* (Lepidoptera: Danaidae). *Z. Tierpsychol.* **31**, 513–530.

Sheppard P. M. (1961) Recent genetical work on polymorphic mimetic Papilios. *Symp. R. ent. Soc. Lond.* **1**, 20–29.

Smith D. A. S. (1976) Phenotypic diversity, mimicry and natural selection in the African butterfly *Hypolimnas misippus* L. (Lepidoptera: Nymphalidae). *Biol. J. Linn. Soc.* 8, 183–204.

Talbot G. (1940) Revisional notes of the genus *Amauris* Hübner (Lep.). *Trans R. ent. Soc. Lond.* **90**, 319–336.

Vane–Wright R. I. (1971) The systematics of *Drusillopsis* Oberthür (Satyrinae) and the supposed Amathusiid *Bigaena* van Eecke (Lepidoptera: Nymphalidae), with some observations on Batesian mimicry. *Trans. R. ent. Soc. Lond.* **123**, 97–123.

Vane–Wright R. I. (1974) Further observations on the occurence and mimicry of *Mycalesis drusillodes* (Lepidoptera: Nymphalidae, Satyrinae). *J. Ent. (B)* **42**, 213–216.

Vane–Wright R. I. (1975) An integrated classification for polymorphism and sexual dimorphism in butterflies. *J. Zool., Lond.* **177**, 329–337.

Vane–Wright R. I. (1976) A unified classification of mimetic resemblances. *Biol. J. Linn, Soc.* 8, 25–56.

Vane–Wright R. I. Towards a theory of the evolution of butterfly colour patterns under directional and disruptive selection. *Biol. J. Linn. Soc.* (In Press-*a*).

Vane–Wright R. I. The identification coloration and phylogeny of *Nessaea* butterflies (Lepidoptera: Nymphalidae). *Bull. Br. Mus. nat. Hist. (Ent.).* (In Press-*b*).

Vane–Wright R. I. Polymorphism, speciation, and the origin of the mimetic butterfly *Papilio dardanus* Brown. *Bull. Br. Mus. nat. Hist. (Ent.).* (In preparation).

Vane–Wright R. I., Ackery P. R. & Smiles R. L. (1977) The polymorphism, mimicry and host-plant relationships of *Hypolimnas* butterflies. *Biol. J. Linn. Soc.* **9**, 285–297.

Willis J. C. (1973) *A dictionary of the flowering plants and ferns*, 8th edn, Cambridge University Press.

Woodhouse L. G. O. (1950) *The butterfly fauna of Ceylon*, 2nd edn (abridged), Apothecaries Company, Colombo.

5 • Diversity of the Sternorrhyncha within major climatic zones

V. F. EASTOP

Department of Entomology, British Museum (Natural History), London SW7 5BD

The Sternorrhyncha consists of the Aphidoidea, Coccoidea, Psylloidea and Aleyrodoidea, that is aphids, greenfly, mealy bugs, scale insects, lac insects, jumping plant lice and whiteflies. All these rather small insects feed on plant sap. Sternorrhyncha are widely, and the group as a whole rather evenly distributed throughout the world, half the world's species (6457 out of 12 782) coming from the north temperate half of the world's land surface. The purpose of this paper is to summarise the distribution of the described species of Sternorrhyncha and to examine the factors which may have led to the differing diversities of the various groups in different localities.

DIVERSITY

Diversity is concerned with the amount of variation in a sample. The variates assessed and the method of taking the sample are all-important. The variates can be any combination of appearance, behaviour, physiology or cytology, all of which are themselves inter-related. One of the variates is usually taxonomic and the most objective taxonomic groups are species and subspecies, as these are real groups of individuals in population genetics terms. Genera, tribes, families and orders are all subjective groupings of species.

STRUCTURAL DIVERSITY

No two individuals are identical and taxonomy is the interpretation of the observed differences between specimens. Morphological diversity is easier to see than it is to score meaningfully. The number of antennal segments can be counted but their shape is assigned to arbitrary classes based on length/width ratios and assessments of curvature. One problem is that the eye only sees what it has been trained to see. We survive in this world by disregarding much of the information gathered by our senses. Observed diversity is based to a considerable extent on what we have already decided is important. Mordvilko (1919; pp. 363–450) described the rose aphid, (*Macrosiphum rosae*) in great detail but did not mention the character later used to define the Macrosiphina (Börner, 1952, pp. 157, 163). It is difficult to interpret the morphological diversity within and between different groups of insects. For instance, considering only adults, psyllids usually have 10 (rarely eight) antennal segments and aleyrodids seven, although in a few species some of the antennal segments are fused. Most aphids have six antennal segments, some have five and a few only three or four. Male coccids have up to 10 antennal segments while in females the antennae vary from 11-segmented in Margarodidae to only unpigmented tubercles in Diaspididae. As neither apterous adult psyllids nor winged adult female coccids have been discovered (recognised) it is particularly difficult to evaluate these ranges of variation. Should the amount of morphological variation observed be corrected for the number of species studied in the group?

The selection of characters is difficult, as individual structures may vary more in one group than in another and the presence of such structures as wings and sensoria is correlated. Environmentally induced diversity (variation) and balanced polymorphisms also provide practical problems when comparing the diversity of different groups. For instance the colour of aphids may be affected both by nutrition and by the species of attendant ant. Much has been written about variation and polymorphism in aphid species, a lesser amount about aleyrodids and coccids and little concerning psyllids which seem to have little variation within species apart from sexual dimorphism and some seasonal differences in pigmentation. No practical, objective way of comparing the structural diversity within the major groups of Sternorryncha has been devised, but adult psyllids and aleyrodids are probably less diverse than adult coccids and aphids, while the reverse may be true of the immature stages.

BEHAVIOURAL DIVERSITY

Behavioural differences are as easy to observe but at least as difficult to quantify as morphological differences. Differences in behaviour within and between cryptic species are probably more important as isolating mechanisms than appearance, and 'appearance' is affected by behaviour. Host-plant recognition and selection is behaviour of particular importance to the Sternorrhyncha, all of which are phytophagous. Hosts may be selected by both appearance (colour, structure) and by taste (chemical composition and physiological state), both of which are likely to vary with age. Different parts of a plant may be selected by different species, or by the same species for different purposes. Polyphagous species tend to prefer one part of their many hosts, e.g. flower stalk or young leaves, while a species specific to an annual herb may live on it from the seedling stage until after it sets seed. Although the populations of many species of aphids increase rapidly on young shoots and leaves during the temperate region spring, other species live under lower leaves and may be most abundant later in the year. Some aleyrodids also occur on mature leaves and are not directly associated with the growth flushes of their host. Psyllids and most aphids are highly host specific, while more species of coccids and aleyrodids have a wider range of host plants. The Myrtaceae and Solanaceae bear many host specific psyllids, but few aphids are specific to members of these families, although polyphagous aphids often do well on them. Selections from the numerous behavioural differences are strongly biassed by the observers' interests.

GENETICAL DIVERSITY

Morphology, physiology and behaviour are all the end products of gene expression, and thus the genetical diversity within a group could be regarded as its 'true' diversity. Some genetical differences influence short-term events and operate in most individuals of a species, while others may not be used for many generations but be of vital importance to a species eventually, e.g. ability to detoxify a pesticide. The six pericentric inversions and one dissociation that distinguish the chromosomes of man from those of chimpanzee (King & Wilson, 1975) seem smaller differences than those between the morphologically inseparable species of the *Anopheles gambiae* complex (Kitzmiller, 1977). Nixon & Taylor (1977) record a similar situation in planarian worms. The 'percentage of polymorphic loci', 'amounts of heterozygosity' or 'gene and genotype frequencies' estimated from enzyme polymorphisms indicate the amount of cytological diversity, but the information is not available from enough species for useful comparisons of diversity. Widespread species are likely to be more variable than species with a more restricted range. Can a given amount of genetical diversity be packaged as either a few variable species or many less variable species? Some diversity is presumably determined by the advantage of having all possible recombinations available each year (one species) or all the 'best' combinations fixed (many species). The former seems more suited to variable environments and the latter to more constant environments. Species which have alternative overwintering methods maintained by a balanced genetic polymorphism (Blackman, 1976*a,b*) present particular problems for the measurement of diversity.

Speciation is the outcome of the variability needed by an organism to survive in a changing world. Different ecological conditions require different numbers of species to give genes the best chance of survival. The number of species in a group is often regarded as a measure of the 'success' of that group, especially by specialists in large groups. Grouchy (1973), however, has suggested that cancer and evolution are similar processes, cancer at the somatic level and evolution at the genetic level, and that speciation is the price which has to be paid for adaptability. The 'price' of complete interfertility is the failure to utilise many available resources. The Endopterygota produce specialised immature forms which may take advantage of resources different from those utilised by their adults; they seem as a result to have been forced into faster speciation than the Exopterygota which in some groups at least have specialised in adaptability, leading to less speciation. Speciation is a snowballing hydra, each new species complicates the environment of its fellows, and thus makes it more suitable for speciation. This exponential increase in species is likely to continue until an abrupt change in the environment causes widespread extinction. Such extinctions not only reduce the number of species but leave a skew distribution, with many insects in the tropics as we have now. The effect of the recent glacials is being accentuated by agricultural man in the present interglacial. The amount of extra diversity induced in a few hundred domesticated species and a few thousand pests is unlikely to compensate for the genetical diversity lost in extinguished species.

SPECIES DIVERSITY

Diversity in its usual ecological sense is the relationship between the number of individuals and the number of taxa in a sample. Table 5.1 gives data and indices of diversity from traps in different parts of the world. It will be seen that diversity increases as climate ameliorates, with the exception of Grove, Tasmania, which lacks the aphid fauna of ornamental plants characteristic of suburban areas. Trapping techniques select species differentially; yellow traps select dicotyledon-feeding species for instance. An outbreak of a dicotyledon-feeder will thus reduce the index of diversity calculated from such trap catches more than a similar outbreak of a grass-feeder. Many aphids are polymorphic, producing winged adults in response to overcrowding. The trap catches thus bear some relationship to past biomass of these species, but a comparison with the catches of species in which all summer adults are

Table 5.1. Diversity of aphid species in trap catches

Locality	Type of trap	Trapping period	Ref.	Number of species	Number of specimens	Index of diversity and standard error
Gordonvale, Qld.	Yellow tray	1960	a	16	448	3.2 ± 0.91
Gordonvale, Qld.	Yellow tray	1961	a	17	1196	2.8 ± 0.74
Brandon, Qld.	Yellow tray	1960	a	16	474	3.2 ± 0.89
Brandon, Qld.	Yellow tray	1961	a	17	579	3.3 ± 0.89
Mackay Qld.	Yellow tray	v–xi, 1960	a	18	2177	2.7 ± 0.69
Mackay Qld.	Yellow tray	1961	a	13	1257	2.0 ± 0.61
Bundaberg Qld.	Yellow tray	v–xi, 1960	a	20	308	4.8 ± 1.22
Bundaberg Qld.	Yellow tray	1961	a	20	675	3.9 ± 0.96
Rydalmere, N.S.W.	Yellow tray	x–xi, 1960	a	23	356	5.5 ± 1.31
Rydalmere, N.S.W.	Yellow tray	1961	a	45	3721	7.2 ± 1.17
Canberra, A.C.T.	Yellow tray	1960	a	31	1716	5.4 ± 1.06
Canberra, A.C.T.	Yellow tray	1961	a	32	2220	5.3 ± 1.02
Merbein, Victoria	Yellow tray	viii–x, 1960	a	14	813	2.4 ± 0.71
Merbein, Victoria	Yellow tray	1961	a	12	404	2.3 ± 0.75
Burnley, Victoria	Yellow tray	1960	a	45	7463	6.4 ± 1.02
Burnley, Victoria	Yellow tray	1961	a	46	6646	6.7 ± 1.06
Victoria	Yellow tray	ii, 1960–ii, 1961	b	48	10 943	6.4 ± 1.00
Victoria	Sticky trap	ii, 1960–ii, 1961	b	36	1535	6.6 ± 1.22
New Town, Tasmania	Yellow tray	vii–xi, 1960	a	38	4486	5.7 ± 1.00
New Town, Tasmania	Yellow tray	1961	a	49	6907	7.1 ± 1.10
Grove, Tasmania	Yellow tray	x–xi, 1960	a	16	1191	2.6 ± 0.71
Grove, Tasmania	Yellow tray	1961	a	29	9734	3.7 ± 0.73
Grove, Tasmania	Yellow tray	1962	a	31	12 037	3.8 ± 0.74
Adelaide, Waite Inst. S.A.	Yellow tray	1961	a	23	487	5.0 ± 1.18
Adelaide, Hills, S.A.	Yellow tray	1961	a	22	518	4.7 ± 1.12
Australia, all sites	Yellow tray	1961–1962	a	78	69 248	8.6 ± 1.04
Davao, P.I.	Yellow tray	1962–1963	c	28	4605	4.0 ± 0.81
Tawau, Borneo	Yellow tray	ix, 1961–iii, 1962	d	14	151	3.8 ± 1.17
Chitedze, Malawi	Suction	1964–1967	e	32	5770	4.5 ± 0.85
Chitedze, Malawi	Yellow tray	1964–1967	e	31	6359	4.2 ± 0.82
Muguga, Kenya	Suction	1952–1955	f	42	2713	7.1 ± 1.19
Muguga, Kenya	Yellow tray	1952–1955	f	44	6879	6.3 ± 1.02
Nachingwea, Tanzania	Yellow tray	xi, 1953–iii, 1954	f	20	1361	3.3 ± 0.81
Ibadan, Nigeria	Yellow tray	1956–1957	g	21	5475	2.8 ± 0.65
Samaru, nr. Zaria, Nigeria	Yellow tray	viii.–xii. 1956	g	16	493	3.2 ± 0.88
Badeggi, nr. Bida, Nigeria	Yellow tray	i, 1957–iv, 1958	h	16	430	3.3 ± 0.92
Jos, Nigeria	Yellow tray	x, xi, 1956	h	12	49	5.1 ± 1.86
Maiduguri, Nigeria	Yellow tray	xi.–xii. 1956	h	5	242	0.9 ± 0.44
Yola, Nigeria	Yellow tray	vii.–viii. 1956	h	7	44	2.3 ± 1.07
Ibadan, Nigeria	Yellow tray	1956–1962	h	30	24 627	3.4 ± 0.65
Umudike, Nigeria	Yellow tray	vii.–viii. 1956	j	16	698	2.9 ± 0.81
Akure, Nigeria	Yellow tray	1956–1957	h	10	69	3.2 ± 1.22
Benin, Nigeria	Yellow tray	iv.–vi. 1958	h	7	97	1.7 ± 0.75
Bamenda, Cameroon	Yellow tray	i.–ii. 1957	h	11	51	4.3 ± 1.63

References: a, Hughes *et al*, 1964; b, O'Loughlin, 1963; c, M. R. Gavarra, unpublished; d, Cole *et al*, 1963; e, Adams & Farrell, 1967; f, Eastop, 1957; g, Eastop, 1958; h, unpublished; j, J. L. Gregory, unpublished.

winged may be misleading. A practical problem with trap catches from a large fauna is the identification of members of large and difficult species groups, for instance *Aphis* in the European fauna. While it may be statistically satisfactory to regard all *Aphis* species as constituting one taxon, this underestimates the diversity. It is much easier to identify all the species in the agricultural areas of Africa and Australia because these areas have a smaller fauna containing many introduced pests; the close relatives of the pests which feed on other plants have usually not been introduced. Many species of *Psylla* have short nymphal and long adult lives while many *Trioza* have long nymphal and short adult lives. Males and females may fly for different proportions of their lives. Crops are traps and the date their leaves appear above ground will determine whether they are colonised by a migrant pest. The only practical way of comparing diversity between different groups is to accept the number of species in each group as an indication of its diversity. The number of species described in each subfamily of Sternorrhyncha is listed by geographical area in Table 5.3.

Diversity in space and on host plants cannot be separated from diversity in time. An ecological sample is usually taken in a short period of time relative to the life of an organism. When the whole fauna of an area is considered, the 'sample' is likely to have been obtained over a number of years, and perhaps even centuries. The number of species of elephant relative to the number of specimens of elephant in Africa during the last few years is meaningful, but the number of species of *Aphis* relative to the number of specimens of *Aphis* in Europe for the last few years is not. The number of aphids fluctuates greatly during the year and the proportion of adults and juveniles, active and inactive stages, is also variable. Is the number of individuals 100 times as great after a female has laid 99 eggs? If not why is an aphid population larger after a parthenogenetic female has given birth to 99 young? The concept of the number of individuals in a sample is relevant to 'instantaneous' sampling. Overwintering eggs, or any other dormant stage, can be counted when the 'instants' are longer. The number of, and mortality among, overwintering stages is often very different from year to year, which, together with alternative methods of overwintering and the difficulty of identifying eggs and pupae, mitigates against the census of dormant populations.

The Sternorrhyncha

The four major groups of Sternorrhyncha each have characteristic biologies apparently favoured by particular climatic conditions. Parthenogenesis is usual in aphids, not uncommon in coccids, occurs in aleyrodids but is very rare in psyllids. Aphids respond to changing environments by producing morphs with different biological functions. Psyllids, adult aleyrodids and male coccids vary little, but female coccids (Takahashi, 1952; Takagi, 1967) and the pupae of aleyrodids (Mound, 1963) may differ when reared on different hosts. The major groups of phytophagous insects appear from the fossil record to be older than the angiosperms, and while some of the more primitive subfamilies of aphids and coccids live on Coniferae, most families probably originated on angiosperms. Angiosperms may be older than the present interpretation of fossil records and some of the insect fragments may be misinterpreted. Short accounts of the biologies of the different groups are given in Gibbs (1973). Evans (1963) has discussed the phylogeny of the Homoptera.

PSYLLOIDEA *(jumping plant lice)*

All adult psyllids are winged, the females are oviparous and sexes usually occur in equal numbers. All psyllids are highly host specific, although some species overwinter as adults on evergreens unrelated to their nymphal hosts. Most psyllids live on woody plants but the few Liviinae live on Juncaceae and Cyperaceae, and many Aphalarinae and some Triozinae feed on herbaceous dicotyledons. Psyllids occur in permanent habitats and are sometimes attended by ants.

Many psyllids are described from Myrtaceae and Leguminosae and the group probably originated from a southern fauna, with only a few recent large genera such as *Psylla* and *Trioza* and some Aphalarinae and Liviinae colonising the northern temperate flora. The Myrtaceae-feeding Spondyliaspinae probably developed from the same fauna that gave rise to the Leguminosae-feeding Psyllinae. The tribe Psyllini and subfamilies Aphalarinae and Liviinae seem to be north temperate specialisations of the same stock. The Triozidae appear to have originated on Moraceae, on which the Carsidarinae still live in the old world tropics. Hodkinson (1974) gives an account of the biology of psyllids.

ALEYRODIDAE *(white flies)*

Aleyrodids are small, soft-bodied insects with winged adults. Males are often present but sex ratios are variable. Most aleyrodids feed on woody plants but *Trialeurodes* often live on herbs, and several genera live on Gramineae and ferns. Aleyrodids seem to be degenerate psyllids morphologically, but many species are much less specific to their hosts. The subfamily Aleurodicinae occurs mostly in South America as do a number of only distantly related genera of Aleyrodinae. The Aleyrodinae are particularly abundant in South-east Asia with 35.8%, and South America with 19.2% of the world's fauna, much higher proportions than for other groups of Sternorrhyncha. More species (96 = 8.5%) have been described from the Mascarenes than from either Europe or North America. The sparse northern fauna is perhaps an indication of late arrival, as many of the species which are present occur in the warmer parts, e.g. *Trialeurodes* in Florida. Mound and Halsey (1978) catalogue the aleyrodids together with their host plants and natural enemies, and Mound (1973) gives an account of their biology.

APHIDOIDEA *(greenfly, blackfly, plant lice, aphids)*

Aphids are noted for their complicated life cycles. Most species produce at least four different forms of female, and the first instar larvae may also be dimorphic, being especially modified for aestivation, distribution or even as soldiers (Aoki, 1977). About 10% of aphids alternate between two only distantly related hosts, about 10 species are polyphagous and most others live on only one or more related genera of plants. Most members of many groups of aphids live on trees. The north temperate region contains 80% of aphid species, but there is a small fauna associated with *Podocarpus*, *Nothofagus* and Epacridaceae in the Southern Hemisphere, and the African Combretaceae and Burseraceae-feeding fauna has a few relatives in South America on Lauraceae and Nyctaginaceae. South-east Asia has 15% of the world's aphid fauna, the subfamilies Hormaphidinae and Greenideinae from Lauraceae, Fagaceae, Moraceae and Myrtaceae perhaps being the remains of the old northern tropical fauna, having colonised Fagaceae from the Lauraceae. Aphidinae are colonisers of the herbs growing in temporary habitats and are thus pre-adapted as agricultural pests. Most aphid pests belong to the Aphidinae, many of which are associated with Rosaceae and/or herbs. Blackman (1974) gives a general account of aphids; van Emden (1972*a*) classifies the sources of information; Harris and Maramorosch (1977) and Lowe (1973) contain accounts of the biologies of aphids; and Eastop and Hille Ris Lambers (1976) list the world fauna with references to recent bibliographies.

The Adelgidae and Phylloxeridae differ from the Aphididae in that the parthenogenetic morphs are oviparous and the alates have a reduced wing venation. The Adelgidae occur only on Coniferae. The Phylloxeridae occur mostly on Juglandaceae and Fagaceae, with a few species on Saliceae, but the best known species of all is exceptional in living on vines.

COCCOIDEA *(mealy bugs, scale insects, lac insects, ground pearls)*

Female coccids are always wingless but some species have winged males. Adult females of Margarodidae, Ortheziidae and Pseudococcidae usually have normally developed appendages but members of some other families have their appendages so reduced as to be barely recognisable as insects. Some coccids are specific to their hosts while many others colonise a wide range of plants. Although coccids are often thought of as a tropical group about one third of the known world fauna occurs in the north temperate region. The Kermesidae occur only on Fagaceae, and the Dactylopiidae on *Opuntia*; this limits their geographical range. Zimmerman (1948), Lepelley (1968) and Entwhistle (1972) give useful accounts of coccids. Morrison and Morrison (1966) supplemented by Russell (1970) give annotated lists of coccid genera, and Morrison and Renk (1957), Morrison and Morrison (1965) and Russell *et al* (1974) give selected bibliographies of coccid literature.

Regional distribution

Table 5.2 summarises the proportion of the world fauna of Psylloidea, Aleyrodoidea, Aphidoidea, Coccoidea, vascular plants and land area occurring in each of the major areas. The most striking feature is that the correlation between the number of species of plants in an area is much better with the number of species of the least host specific groups, the aleyrodids and coccids, than with the more host specific groups, aphids and psyllids. The reason may be that specificity is a disadvantage in the more diverse flora of the tropics.

The distribution of organisms adapted to temperate climates must also be influenced by the distribution of the land; 96% of the land surface between 40° and 60° (i.e., temperate up to 1000 metres altitude) is in the Northern Hemisphere and only 4% in the Southern. The temperate regions of Australia, South America and South Africa are thought to have undergone considerable changes in recent geological time, which probably accounts for the poor soil in much of the area; this in turn may slow the replacement of the flora.

The north temperate regions have few species of plants relative to their area, but many aphids and psyllids and few aleyrodids. Tropical Africa has many coccids and aleyrodids but only a small aphid fauna. Temperate South Africa has a large flora relative to its area but few Sternorrhyncha. The Mascarene area is very small but has many aleyrodids. The tropical climate and partial isolation from an adjacent fauna appear to favour aleyrodids, which also have a characteristic fauna in the Pacific apparently derived from that of South-east Asia. Australia has a large psyllid and coccid fauna, but a small aphid fauna, part of which has South-east Asian affinities and the other has relatives in New Zealand and South America. Temperate South America has a relatively large flora bearing few Sternorrhyncha. Tropical South America contains 40% of the world's vascular plants, a large aleyrodid and a small aphid fauna, part of which is related to the North American fauna and part to the African fauna. The psyllid tribe Ciriacremini is also mostly African with a few tropical American species which probably reflects the Cretaceous position of Brazil in the Gulf of Guinea.

Table 5.2. Fauna, flora and land area as a percentage of the world total

	Central & Western Palaearctic	Eastern Palaearctic Northern Oriental	Ethiopian		Mascarene	S.E. Asia	Australia	New Zealand	Pacific	Neotropical		Nearctic	Unknown origin	Total
			Tropical	Temperate						Temperate (Argentina Chile)	Tropical			
Psylloidea	23.7	9.0	4.9	1.4	0.1	12.8	13.9	2.5	4.6	0.9	10.0	15.6	0.5	1728 species
Aleyrodoidea	5.1	5.1	11.6	0.1	8.5	35.8	1.7	0.7	3.2	0.9	19.2	7.0	1.2	1127 species
Aphidoidea	39.3	14.5	1.8	0.1	0.1	15.0	0.3	0.1	>0.1	0.9	0.8	26.7	0.4	4064 species
Coccoidea	18.6	7.2	12.1	2.1	2.8	16.2	8.4	3.6	1.9	1.0	13.5	11.0	1.8	5853 species
Vascular plants	5.4	2.6	9.5	4.3	3.9	16.2	5.6	0.9	2.4	4.5	40.1	4.1	–	232 000 species
Land area	23.5	6.4	19.6	0.6	0.5	9.8	6.1	0.2	0.1	2.8	13.3	17.1	–	48.6 million square miles

Diversity on hosts

Diversity is measured by the number of different sorts of thing in a sample. The sample can be either individuals in a trap catch, the species in a 'habitat', or in a larger geographical area. One 'habitat' that can be studied is the species, genus or family of plant inhabited. In any area some groups of plants can be seen to have many insects feeding on them while others are comparatively free from insects. The relationship between plants and insects was the subject of the sixth symposium of this Society (van Emden, 1972*b*) in which attention was drawn (Eastop, 1972) to the very different numbers of aphids and psyllids colonising different families of plants. Since the different major groups of plants are not evenly distributed around the world it can be difficult to distinguish between 'host plant' factors and geographical factors affecting distribution.

Most phytophagous insects have evident food plant preferences. This is often expressed as specificity to members of a particular species, genus or family of plants. For instance the coccid family Kermesidae is found only on Fagaceae, and the Dactylopiidae only on *Opuntia* (Cactaceae). Tables 5.4 and 5.5 summarise the host plants known for psyllids (1022 species), host-specific aphids (3523 species), 702 apparently host-specific aleyrodids and 697 apparently host-specific Diaspini (Coccoidea). Insects are in most cases regarded as 'host specific' if all their hosts belong to one family of plants, with the exception that species feeding both on Polypodiaceae and Aspidiaceae were regarded as 'specific to ferns'. The criteria for host-alternating aphids were those used previously (Eastop, 1972; pp. 157–8).

It is evident that the number of insects specific to members of a particular family is not correlated with the number of species of plants in that family, but the large number of psyllids and aleyrodids described from Leguminosae (12 500 species of plants) and Myrtaceae (3000) suggests that specificity may act as habitat-isolating mechanisms for psyllids and aleyrodids. The relationship between morphological and cytological differences in insects, relationships of their host plants and presumed periods of genetical isolation are outside the scope of this discussion, except in as far as they have already influenced the classification of the insects.

The 400 species of Coniferae bear 363 species of aphids, 45 Diaspini, three psyllids and no aleyrodids. The 20 000 species of orchids have only nine aphids, eight diaspids, three aleyrodids and no psyllids specific to them. Some rather small families of plants are colonised by many host-specific insects. The 150 species of Betulaceae bear 108 species of aphids and 16 psyllids but only one aleyrodid and one diaspid. The 750 species of Fagaceae bear 211 species of aphids, 37 diaspids, 20 aleyrodids but only four psyllids.

The six groups of plants with most host-specific species of the groups of insects in Table 5.4 are:

Aphids		*Psyllids*		*Aleyrodids*		*Diaspini*	
Compositae	605	Myrtaceae	212	Leguminosae	58	Leguminosae	74
Coniferae	363	Leguminosae	125	Myrtaceae	43	Myrtaceae	46
Rosaceae	293	Compositae	84	Lauraceae	42	Coniferae	45
Gramineae	242	Saliceae	71	Moraceae	42	Fagaceae	37
Saliceae	216	Anacardiaceae	32	Gramineae	38	Euphorbiaceae	32
Fagaceae	211	Moraceae	29	Euphorbiaceae	25	Lauraceae	22

The Leguminosae (12 500 species of plants), and the Myrtaceae (3000) feature in three of the lists, the Coniferae (400 species), Lauraceae (2200), Moraceae (1450), Fagaceae (750), Saliceae (530), Euphorbiaceae (6000), Compositae (19 000) and Gramineae (8000 species)

Table 5.3. Geographical distribution of the species of Sternorrhyncha

	Palaearctic		Ethiopian							Neotropical				
	Western & Central	Eastern also Northern Oriental	Tropical Africa	Temperate (Cape)	Mascarene	S.E. Asia	Australia	New Zealand	Pacific	Temperate (Argentina Chile)	Tropical	Nearctic	Unknown origin	Total species
PSYLLOIDEA	409	156	85	25	2	222	240	43	80	15	172	270	9	1728
incertae sedis										1	2			3
PSYLLIDAE														
Aphalarinae	134	37	12	4	–	24	1	2	3	2	9	62	3	293
Liviinae	4	3				2						16	2	27
Spondyliaspinae						4	201	4	3					212
Psyllinae (incertae sedis)	1	2				7	3	1		1	12	20		47
Arytainini														
Arytainina	47	7	5	5		18	16	1	10	2	42	21	1	175
Ciriacremina			29	1							3			33
Psyllini	94	65	3	1	1	22	5	1		2	11	69	1	275
Diaphorinini	14	–	20	13		11								58
Calophyini	1	5				2				3	4	13		28
TRIOZIDAE														
Carsidarinae	4		13			64	4	1	5		13			104
Triozinae	110	37	3	1	1	68	10	33	59	4	76	69	2	473
ALEYRODIDAE	57	58	131	1	96	403	19	8	36	10	216	79	13	1127
Aleyrodicinae						6	1		1	2	83		1	94
Aleyrodinae	57	58	131	1	96	397	18	8	35	8	133	79	12	1033
APHIDOIDEA	1599	589	72	3	6	611	12	5	1	35	32	1084	15	4064
APHIDIDAE														
Lachninae	104	37	1			33						171		346
Chaitophorinae	76	28				10					1	35		150
Drepanosiphinae	107	67	26	1	1	53	6	2		9	9	126		407
Pterocommatinae	16	6										22		44
Aphidinae	1121	329	42	2	2	254	3	3	1	26	20	598	11	2412
Greenideinae		15	1		3	99	3				1			122
Phloeomyzinae	1													1
Anoeciinae														
Anoeciini	13	1										5		19
Aiceonini		2				10								12
Hormaphidinae	3	41				119						3	1	167

ADELGIDAE	21	8				2						17		48
PHYLLOXERIDAE	20	3										44		67
COCCOIDEA	1089	421	706	122	116	949	482	210	110	56	792	646	104	5853
MARGARODIDAE	22	3	52	1	1	42	17	5	1	1	35	17		196
ORTHEZIIDAE			1		2	1		2	3	4	37	26		76
ERIOCOCCIDAE	83	16	5		2	25	132	90		7	38			378
DACTYLOPIIDAE											9			9
KERMESIDAE	21	12				8				2	26			69
PSEUDOCOCCIDAE	442	105	159	28	50	144	33	– 42	72	8	129	334	57	1603
ACLERDIDAE	9	4	7		1	6	1		2	10	13			53
ASTEROLECANIIDAE	18	29	5	4	2	72	15	1		1	22	2	1	172
LECANODIASPIDIDAE	2	2	10	2		23	20			1	8	6		74
COCCIDAE	215	80	181	56	15	185	60	22	3	17	222	83	23	1162
STICTOCOCCIDAE			15											15
LACCIFERIDAE														
Lacciferinae						19	6			3	9	9		46
Tachardininae			8	4		6	1							19
BEESONIIDAE						2								2
CONCHASPIDIDAE			1	1	7	1					8			18
DIASPIDIDAE	276	169	260	26	84	412	207	48	26	18	253	130	23	1932
HALIMOCOCCIDAE		2	2		2	3			3	1	5			18
PHOENICOCOCCIDAE	1													1
ALL STERNORRHYNCHA	3154	1224	994	151	270	2185	753	266	227	116	1212	2079	141	12 782
Number of species of vascular plants	12 500	6000	22 000	10 000	9000	37 600	13 100	2100	5600	10 400	93 000	9600		232 000
Land area in thousands of square miles	11 446	3099	9546	277	229	4784	2971	104	25	1359	6473	8307		48 620

Table 5.4. Food-plant preferences of host-specific Sternorrhyncha

	Number of species				
	Plants	Aphidoidea	Psylloidea	Aleyrodidae	Diaspini (Coccoidea)
'Vascular' Bryophytes	1000	11	0	0	0
Equisetales	23	2	0	0	0
Filicales	10 000	29	0	11	4
Cycadales	100	0	0	0	1
Coniferae	400	363	3	0	45
Gnetales	71	3	0	0	5
Other Gymnosperms and vascular Cryptogams	538	0	0	0	0
Annonaceae	2100	3	2	8	3
Lauraceae	2200	24	22	42	22
Ranunculaceae	800	38	5	1	0
Other Magnolidae	5570	24	5	17	10
Hamamelidaceae	100	21	1	4	2
Ulmaceae	175	54	9	8	8
Moraceae	1450	16	29	42	13
Urticaceae	550	16	3	9	1
Juglandaceae	50	55	0	0	1
Fagaceae	750	211	4	20	37
Betulaceae	150	108	16	1	1
Other Hamamelidae	120	13	2	0	12
Caryophyllaceae	1800	28	0	0	0
Chenopodiaceae	1450	20	16	1	9
Polygonaceae	800	57	17	18	2
Other Caryophyllidae	7220	9	3	2	7
Guttiferae	950	3	4	4	7
Tiliaceae	425	7	4	2	5
Sterculiaceae	850	2	5	1	4
Bombacaceae	190	0	5	3	0
Malvaceae	1250	7	9	3	1
Flacourtiaceae	1150	0	4	5	1
Tamaricaceae	120	7	22	0	7
Saliceae	530	216	71	4	7
Cruciferae	3000	38	1	1	0
Ericaceae	1900	43	21	18	10
Epacridaceae	400	2	1	9	0
Sapotaceae	800	0	9	8	6
Styracaceae	150	17	12	3	0
Symplocaceae	400	0	8	1	0
Other Dilleniidae	10 175	35	16	27	22
Grossulariaceae	240	43	1	2	1
Rosaceae	2500	293	48	18	17
Leguminosae	12 500	106	125	58	74
Myrtaceae	3000	7	212	43	46
Onagraceae	650	26	4	0	1
Elaeagnaceae	50	7	12	1	3
Cornaceae	95	14	0	1	1

Table 5.4. (cont.)

	Number of species				
	Plants	Aphidoidea	Psylloidea	Aleyrodidae	Diaspini (Coccoidea)
Loranthaceae	1300	6	10	12	12
Euphorbiaceae	6000	29	10	25	32
Rhamnaceae	900	20	27	7	2
Sapindaceae	1750	4	4	6	5
Aceraceae	175	67	4	7	4
Anacardiaceae	600	55	32	5	11
Rutaceae	1200	8	21	22	14
Meliaceae	1400	0	11	3	4
Araliaceae	700	8	6	5	3
Umbelliferae	2900	107	8	1	1
Other Rosidae	22 150	92	30	47	42
Gentianaceae	1100	1	0	0	0
Apocynaceae	1750	6	3	5	5
Asclepiadeceae	2100	5	0	1	0
Solanaceae	2300	5	7	5	4
Convolvulaceae	1500	3	1	0	0
Boraginaceae	2200	13	3	0	0
Verbenaceae	2750	7	2	7	1
Labiatae	3300	68	0	1	0
Oleaceae	600	12	11	11	10
Scrophulariaceae	2800	27	3	1	0
Acanthaceae	2500	0	1	1	1
Campanulaceae	2000	12	1	2	0
Rubiaceae	6200	21	1	24	8
Caprifoliaceae	400	45	3	2	2
Valerianadeae	380	10	1	0	0
Compositae	19 000	605	84	14	13
Other Asteridae	5855	29	7	12	2
Juncaceae	300	5	5	0	1
Cyperaceae	4000	75	1	0	2
Gramineae	8000	242	0	38	75
Palmae	3500	6	3	18	16
Liliaceae	4000	31	1	3	20
Orchidaceae	20 000	9	0	3	8
Other Liliatae (= monocotyledons)	11 900	19	0	10	0
All vascular plants	232 132	3523	1022	702	697

each feature twice. Many large groups of plants such as the Orchidaceae (20 000 species), Filicales (10 000), Rubiaceae (6200), Cyperaceae and Liliaceae (each 4000), Palmae (3500), Labiatae (3300), Cruciferae (3000), Umbelliferae (2900), Scrophulariaceae (2800), Verbenaceae (2750), Acanthaceae (2500), Solanaceae (2300), Boraginaceae (2200), Annonaceae and Asclepiadaceae (each 2100) and Campanulaceae (2000) do not appear among the six most favoured hosts of any of these groups of Sternorrhyncha. This is not simply due to toxic substances or other nutritional unsuitability. Most of the few species of polyphagous aphids do well on Solanaceae for instance, and the aphid-borne potato viruses are

Table 5.5. Proportion of the world's fauna specific to groups of plants

	% of species of vascular plants	Specific species of			
		Aphids	Psyllids	Aleyrodids	Diaspini
Coniferae	0.17	10.3	0.3	0	6.5
Lauraceae	0.95	0.7	2.2	6.0	3.2
Moraceae	0.62	0.5	2.8	6.0	1.9
Fagaceae	0.32	6.0	0.4	2.8	5.3
Saliceae	0.23	6.1	6.9	0.6	1.0
Rosaceae	1.08	8.3	4.7	2.6	2.4
Leguminosae	5.38	3.0	12.2	8.3	10.6
Myrtaceae	1.29	0.2	20.7	6.1	6.6
Euphorbiaceae	2.58	0.8	1.0	3.6	4.6
Compositae	8.18	17.2	8.2	2.0	1.9
Gramineae	3.45	6.9	0	5.4	10.8

transmitted by species which had not encountered potato until 400 years ago. *Artemisia campestris* has eight species of aphids specific to it, more than are specific to the 3000 species of Solanaceae.

Some groups of plants support a far more diverse fauna of Sternorrhyncha than others, and probably for different reasons. The large number of aphids on Coniferae and Fagaceae probably reflects their ability to feed on plants well armed with insect repellants, while the large number of species on Compositae and particularly the Anthemidae perhaps indicates that the aromatic oils of these plants make useful arrestants for indicating habitats. The large number of psyllids, aleyrodids and Diaspini on Leguminosae and Myrtaceae probably reflects the abundance of these plants during the development of the Southern Hemisphere fauna. The percentage of the world's fauna specific to the 11 most favoured groups of plants is given in Table 5.5. The absence of aleyrodids from Coniferae is perhaps due to infrequent encounters, as aleyrodids are mostly tropical and Coniferae mostly cool temperate. The absence of psyllids from Gramineae and their rarity on Coniferae and Fagaceae is perhaps due to the southern origin of psyllids and their recent colonisation of the north temperate zone, but species are neither described from *Nothofagus*, which may actually be colonised by psyllids, nor *Podocarpus* which probably is not. The abundance of psyllids on Saliceae and the rarity of aleyrodids has not been explained, perhaps *Salix* is a recent invader of the tropics. The relative rarity of aphids on Lauraceae, Moraceae, Myrtaceae and Euphorbiaceae is probably due to the great speciation of aphids in the north temperate region after the Lauraceae flora had been replaced by Rosaceae, Compositae, Gramineae etc. Many of the psyllids described from Myrtaceae belong to the Australasian subfamily Spondyliaspinae. These appear to have colonised Myrtaceae from leguminous-feeding stock, probably reflecting a change in the vegetation of Australia.

Discussion

Diversity in form, in genetical constitution, in different localities, on different plants and through time are interlinked phenomena. For instance, where a physical barrier such as the sea forms the edge of a species range, many genotypes are likely to occur there, but when the barriers are more gradual (relative to the distances moved by individuals), such as shortage of suitable hosts or intolerable temperatures, then only a few genotypes are likely to be able to survive at the mobile edges of the species range. Seasonal and climatic changes will affect both the distribution of species, and thus the diversity of the group, at any one place and the genetical diversity within any species at that place. Temperate climates select for diversity within species more than extremes, as the selection pressures are more evenly balanced, while extreme climates have a strongly skew distribution of pressures. Sudden changes of climate are one of the ways the multiplication of species is checked (p. 73).

The rarity of parthenogenesis in psyllids and its more frequent occurrence in coccids and aleyrodids may be associated with the polyphagy of the latter insects. Finding a specific plant may be part of the aggregation behaviour conducive to sexual reproduction. Colonisers of temporary habitats (opportunists) need vagility, and this, combined with polymorphism and cyclical parthenogenesis, produces groups in which diversity can be difficult either to assess or interpret. Genetically variable species must have a high reproductive rate to capitalise on the individual which is well adapted to its particular circumstance and to compensate for the many failures. There is the constant balance of advantages in rearrangement of genetical potential against the disadvantage of needing to package it to meet precise requirements. Almost by definition, temporary habitats are more variable in temperate regions. The climatic extremes, the tropics and tundra, provide a more regular (reliable) environment, to be colonised by specialists with little genetical variation, which implies speciation. As far as most living things are concerned, tropical deserts can only get cooler and wetter; if they get hotter and drier it is of no interest except in as far as it increases the uninhabitable area. The few species present in deserts and tundra compared with tropical forest is due to the closeness to the limits of life, rather than the predictable environment. Highly specialised groups are more likely to find difficulty in speciating on to new hosts. Specialisation on different parts of the same species of host plant is probably a sign of an old association, e.g., *Cinara* on conifers. This is part of a tendency for the number of species to increase with time. In temperate regions, longer or shorter, warmer or cooler, drier or wetter summers or winters will select different genotypes of both plants and insects. Aphids have produced comparatively few polymorphic species in response to this situation, but are less well adapted to extreme conditions where few of their morphs are produced. Where sexual reproduction is linked to one of the inhibited morphs, speciation is also likely to be inhibited. The 'built in' ability of aphids to respond to change by producing different forms may have slowed speciation, but another explanation must be found for the even smaller number of species of psyllids. The presence of predator complexes of related groups may deter the establishment of late comers to an area, but both *Psylla* and *Trioza* (Psylloidea) and *Pterocomma*, *Chaitophorus* and *Cavariella* (Aphidoidea) have many species on *Salix* in the Northern Hemisphere.

Numerous species of aphids can be collected from a small area with a rich flora in Europe. Some plants have several species of aphids specific to them. The different aphids may feed on different parts of the plant or on herbs growing in wetter or drier situations, but some trees regularly bear several species of aphids on the same leaf. Guava and cinnamon leaves often bear several species of aleyrodids in the tropics (teste L. A. Mound) and up to 11 species of coccid have been found on one piece of palm frond (teste D. J. Williams). Aphids are sporadic in both time and space. The number of individuals present in a locality is variable both seasonally and from year to year. A species may be common for a few years and then rare for many years. The commonest species in a trap catch one year may be barely represented the following year. The reproductive rate of aphids depends on the precise conditions, and when conditions are just right (wrong), a species multiplies rapidly. Several species have extended and then contracted their range in Europe during the last few years.

Taylor and Taylor (1977) have suggested a mechanism that accounts for the observed spatial behaviour.

Conclusions

The present diversity of Sternorrhyncha at any site is the result of the interaction of a number of processes. This is only a restatement of the opening sentence in Willis' (1922) *Age and Area*, 'The existing distribution of a plant (or animal) upon the surface of the globe, which is often a very complex phenomenon, is due to the interaction of many factors'. The present climatic conditions limit the type of vegetation which limits the fauna of phytophagous insects. Geological history has influenced the groups of plants to be found in any one area. Inherent factors in the organisms themselves, such as activity, resting stages, reproductive rates and the occurrence of parthenogenesis have also influenced their distribution and adaptability may retard the rate of speciation. In the last 400 years man has made enormous differences by transporting species from one continent to another and by speeding up the transition of vegetation from trees to herbs. Such introductions reduce the number of species in a continent by destroying habitats; however, agriculture tends to increase the diversity within the few species able to exist with and capitalise on it. In areas with a small native fauna (Australian aphids), the introduction of agriculture may actually increase the diversity by the introduction of agricultural and horticultural pests. Suburban gardens provide a wide range of additional exotic hosts. However, man almost inevitably reduces the numbers of species other than those he calls pests.

Diversity like taxonomy is fossilised ecology and population genetics. It is the result of the interaction of the nature of the founding species with the habitat. Diversity is a matter of history and habitat, and habitat itself is historically determined. The nature of the founding species is also important; for instance a group can hardly consist of numerous, closely related, vagile, polyphagous insects. The combination of any three is possible, but not all four, and it would be difficult to recognise the species if they did exist. Coccids are not vagile, all females being wingless, and the first instar larvae, the 'crawlers' which may be blown by the wind, are the distribution form. All adult psyllids are winged but they are not usually regarded as vagile insects, being rather infrequent in traps in the temperate region. They occur more frequently in traps in the tropics and may be more active there. Perhaps psyllids are so host specific, and coccids are so polyphagous, because winged psyllids can indulge in positive host finding while the 'crawlers' of coccids must often land on a species of plant different from that on which their parents developed.

Present diversity depends upon the quality of the taxonomy; the actual population genetics depends on the history of the area. There are a great many reasons why any species does not occur in a particular place and diversity is the reciprocal of the sum of all these reasons. In other words, the amount of diversity in a locality is dependent upon the number of opportunities offered and difficulties presented by the environment, the accessibility of the locality to neighbouring populations, the nature of these neighbouring populations and the length of time the locality has been available to them. The behavioural versatility of Sternorrhyncha has slowed speciation compared with endopterygote insects, but Sternorrhyncha are diverse in old, available, suitable habitats. When the ecology of Sternorrhyncha is properly understood, the experts will be pest control consultants, not salaried teachers and museum staff.

Acknowledgements

The data on insects have been obtained from the collections, catalogues and indexes of the British Museum (Natural History). Thanks are due to all past collectors and specialists as well as to the present incumbents who maintain catalogues and curate specimens so that the associated data are readily accessible. Dr L. A. Mound and Miss Sheila Halsey provided a galley proof of their catalogue of world aleyrodids which contained all the information needed in readily available form; David Hollis provided constant advice on psyllids and is responsible for the classification adopted; Dr D. J. Williams of the Commonwealth Institute of Entomology was a source of information on many aspects of Coccoidea and is responsible for the classification adopted and Mrs Linda Huddleston provided the information about the distribution of Coccidae. Dr R. L. Blackman read and improved the typescript and is a fount of information about the genetics and other aspects of aphids. Dr C. J. Humphries and John Lewis provided information about the numbers and distribution of flowering plants and Messrs A. C. Jermy and J. A. Crabbe about the ferns. Dr. L. R. Taylor explained diversity and Mr. I. P. Woiwod derived the indices of diversity.

References

Only references to aphids and coccids not included in Smith (1972), Morrison & Morrison (1965), Morrison & Renk (1957) or Russell *et al* (1974) are cited in list of references.

Adams A. N. & Farrell J. A. K. (1967) The seasonal occurrence of aphids in traps at Chitedze, Malawi. *Rhodesia Zambia Malawi J. agric. Res.* **5**, 153–159.

Aoki S. (1977) *Colopha clematis* (Homoptera, Pemphigidae), an Aphid Species with 'Soldiers'. *Kontyû* **45**, 276–282.

Blackman R. (1974) *Aphids*, 175 pp. Ginn., Aylesbury.

Blackman R. L. (1976*a*) Biological approaches to the control of aphids. *Phil. Trans. R. Soc. Lond. B.* **274**, 473–488.

Blackman R. L. (1976*b*) The puzzle of the adaptable aphid. *Spectrum* **144**, 2–4.

Cronquist A. (1968) *The evolution and classification of flowering plants*, 396 pp. Nelson, London.

Eastop V. F. (1972) Deductions from the present day host plants of aphids and related insects. *Symposia R. ent. Soc. Lond.* **6**, 157–178.

Eastop V. F. & Hille Ris Lambers D. (1976) *Survey of the World's Aphids*, vi + 573 pp. W. Junk, The Hague.

van Emden H. F. (ed.) (1972*a*) *Aphid technology*, xiv + 344 pp. Academic Press, London.

van Emden H. F. (ed.) (1972*b*) Insect/Plant Relationships. *Symposia R. ent. Soc. Lond.* **6**, viii + 213 pp.

Entwhistle P. F. (1972) *Pests of Cocoa*, xxiv + 779, Ch. 9 Coccoidea, pp. 121–168. Longman, London.

Evans J. W. (1963) The phylogeny of the Homoptera. *A. Rev. Ent.* **8**, 77–94.

Gibbs A. J. (ed.) (1973) *Viruses and Invertebrates*, (aphids pp. 112–128, psyllids pp. 129–136, coccoids pp. 178–191, whitefly pp. 237–242). North Holland publ. Co., Amsterdam.

Grouchy, J. G. de (1973) Cancer and the evolution of species: a ransom. *Biomedicine* **18**, 6–8.

Harris K. F. & Maramorosch K. (Eds.) (1977) *Aphids as Virus Vectors*, xvi + 559 pp. Academic Press, London.

Hodkinson I. D. (1974) The biology of the Psylloidea (Hemiptera): a review. *Bull. ent. Res.* **64**, 325–339.

King M.-C. & Wilson A. C. (1974) Evolution at two levels in humans and chimpanzees. *Science* 188, 107–116.

Kitzmiller J. B. (1977) Chromosomal Differences among species of *Anopheles* Mosquitos. **Mosquito Syst. 9**, 112–122.

LePelley R. H. (1968) *Pests of Coffee*, xii + 500 pp. (Coccoidea, 316–380). Longmans, London.

Lowe, A. D. (ed.) (1973) Perspectives in aphid biology. *Bull. ent. Soc. N.Z.* **2**, 123 pp.

Morrison H. & Morrison E. R. (1965) A selected Bibliography of the Coccoidea. First Supplement. *Misc. Publs U.S. Dep. Agric.* 987, 44 pp.

Morrison H & Morrison E. R. (1966) An annotated list of the generic names of the Scale Insects (Homoptera: Coccoidea). *Misc. Publs U.S. Dep. Agric.* **1015**, 206 pp.

Morrison H. & Renk A. V. (1957) A Selected Bibliography of the Coccoidea. *Misc. Publs U.S. Dep. Agric.* **734**, 222 pp.

Mound L. A. (1963) Host correlated variation in *Bemisia tabaci* (Gennadius) (Homoptera, Aleyrodidae). *Proc. R. ent. Soc. Lond.* (A) **38**, 171–180, 17 figs.

Mound L. A. (1973) In *Viruses and Invertebrates*, Ed. A. J. Gibbs, 673 pp. (Whitefly, pp. 237–242). North Holland. publ. Co., Amsterdam.

Mound L. A. & Halsey S. H. (1978) *Whitefly of the World*, 450 pp. John Wiley and British Museum (Natural History), London.

Nixon S. E. & Taylor R. J. (1977) Large genetic distances associated with little morphological variation in *Polycelis coronata* and *Dugesia tigrina* (Planaria). *Syst. Zool.* **26**, 152–164.

Russell L. M. (1970) Additions and corrections to An Annotated List of the Generic Names of the Scale Insects (Homoptera: Coccoidea). *Misc. Publs U.S. Dep. Agric.* **1015**. (Supplement): 13 pp.

Russell L. M., Kosztarab M. & Kosztarab M. P. (1974) A selected Bibliography of the Coccoidea. Second Supplement. *Misc. Publs U.S. Dep. Agric.* **1281**, 122 pp.

Smith C. F. (1972) Bibliography of the Aphididae of the World. *Tech. Bull. N. Carol.agric. Exp. Stn* **216**, 717 pp.

Taylor L. R. & Taylor R. A. J. (1977) Aggregation, migration and population mechanics. *Nature, Lond.* **265**, 415–421.

Willis J. C. (1922) *Age and Area*, x + 259 pp. Cambridge University Press, Cambridge.

6 • Determinants of local diversity in phytophagous insects: host specialists in tropical environments

LAWRENCE E. GILBERT & JOHN T. SMILEY
Department of Zoology, University of Texas, Austin, Texas 78712

This symposium illustrates that insect diversity may be viewed from various taxonomic and geographical perspectives. An aspect of insect diversity receiving considerable attention recently involves the study of taxonomically heterogeneous faunas delimited by sampling methods (see Taylor, this symposium, Chapter 1), by host plant taxa (Strong, 1974; Lawton, Chapter 7), or by geographical boundaries (Simberloff, Chapter 9 and references therein). The primary goal of all such studies has been to explain observed variation in overall insect faunal diversity between different traps, plant species, or islands.

Simberloff (this volume) has elegantly shown that rates of build-up, turnover and equilibrium numbers of species on small mangrove islands are predicted by island biogeography theory. Strong (1974) likewise, has rekindled interest in the relationship, first explored by Southwood (1961), that area occupied by a plant taxon predicts the size of the total insect fauna associated with it over its range.

In this paper we are concerned with explaining patterns of within-habitat (or point) diversity in host-specialist insects. We further limit our scope by emphasising recent work in the neotropics with special reference to our own studies of neotropical butterfly communities. Unfortunately, at this taxonomic and geographic level of resolution, published host records, so useful to broad scale comparisons (e.g. Strong, 1974) are worthless. Most publications giving host records for butterflies and other specialist insects fail to specify precise locality or other potential hosts and herbivores in sympatry. Gilbert and Singer (1975) and Gilbert (1978) provide several examples of extreme discrepancies between the local versus the cumulative regional pattern of host usage by particular butterfly species. Obviously, to distinguish among hypotheses which explain variation in diversity between sites we must know details of interactions between herbivore and host plants locally.

It should be emphasised that the significant correlations between total herbivore diversity and host-plant range is probably not relevant to the very different problem which is of interest to us; i.e., the local diversity of specialist or coevolved herbivores. Gilbert and Singer (1975) provide a general discussion of the various niche dimensions which appear relevant to explaining the diversity of coexisting species of butterflies, and Gilbert (1978) discusses the extension of equilibrium biogeography to the analysis of host plant 'islands' with special reference to butterflies.

While 'small-scale' island biogeography may explain the diversity of some insect groups on local patches, this is undoubtedly not the case with butterflies. Since deliberate egg placement is the only 'parental care' which these insects exhibit, many cues other than patch size are used in assessing host plant. Females of some *Euptychia* butterflies actually prefer small isolated patches of host plant for oviposition (Singer, personal communication), possibly as a predator escape tactic. Female *Heliconius*, which oviposit on shoots of *Passiflora* vines, can revisit the same patches daily for several months (Gilbert, 1975; Benson *et al*, 1976).

As we have been primarily involved with the behavioral ecology and population biology of neotropical heliconiine butterflies, a group confined to, and coevolved with, the Passifloraceae (Benson *et al*, 1976), we automatically approach community-level problems in a somewhat reductionist fashion. We are specifically concerned with the diversity within local host-plant guilds over areas ideally defined by the size of local heliconiine demes. Some explanations of community diversity in studies of this kind are inspired by direct observation (sometimes involving experiments) of individual insects interacting with their resources, predators, and competitors.

We therefore attempt to account for the host-specialist fraction of phytophagous insect diversity by dissecting coevolved subsystems in which the relevant interactions take place (Gilbert, 1977). Admittedly we do not know what fraction of overall within-habitat diversity is accounted for by specialists, but the increased intensity of biotic selection in tropical regions should shift the balance toward monophagous species.

We know of few analyses of host-specialist tropical insects which would provide adequate information on geographical patterns of local diversity as well as all the necessary ecological information required to explain them. An interesting example from the temperate zone is the study by Opler (1974) on the leaf-mining microlepidoptera of oak in California. According to Opler, *within-habitat diversity* of oak leaf-miners is predicted by the overall range of the host. No similar tropical study to our knowledge leads to this conclusion. Admittedly, none involve leaf-mining lepidoptera.

STUDIES OF TROPICAL HOST SPECIALISTS

HISPINE CHRYSOMELIDS ON ZINGIBERALES

Strong (1977*a*, 1977*b*) has conducted population and community studies of hispine beetles which feed on the Zingiberales. In addition to local studies at the Organization for Tropical Studies 'La Selva' site, Costa Rica, he has sampled several geographically separate habitats of Central America. Strong shows that widely distributed species of Zingiberales have more associated hispines over their range (because of geographical replacement of hispine species). However, the numbers of beetles coexisting locally depend on host size, abundance, and the species diversity of coevolved host families, but not the extent of host distribution outside of the sample area.

BEETLES ON TREE SEEDS

A detailed and long-term study of tropical host-specialist insects is that of Janzen (1977) on the seed-feeding beetles in Guanacaste Province, Costa Rica. Ninety-five beetle species have been identified from seeds of 83 tree species; of these, 73% of the bruchids ($N = 78$) were restricted to a single host and 100% of cerambicids and curculionids ($N = 17$) were on only one host. The maximum number of hosts used by a single bruchid species was only six. In addition, the number of bruchids coexisting on a host ranged from one to three.

Janzen's results rule out both local and regional distribution as factors involved in the number of beetles per tree species. Janzen believes that chemical diversity in tree seeds has been generated by coevolutionary interaction with the beetles. Reciprocally, chemical diversity of these resources leads to a diversity of digestive specialisations and speciation by the beetles. Thus, the number of tree species determine the number of beetles. Ehrlich and Raven (1965) proposed just this kind of feedback system for the generation of tropical insect and plant diversity. It may be argued that Janzen's view of community structure is biassed by his work on such extreme specialists, and that they are specialised partly because they feed internally. However, at least one group of foliage feeders illustrates a similar pattern.

ITHOMIINE BUTTERFLIES ON SOLANACEAE

Ithomiines (Nymphalidae) are conspicuous neotropical insects most species of which specialise on the Solanaceae. A preliminary study in Costa Rica by Gilbert (1969) suggested a one-to-one, host-plant species to butterfly species pattern. More recently Drummond (1976) has completed a detailed analysis of a local ithomiine community in Ecuador. He found 53 ithomiines co-occurring with 44 potential host plants. Of 27 species whose life histories were established, 81% ($N = 22$) were restricted to a single host plant while 19% ($N = 5$) used two to five hosts. Conversely, 85% ($N = 28$) of host plants known to be utilised supported only one ithomine species, the remaining 15% ($N = 5$) supporting two to three species. Even without knowing the ranges of all host plants we can say with certainty that plant geographical distribution does not account for variance in numbers of ithomiines per host within habitats. Moreover, there is no relationship between local abundance of hosts and the number of butterfly species associated with them. As with Janzen's beetles there seems to be almost no opportunity for larval competition on host plants, presumably an outcome of the fact that observed resource partitioning is generated by coevolutionary interaction of host chemical defense and parasite digestive specialisation (see Gilbert and Singer, 1975, for a discussion of this aspect of butterfly community interactions).

Gilbert (1969) noted a positive relationship between local abundances of individual ithomiine species and the corresponding host-plant species. This seemed remarkable given the fact that heavy parasitoid and predator mortality of early stages keeps ithomiines well below food limitation (also noted by Drummond, 1976). Gilbert (1969) suggested an 'indirect food limitation hypothesis' to account for this pattern. Since the effective density of larvae available to foraging ants and wasps is probably measured in number of larvae per unit of leaf area, species using an abundant host should be expected to reach greater absolute numbers before reaching densities which elicit functional response by resident hymenopteran predators and parasitoids. Rare ithomiines may have similar larval densities measured in terms of leaf area but, having rare host plants, they are limited well before host leaves are depleted or a host becomes difficult for females to locate. For example *Pteronymia notilla* is consistently present but in low density (less than 5% of any ithomiine sample) at Finca Las Cruces near San Vito, Costa Rica. The host plant, a forest-understory shrub, *Solanum brenesii* is extremely rare and scattered. Only one small individual of this host plant has been located in the study site (by following ovipositing *P. notilla*). Observations in 1969, 1970, 1971, and 1977 by one of us (L.E.G.) indicate low density persistence of the butterfly, relatively little change in the size or leaf area of the plant, few *P. notilla* eggs or larvae present at each observation period, and evidence of some larval damage on most older leaves (which persist for years).

This example illustrates a pattern of interaction consistent with the 'indirect host limitation' idea. The plant occurs in scattered, small patches but is consistently located by its specialist herbivore. The plant is consistently damaged but no evidence of defoliation has been seen, and it is clear that the butterfly is generally not converting leaves into butterfly as fast as amount of host would allow. It thus remains at a lower relative abundance than an ithomiine specialised on a locally more common host. Strong (1977*b*) has suggested a similar process to account for correspondence between abundances of hispines and abundances of their specific hosts.

It is relevant to mention that ithomiine species richness, equitability, and overall abundance increases as one proceeds from flat lowland to middle elevation montane habitats. Gilbert (1969) suggested that in steep terrain, abundant stream-cuts and land-slips maintain a greater fraction of the available habitat in early successional stages. Solanaceae thrive well along forest edges and in early succession, and are themselves more diverse and on average more abundant than in less disturbed forest. Janzen (1973*b*) noting increases in beetle and moth density at mid-elevations, suggested that plants there are able to increase net productivity because cool nights reduce respiratory losses. Thus without changing number of plant species or climatic predictability, each plant represents a larger resource base allowing more specialists to coexist per plant. Scott (1976) provides good evidence for such a process affecting diversities of the litter herpetofaunas in tropical regions. However, available data on ithomiines from mid-elevation Costa Rica (Gilbert, unpublished; Haber, unpublished) support the hypothesis of increased host diversity rather than that of increased energy available per plant, since numbers of specialists per plant appear not to be different from lowland habitats. Study sites along elevational transects in several areas are required to verify this apparent trend.

SATYRINE BUTTERFLIES ON GRAMINEAE

Singer and Ehrlich (unpublished data) have studied local diversity and host utilisation by 14 species of *Euptychia* (Satyrinae) in Trinidad. Eleven (or 79%) were found to be polyphagous (within the Gramineae) while the remaining three (21%) were found to be host specialists either on sedges or grasses.

In marked contrast to the other examples discussed in this paper, *Euptychia* population and community structure is highly variable in space and time, even in areas such as Andrews Trace, Trinidad where *Heliconius ethilla* exhibits extreme numerical constancy (Ehrlich & Gilbert, 1973). In most *Euptychia* study sites, specifying numbers of host-plant species present does not predict which or how many *Euptychia* species will be present. A graph of numbers of host species against numbers of *Euptychia* species found (lumping several sampling periods) in 10 Trinidadian study sites (Singer & Ehrlich, unpublished data) would show all points on or below the 45 degree line. Thus no local area rich in host plants had fewer than one butterfly species per plant species. However, one of the richest sites for *Euptychia* (10 species) had only one grass species present and a total of four (40%) sites had only half as many host species as butterfly species.

In contrast, the occurrence in an area of a plant species used by a monophagous *Euptychia* practically assures that the butterfly will be found there. For example, in 12 instances in which such host plants were found among 10 study sites, the appropriate *Euptychia* species was found to co-occur 11 times. Thus, the component of *Euptychia* diversity due to the host-specialist species (an average of 15% of the diversity of any site) is, in effect, determined when the host plant survey is completed. A similar pattern was found in Costa Rica where 79% ($N = 14$) of species studied were relatively polyphagous on grass species, and the remaining 21% ($N = 3$) were specialists on palm seedlings (one) and *Selaginella* (two) (Singer *et al*, 1971).

Singer's observations of numerous bizarre oviposition behavior patterns by *Euptychia* species (e.g., preferences by several species for the same isolated clump of grass over extended periods!) combined with the observation that *Euptychia* larvae have greater phenotypic diversity than do adults (the reverse of European satyrines) suggest that structure of the grass-feeding component of *Euptychia* communities may be affected much more by predation and unknown aspects of microsite heterogeneity than by direct competitive interaction (or coevolutionary interaction with host).

As species-rich *Euptychia* communities overlap in space and time with ithomiines, heliconiines, and hispine chrysomelids, they provide an interesting contrast and should be studied more intensively. Their distinct pattern of host utilisation at the species level may reflect a relative lack of chemical distinctiveness among grass species. This is also indicated by the lower incidence of species-specialists among grass-feeding, as opposed to Umbelliferae-feeding agromyzids (Price, 1977).

Further study may well show *Euptychia* communities to be structured along the lines of the stem boring guild studied by Rathcke (1976).

THE HELICONIINE-PASSIFLORACEAE SYSTEM

Heliconiine butterflies are conspicuous and constant components of neotropical butterfly communities. In any area they account for only about 3–7% of total butterfly diversity. Yet, even more than the similarly aposematic ithomiines, heliconiines are one of those conspicuous attributes of neotropical forests which invariably have caught the attention of tropical naturalists since H. W. Bates. Consequently, accumulated information on behavior (Gilbert, 1976), reproduction (Dunlap-Pianka *et al*, 1977), genetics (Turner, 1976), biogeography (Brown *et al*, 1974), systematics (Emsley, 1965; Brown, 1976), host relations (Benson *et al*, 1976), population biology (Ehrlich & Gilbert, 1973; Gilbert, 1975) and community ecology (Gilbert, 1977; Benson, 1978) of this group rivals that available for any group of insects.

All heliconiines are restricted in larval feeding to the family Passifloraceae, which contains one major New World genus, *Passiflora*, and several minor genera including *Tetrastylis*, *Dilkea*, and *Mitostemma*, which occasionally serve as hosts to heliconiines (Benson *et al*, 1976).

Since we have been studying the entire food web based on *Passiflora* (Gilbert, 1977), we know that flea beetles (e.g. *Altica*, *Disonycha*, *Monomacra*, *Strabala*) are the principal host specialists aside from the heliconiines which are significant herbivores of these plants. For particular *Passiflora* species coreid bugs of the genera *Anisocelis* and *Holymenia* can be important locally. Our data on flea beetle communities is less extensive than that for heliconiines, yet adequate beetle data exist from our primary study site (the O.T.S. 'La Selva' field station in Costa Rica) to allow useful comparisons between host specialists of rather distinct life histories which live on the same host-plant family.

PATTERNS OF HELICONIINE DIVERSITY

Of the 66 heliconiine species, up to 30 have been found in an area along the Rio Negro, Meta, Colombia (Brown, 1972), but this figure may involve some between-habitat component. Local site diversity rarely exceeds 20 species and drops to eight in the northernmost semi-evergreen neotropical forests in Tamaulipas, Mexico (L.E.G., personal observation). Brown's (1972) remarkable data based on intensive daily sampling of butterfly faunas in numerous Brazilian localities reveal a high degree of consistency in the species richness of particular butterfly taxa from region to region within Brazil. This is particularly true of heliconiines in that they are represented by 12, 14, 16, 13, and 17 species respectively for five regions sampled in extra-Amazonian Brazil (see Brown's Table 1, p. 191). Region totals for heliconiines equal 60–81% of the 21 species available in all of extra-Amazonian Brazil. In contrast, riodinid butterflies with more local endemism are represented at a regional level by only 24–44% of the total of 350 species known to exist in extra-Amazonian Brazil.

A similar comparison may be made between local diversity and regional diversity. Brown's data show that cumulative sampling period totals of 15–40 days over a year or more at a locality typically reveal 80–100% of heliconiines known from the region. Over a period of years, the number of heliconiine counted daily samples are remarkably constant at locations such as Sumaré (Brown's (1972) Table 1), where 14 daily counts from 1967 to 1971 yielded from eight to 10 species of the total of 16 known from Sumaré. These data are superficially consistent with the hypothesis that eight to 10 species represents a dynamic balance between colonisation and extinction in the Sumaré study site. However, observations of single species populations (Ehrlich & Gilbert, 1973) and our experience in Trinidad and Central American localities suggest that such a pattern is more likely the outcome of the presence of six to eight constant, conspicuous species, plus six to seven species which, because of rarity, arboreal flight habits, or mimicry association are less likely to be 'counted' on a particular day, but have a constant probability of being counted. Finally there are likely to be two to four species which would not persist in the area if it were cut off from adjacent habitats (of a different elevation) from which they disperse at some regular flow rate.

If any collection data is to reflect some underlying reality about butterfly faunas, that of Brown should. However, the highly consistent patterns in space and time revealed in his samples of butterfly communities, heliconiines in particular, cannot be explained without further knowledge of the fine structure of these communities including ecological interactions among component species and their resources.

PATTERNS OF PASSIFLORACEAE DIVERSITY

In contrast to heliconiines, the Passifloraceae contain vastly more species in the neotropics than would ever be found in one locality. Killip (1938) describes about 350 *Passiflora* species and there may be as many as 500. General floristic surveys are of little value in estimating local species densities because few individuals and species of *Passiflora* are likely to be in flower at any given time and are thus not well sampled by botanists. Therefore we must rely on data from those areas where we and/or our colleagues have followed ovipositing heliconiines and have located most of the non-flowering *Passiflora* species. In a dozen or so sites intensively sampled by ourselves or Benson and Brown in South America, nowhere to our knowledge has the number of locally coexisting *Passiflora* species exceeded 15, and 10–12 is a more typical figure for moist to wet habitats favored by heliconiines.

In spite of the large number of species in the genus *Passiflora,* there are reasons to believe that the number to heliconiines may be much less. It may well be that *Passiflora* taxonomy, like that of *Heliconius* 80 years ago, contains many 'species' which will turn out to be racial variants of widespread species. For example, what were about 70 'species' of silvaniform *Heliconius* in 1894, are now known to be about seven species (Brown, 1976). Also, a substantial fraction of *Passiflora* species occur at higher elevations of Central and South America so that they are not potential invaders of the mid to low elevation habitats occupied by most heliconiines. We can also eliminate the few dry-adapted and temperate zone species. Even so, it is reasonable to estimate that 100–150 *Passiflora* species exist which are potential colonisers of any moist neotropical habitat below 1500 meters. Yet the maximum number which locally coexist is consistently 10% or less of that number. The relationship of within habitat diversity of *Passiflora* to regional diversity is not adequately understood because local diversity is accurately known in so few cases. Since *Passiflora* diversity appears to be a major factor regulating heliconiine diversity at a local level (see below), it is crucial to learn more concerning the geographical ecology of the group.

LOCAL HERBIVORE DIVERSITY ON *PASSIFLORA* SPECIES

We have examined the heliconiine and flea beetle communities on nine species of Passifloraceae which occur at the La Selva field station in Costa Rica, and have used strictly heliconiine data from Trinidad, Costa Rica and Brazil summarised by Benson, 1978. Host records are obtained slowly and interactions between rare herbivores and rare plants are probably still to be made. However, the La Selva data represent our accumulated observations over a four year period and we feel confident that the major interactions of *Passiflora* species and the herbivore community are known. At least one of us has conducted field studies in Arima Valley, Trinidad, and/or Rincon, Costa Rica, and have supplemented Benson's data for both areas in constructing graphs for this paper. The data collected by Benson and Brown near Rio de Janeiro are likely to represent a nearly complete picture of *Passiflora*-heliconiine interaction there, as their studies were specifically directed to host utilisation by heliconiines.

We ask the following question: How do geographical distributions, local abundance, plant size, and successional status of a *Passiflora* species correlate with the number of heliconiine and/or flea beetles which will be associated with it?

Geographical distribution

It can be hypothesised that plant species with larger distributions are larger targets for evolutionary colonisation by host specialists and that these subsequently spread throughout the total range of the plant. It could be predicted then, that local diversity on a host would be an indirect result of this more global process. This effect might be particularly likely if competitive interactions were unimportant in the local herbivore guild as was probably the case in Opler's (1974) study.

We examined this possibility by plotting numbers of heliconiines found on each *Passiflora* species (including *Tetrastylis*) against the area occupied by the plants. Areas were estimated from rough distributional ranges given in Killip (1938). Fig. 6.1 shows data plotted from the four areas. Since many plants in the Rio de Janeiro locality were endemic and of uncertain distribution we simply distinguish endemics (E) from widespread species (W). Correlation coefficients calculated for the remaining three areas were not significantly different from zero.

It is worth noting that the open circle in the La Selva graph is *Tetrastylis lobata,* a species possessing hooked trichomes (Gilbert, 1971) and avoided by all heliconiines but *H. charitonius.* Thus its position reinforces the weak positive trend in the La Selva data not

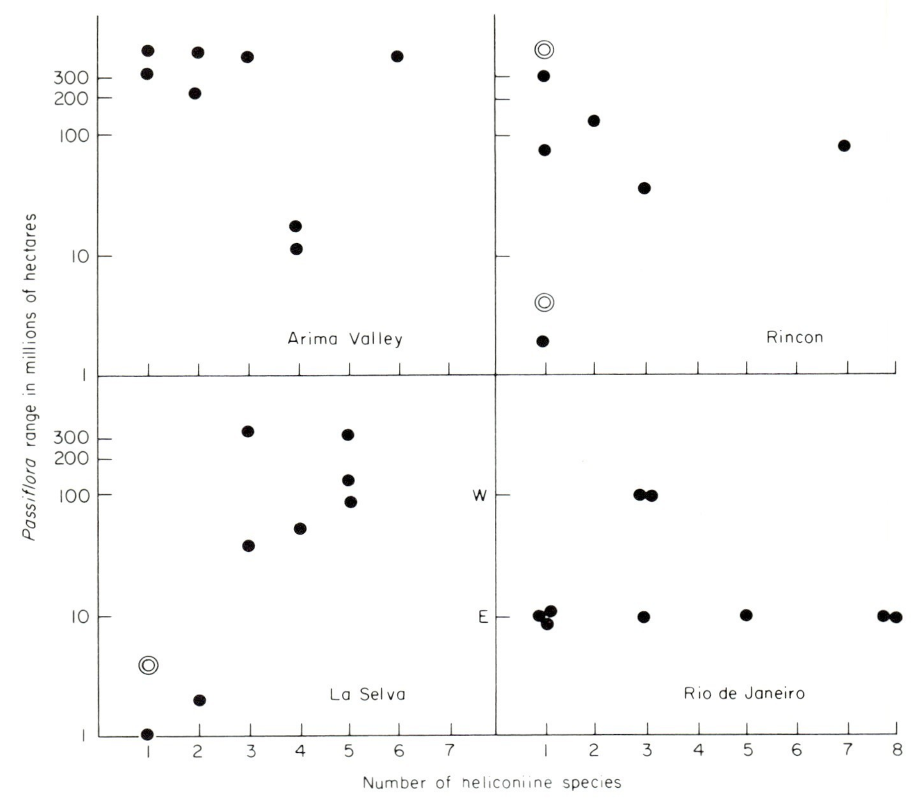

FIG. 6.1. Number of heliconiine species found locally using species of Passifloraceae as a function of their total geographical range. Data for Arima Valley, Trinidad; Rincon, Costa Rica; and Rio de Janeiro, Brazil, taken from Benson (1978); that of La Selva, Costa Rica, from J. Smiley (1978*b*). Janzen (1973*c*) provides descriptions of the two Costa Rican sites.

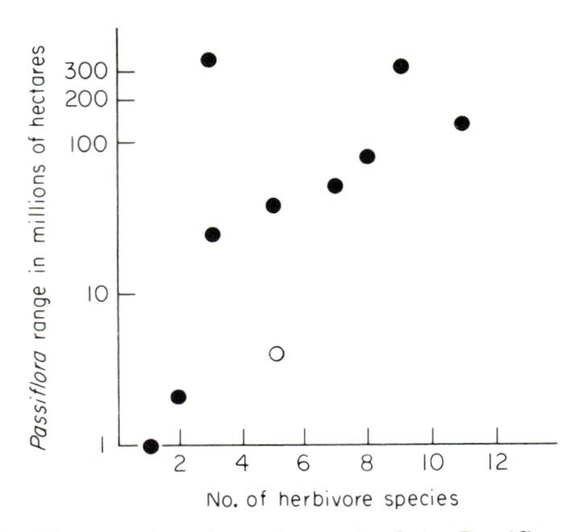

FIG. 6.2. Number of herbivorous insect species on La Selva Passifloraceae as function of plant geographical range. La Selva heliconiine data (Fig. 6.1) are combined with La Selva flea beetle host records (J. T. Smiley, 1978*b*).

because it is narrowly distributed, but because of its heliconiine deterrents. These trichomes do not deter flea beetles. The *Tetrastylis* point moves to the right when La Selva flea beetle data is combined with La Selva heliconiine data (Fig. 6.2), but significant correlation is not obtained ($r = 0.32$).

Local abundance

In cases where strong chemical coevolution has resulted in extreme monophagy at the level of host species (as in seed beetles and many ithomiines) changing the local abundance of a host is unlikely to increase the diversity of specialist herbivores associated with it. Rather, those that already use it simply become more abundant as was suggested above for ithomiines.

At another extreme are family-restricted herbivores which are relatively unconcerned with chemical distinctiveness, if any, among potential host species (within the family). In this case any host species which becomes abundant is likely to attract many of the herbivore species available. *Euptychia* and grasses probably involve this type of interaction.

Heliconiines and *Passiflora* flea beetles fall between the two above extremes. Depending on the area, 40–80% of these specialists utilise two or more host plants. Even some host specialists such as *H. melpomene* at La Selva turn out to be specialist for ecological not chemical reasons (Smiley, 1978*a*), and this has been termed 'ecological monophagy' as opposed to 'coevolved monophagy' (Gilbert, 1978).

Increasing the local abundance of a particular host plant should, in the case of *Euptychia,* and to a lesser extent the heliconiines, increase the number of species which feed on it since many species of these groups are not chemically restricted to a single host. A correlation between numbers of herbivores per *Passiflora* species and local *Passiflora* abundance (measured in numbers of new shoots per hectare) at La Selva (Fig. 6.3) is significant ($r = 0.69, P < 0.05$).

Thus we tentatively suggest that the local abundance of a *Passiflora* measured in terms which reflect productivity (e.g. new shoots available) determines the number of heliconiines and flea beetles found on it within a habitat, even if a substantial part of the herbivore guild is strictly monophagous.

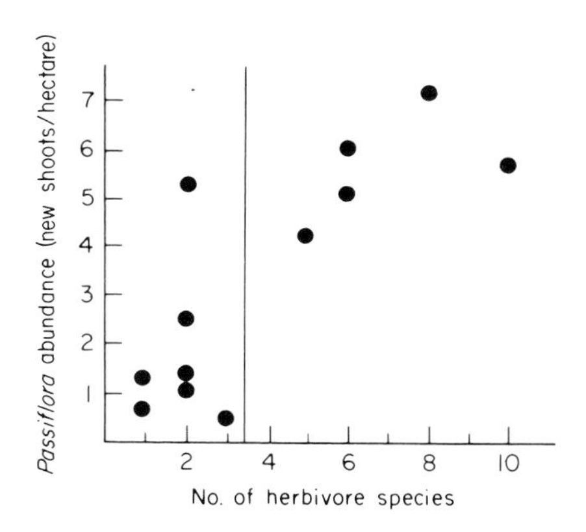

FIG. 6.3. Number of heliconiine plus flea beetle species per host-plant species at La Selva as a function of local abundance of host. Data from J. T. Smiley (1978*b*).

Plant size

Large *Passiflora* vines attract a complex fauna of omnivorous and predaceous insects which exploit their extra-floral nectar production. Field observations and time-lapse cinema photography of leaf or petiolar nectaries of several species (Smiley and Gilbert, in prep.) leaves little doubt that the larger species, and larger individual plants within a species, are occupied by larger general assemblages of insects.

However, the impact of size on the diversity of heliconiines and flea beetles is not so clear. For shoot-feeding *Heliconius* (Benson *et al*, 1976), smaller plants are frequently preferred over adjacent large plants. Females can assess egg loads (Williams and Gilbert, in prep.) and presumably predator activity. Old-leaf feeders such as *Eueides aliphera,* or the cluster-laying *Heliconius doris,* prefer the large plants of their host species, while *H. cydno* might be found only on small plants of the same species.

Thus *Passiflora* species which mature as large woody lianas may support a longer list of herbivores because juvenile and adult plants are effectively different resources from a herbivore viewpoint. However, our data suggest that population productivity measured as active meristems per unit area is more significant than maximum size attainable in predicting heliconiine and flea beetle diversity on a plant species. Better sampling of canopy emergents of the lianas *P. vitifolia, P. ambigua,* and *P. auriculata* might alter this conclusion.

Successional status

Most *Passiflora* species begin as seedlings in early successional areas (stream banks, road-cuts, land slips) or in light-gaps caused by tree falls. While some species such as *P. biflora* drop out as succession proceeds, others grow up with the forest and become emergent lianas (*P. ambigua, P. vitifolia, P. auriculata*). In Fig. 6.3, the five points to the right of the line are all from secondary successional areas at La Selva, while six to the left are from the forest. We have separated juveniles (successional area) from adults (in forest) for *P. vitifolia* and *P. auriculata* in this graph. It appears that *Passiflora* populations which occupy later succession have fewer herbivores because they are less abundant and less productive per unit area.

Implications for total diversity of herbivore guild

Given the interest in the diversity of herbivores on particular host-plant species (e.g. Southwood, 1961; Strong, 1974; Lawton, Chapter 7) it is worth asking the following question: Does an understanding of the factors which determine the number of family-specialist herbivores on a plant species help our understanding of the determinants of diversity in a local herbivore guild sharing many species of the host family?

For example, if geographical range of a host determined the local faunal diversity of specialists on that host, then the total specialist herbivore diversity of habitats dominated by widespread plant taxa would be greater than those dominated by relatively endemic plants. We see no evidence in our studies, or in those of others for such a pattern.

As we have indicated, the local abundance of productive biomass explains most of the variance in family specialist faunas *between plant species* within a habitat; but it explains little of the between site variance in local diversity of a family-specialist guild unless: (1) little overlap occurs between faunas of each host species, and (2) there exists regionally a pool of species for each host capable of invading when local concentrations of particular hosts are high (this is what is not happening in Janzen's system). Note that if only one plant species of a family is available locally, and if (2) above holds, we have a special case exemplified by Simberloff's islands. Even if the local diversity of coevolved mangrove specialists correlates nicely with the area of the island, this result would be relevant only to the problem of herbivore loads *on individual plant species* in more botanically complex communities. It fails to provide any general explanation for pattern of diversity in local specialist herbivore guilds based on species rich host assemblages. Likewise, as we deal with increasingly generalist faunas, we expect island biogeography theory to predict diversity more accurately from plant abundance or habitat area alone since biotic interactions are of lesser significance.

DETERMINANTS OF LOCAL HELICONIINE DIVERSITY

The wealth of taxonomic, ecological and geographical information available on heliconiines allows us to make an attempt at factoring out major causes of local diversity. However, these conclusions must be tentative until many additional local communities are studied in detail.

Historical factors: biogeography and evolution

Heliconiines and the genus *Heliconius* in particular underwent considerable adaptive radiation, in terms of aposematic color patterns, while isolated in the glacial forest refuge areas of South America (Turner, 1976).

Species pairs such as *Heliconius melpomene* and *H. cydno* probably diverged as races on different habitat islands. They now coexist over a large region where *H. cydno* is restricted to forest and *H. melpomene* occupies forest edges and successional areas.

Not all heliconiine species generated by this process of cyclic refuge isolation (Turner, 1976) have spread over the entire range of potential habitats. Thus, while species like *H. erato* range from northern Mexico to southern Brazil, most others occupy a fraction of that range, and historical factors come into play in accounting for between site differences in diversity (if sites are compared between regions differing in the total available pool of heliconiine species). In particular, local sites proximate to numerous Quaternary refugia (see maps in Brown *et al*, 1974) have many more potential species available for colonisation. For example, the previously mentioned Rio Negro site in Colombia (Brown, 1972) as well as other extremely rich heliconiine sites in the Manaus, Brazil, area (Brown, personal communication) are in the overlap zones for the geographically expanded faunas of four to five surrounding refugia. Since forest plants also underwent differentiation in Quaternary refugia (see references in Brown *et al*, 1974) it is possible that sites in zones receiving input from several previously separate refugia may well have higher *Passiflora* diversity than sites in more homogeneous regions.

Clearly, to minimise historical factors in seeking explanations for local diversity we should compare local sites within regions homogeneous for the influence of refugia.

Ecological factors: resource variety

At a first level of approximation, the taxonomic diversity of Passifloraceae locally accounts for much of the variation in heliconiine species richness between 10 sites (Fig. 6.4) for which reasonable data exist ($r = 0.82$, $P < 0.05$). The number of plant species is an even better indication (Fig. 6.4) of how many *Heliconius* species can be expected ($r = 0.87$, $P < 0.01$).

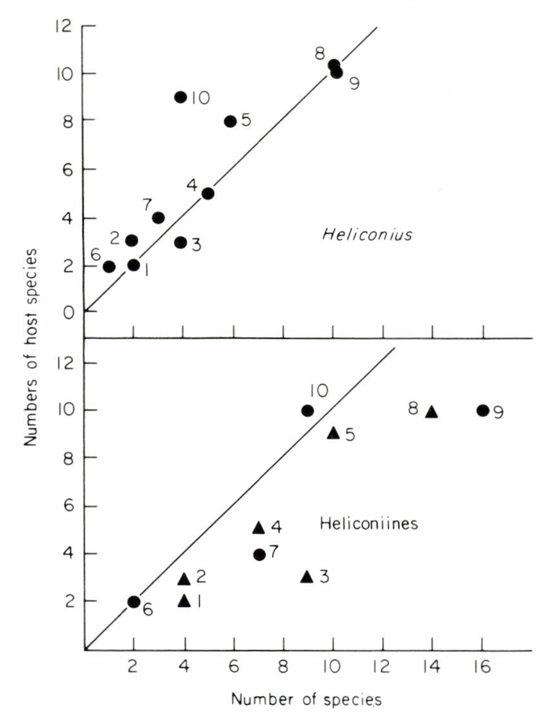

FIG. 6.4. Number of *Heliconius* (above) or heliconiine (below) species as a function of host plant diversity for 10 different locations. Triangles indicate areas for which the most abundant heliconiines are clearly associated with the most abundant *Passiflora* species. Localities are: 1, Headquarters, Santa Rosa Park, Costa Rica; 2, Airstrip hill, Palo Verde Field Station, O.T.S., Costa Rica; 3, Andrews Trace, Arima Pass, Trinidad; 4, Guanapo Dump, near Arima, Trinidad; 5, Monteverde, Costa Rica; 6, Rancho Cielo, Tamps., Mexico; 7, Gomez Farias, Tamps., Mexico; 8, La Selva Field Station, nr. Puerto Viejo, Costa Rica; 9, Rincon, Costa Rica; 10, Rio de Janeiro, Brazil. Sources: Sites 1–7, unpublished field observation of L. E. Gilbert; site 8 unpublished field observations of J. T. Smiley and L. E. Gilbert; sites 9 and 10, data from Benson (1978).

The fact that 90% ($N = 10$) of the *Heliconius* data points fall on or above the 45° line, while 90% ($N = 10$) of heliconiine points fall below the line (Fig. 6.4) reflects the partitioning of *Passiflora* species into two different resources: new shoots (most *Heliconius*) and older leaves (other heliconiines). Benson *et al* (1976) and Benson (1978) give detailed discussion of how heliconiines subdivide *Passiflora*.

The observed correlation between numbers of *Passiflora* and numbers of heliconiines raises two important questions: (1) What determines the number of *Passiflora* species which coexist locally? (2) What determines their distinctiveness from a herbivore point of view?

Janzen's (1973*a*) answer to the first question would be that host-plant diversity is locally regulated by counter-balancing mortality due to density-dependent response by host specialists. This model, like the strictly stochastic models of community structure, is attractive for its convenient lack of complexity. As the extent to which heliconiines kill or reduce the growth of *Passiflora* plants is not known, it is not possible to evaluate their role in regulating the diversity of their hosts. However, the regular patterns of *Passiflora* species locally (see above) suggest some sort of biotic limitation on diversity.

A solution to this problem may be suggested by Gentry's (1976) study of tropical Bignoniaceae. This family is remarkably like Passifloraceae in containing many vine species, having a variety of pollinators, and in exhibiting regular patterns with respect to local site diversity. Gentry was able to account for most variation in local diversity of Bignoniaceae by establishing: (1) the number of pollination niches available, (2) rainfall and edaphic features of the site, (3) the degree of seasonality and climatic predictability, and (4) the balance of successional stages locally. Gentry's study exemplifies the sort of work needed to help unravel a major aspect of resource diversity which contributes to heliconiine diversity.

In spite of the significant correlation between taxonomic diversity of hosts and numbers of heliconiines, we know of other family specialists (e.g. *Euptychia*) for which no such correlation exists. Clearly a more important aspect of resource diversity is the degree of distinctness between available host species.

Gilbert (1975) suggested that the behavioral sophistication of foraging female *Heliconius* accounts for the remarkable leaf-shape diversity of *Passiflora* species. Gilbert (1975) and Benson *et al* (1976) discuss other traits of *Passiflora*, apparently evolved in response to heliconiine herbivory, which increase differences among *Passiflora* species and, in effect, generate new niches for specialists. For example, the hooked trichomes which defend some *Passiflora* and *Tetrastylis* species from heliconiine attack (Gilbert, 1971) make these species resources for *Dione moneta* and *Heliconius charitonia,* both of which are able to utilise these plants in spite of the trichomes. Benson (1978) notes that this feeding specialisation by *H. charitonia* allows it to coexist with similar species such as *H. erato*. This observation indicates that some combinations of *Passiflora* species might maintain higher heliconiine diversity than others, and that qualitative as well as quantitative differences in *Passiflora* communities between local areas may affect differences in heliconiine species density between them.

Gilbert (1975) has argued that for *Heliconius,* which as adults utilise pollen resources for maintaining a long reproductive life (Gilbert, 1972; Dunlap-Pianka *et al,* 1977), the presence of abundant pollen allows particular *Heliconius* to persist in a local habitat at lower densities than would non-pollen feeders. Gilbert (1975) has suggested that selection for improved pollen foraging led to many of the individual and population traits of *Heliconius* which allow them to exploit other scattered, low density resources such as new shoots of *Passiflora*. In ecological time, areas which differ only in pollen abundance and/or predictability might differ in numbers of *Heliconius* present because lacking pollen, no *Heliconius* would be able to specialise on lower density, less predictable plants. It is difficult to imagine, for example, a non-pollen feeding heliconiine specialised on the low density *Passiflora pittieri* at La Selva. However, a pollen feeder, *H. sapho*, is able to persist at low densities using only this host plant.

Ecological factors: habitat structure

Heliconiine species have definite preferences for particular successional stages (Benson, 1978) and for particular levels within the forest (Papageorgis, 1975). Papageorgis builds a case for the vertical structuring of mimicry complexes (ithomiines and heliconiines) being related to different color patterns being optimum (from the standpoint of escape from predation) at different levels in a forest. Unfortunately, how these species differences in average height of flight relate to utilisation of host plants is unknown.

Smiley (1978*b*) has studied similar habitat selection by members of mimicry complexes at La Selva. He has demonstrated that the closely related *H. melpomene* and *H. cydno* differ in host preferences (at the ovipositional level) because they encounter different levels of *Passiflora* abundance in the habitats they choose. *H. melpomene* flies with its co-model *H.*

erato along forest edges and in second growth. *H. cydno* is restricted to the forest where its co-model *H. sapho* flies. Smiley has shown that while *H. melpomene* only uses a single host plant, *P. oerstedii,* its larvae are fully capable of using the same array of *Passiflora* species used by the generalist, *H. cydno.* Thus, vertical or horizontal partitioning of habitat between members of different mimicry complexes can increase *Heliconius* diversity without an increase in host-plant diversity. The butterflies simply adjust host preferences to fit with diversity and abundance of host species encountered in habitat chosen for predator escape reasons.

The extent to which structural differences (balance of successional stages, etc.) between sites accounts for differences in heliconiine species richness between them cannot be estimated from existing data. However, what is known about mimicry, habitat preference and host-plant ecology (Smiley, 1978*b*) emphasises the need to control for habitat structure in diversity studies.

Ecological factors: climatic predictability

Benson (1978) has compared sites having similar numbers of coexisting *Passiflora* species and exposed to similar regional pools of heliconiine species, but differing in the predictability of edible foliage on host plants. He found that in the least predictable site (Rio de Janeiro) average niche breadths with respect to host plants were twice those of sites with more predictable resource availability (Rincon, Costa Rica; Arima Valley, Trinidad). Benson also found fewer heliconiines than predicted from number of host-plant species at the Rio de Janeiro site (see point No. 10 of Fig. 6.4). He points to probable cases of competitive exclusion as accounting for the fewer species.

Ecological factors: topography

As one proceeds from lowlands to the highest heliconiine habitats, ecological replacement occurs for both *Passiflora* and heliconiines. Benson (1978) has documented several examples where heliconiines replace one another over a narrow elevational zone (see Gilbert and Singer, 1975, for other examples in butterflies).

Therefore, sites chosen near transition zones are likely to have higher diversity than sites well within an elevational zone. For example, *Heliconius hecalesia*, a Costa Rican species common in mid-elevation where its primary host, *P. lancearia,* is abundant, occurs in low density at La Selva. As La Selva is within a few miles of montane habitat, and has appropriately cool and wet conditions, some *P. lancearia* and *H. hecalesia* occur there. However, as La Selva becomes increasingly isolated by ongoing forest clearing from surrounding habitats of higher elevation, we expect to see *H. hecalesia* and *P. lancearia* disappear from the site.

Conclusion

It is likely that most higher taxa of tropical plants support at least some tightly associated and coevolved insect groups. It is difficult to understand community properties of either the plants or the insects without specifying how they interact with one another and with other organisms such as predators and adult resources of the insects, and pollinators of the plants (Gilbert, 1977).

While manipulations in simple systems such as tide-pools and mangrove islands have been most useful in the development of community ecology, the complexity, and specificity of interaction as described in the heliconiine-Passifloraceae system make suspect any experimental manipulation of a tropical forest community done in ignorance of the fundamental natural history and geographical ecology of its component coevolved food-webs. In addition to many more field observations of such systems, we must develop general theory (e.g. Southwood, 1977) which will apply to phytophagous insects if we are to understand how to conserve the bulk of tropical diversity.

Acknowledgements

We thank D. R. Strong, M. C. Singer, and P. R. Ehrlich for access to unpublished manuscripts and data. The paper would not have been possible without the work of K. S. Brown and W. W. Benson who, over the years, have provided data and manuscripts concerning heliconiine biology. This paper is as much an extension of their work as it is ours. We thank M. C. Singer and J. Waage for critically reading parts of the manuscript and N. K. Smiley and C. M. Gilbert for doing the typing. Insectary facilities and field expenses were supported in part by N.S.F. grant GB4074X-P. One of us (J. T. S.) has been supported by an N.S.F. predoctoral fellowship. We appreciate the logistic help given by the staff of the Organisation for Tropical Studies.

References

Benson W. W. (1978) Resource partitioning by passion vine butterflies. *Evolution, Lancaster, Pa.* (in press).

Benson W. W., Brown K. S. Jr., & Gilbert L. E. (1976) Coevolution of plants and herbivores: passion flower butterflies. *Evolution, Lancaster, Pa.,* **29**, 659–680.

Brown K. S. (1972) Maximising daily butterfly counts. *J. Lepid. Soc.* **26**, 183–196.

Brown K. S. Jr. (1976) An illustrated key to the silvaniform *Heliconius* (Lepidoptera Nymphalidae) with descriptions of new subspecies. *Trans. Am. ent. Soc.* **102**, 373–484.

Brown K. S. Jr., Sheppard P. M. & Turner J. R. G. (1974) Quaternary refugia in tropical America: evidence from race formations in *Heliconius* butterflies. *Proc. R. Soc. B.* **187**, 364–378.

Drummond B. A. III (1976) Comparative ecology and mimetic relationships of ithomiine butterflies in eastern Ecuador. Ph.D. Dissertation, University of Florida, Gainesville.

Dunlap-Pianka H., Boggs C. L. & Gilbert L. E. (1977) Ovarian dynamics in heliconiine butterflies: Programmed senescence versus external youth. *Science* **197**, 487–490.

Ehrlich P. R. & Gilbert L. E. (1973) Population structure and dynamics of the tropical butterfly *Heliconius ethilla. Biotropica* **5**, 69–82.

Ehrlich P. R. & Raven P. H. (1965) Butterflies and plants: a study in coevolution. *Evolution, Lancaster, Pa.* **18**, 586–608.

Emsley M. G. (1965) Speciation in *Heliconius* (Lep. Nymphalidae): morphology and geographical distribution. *Zoologica, N.Y.* **50**, 191–254.

Gentry, A. H. (1976) Bignoniaceae of Southern Central America: distribution and ecological specificity. *Biotropica* 8, 117–131.

Gilbert L. E. (1969) Some aspects of the ecology and community structure of ithomid butterflies in Costa Rica. Mimeo. Research Report, *In: Advanced Population Biology* (course volume). Organization for Tropical Studies, Universidad de Costa Rica, Ciudad Universitaria. Jul.–Aug., 1969, pp. 68–92.

Gilbert L. E. (1971) Butterfly-plant coevolution: Has *Passiflora adenopoda* won the selectional race with heliconiine butterflies? *Science, N.Y.* **172**, 585–586.

Gilbert L. E. (1972) Pollen feeding and the reproductive biology of *Heliconius* butterflies. *Proc. natn. Acad. Sci. U.S.A.* **69**, 1403–1407.

Gilbert L. E. (1975) Ecological consequences of a coevolved mutualism between butterflies and plants. In *Coevolution of Animals and Plants*, Eds. L. E. Gilbert & P. R. Raven, pp. 210–240. University of Texas Press, Austin and London.

Gilbert L. E. (1976) Postmating female odor in *Heliconius* butterflies: a male contributed antiaphrodisiac? *Science, N.Y.* **193**, 419–420.

Gilbert L. E. (1977) The role of insect-plant coevolution in the organization of ecosystems. In: Labrie V. (Ed) *'Comportement des insectes et milieu trophique. Coll. Int. C.N.R.S.* **265**, 399–413.

Gilbert L. E. (1978) Development of theory in the analysis of insect-plant interactions. In *'Analysis of*

Ecological Systems, Ed. D. Horn, R. Mitchell & G. Stairs. Ohio State University Press, Columbus. (in press).

Gilbert L. E. & Singer M. C. (1975) Butterfly ecology. *Ann. Rev. Ecol. Syst.* **6**, 365–397.

Janzen D. H. (1973*a*) Comments on host-specificity of tropical herbivores and its relevance to species richness. In (Ed) *Taxonomy and Ecology*, Ed. V. H. Heywood, pp. 201–211. Academic Press, London and New York.

Janzen D. H. (1973*b*) Sweep samples of tropical foliage insects: effects of seasons, vegetation types, elevation, time of day, and insularity. *Ecology* **54**, 687–708.

Janzen D. H. (1973*c*) Sweep samples of tropical foliage insects: descriptions of study sites, with data on species abundance and size distributions. *Ecology* **54**, 659–686.

Janzen D. H. (1977) The interaction of seed predators and seed chemistry. In: Labyrie V. (Ed) *'Comportement des insectes et milieu trophique.' Coll. Int. C.N.R.S.* **265**, 415–427.

Killip E. P. (1938) The American species of Passifloraceae. *Publs. Field Mus. nat. Hist.* (*Bot.*) **19**, 1–613.

Opler P. A. (1974) Oaks as evolutionary islands for leaf-mining insects. *Amer. Sci.* **62**, 67–73.

Papageorgis C. (1975) Mimicry in neotropical butterflies. *Am. Scient.* **63**, 522–532.

Price P. W. (1977) General concepts on the evolutionary biology of parasites. *Evolution* **31**, 405–420.

Rathcke B. J. (1976) Competition and coexistence within a guild of herbivore insects. *Ecology, Lancaster, Pa.* **57**, 76–87.

Scott N.. Jr. (1976) Abundance and diversity of the herpetofaunas of tropical forest litter. *Biotropica* **8**, 41–58.

Singer M. C., Ehrlich P. R. & Gilbert L. E. (1971) Butterfly feeding on lycopsid. *Science, N.Y.* **172**, 1341–1342.

Smiley J. T. (1978*a*) Plant chemistry and the evolution of host specificity: new evidence from *Heliconius. Science N.Y.* (in press).

Smiley J. T. (1978*b*) Host plant ecology of *Heliconius* butterflies in North-eastern Costa Rica. PhD dissertation, University of Texas, Austin.

Southwood T. R. E. (1961) The numbers of species of insects associated with various trees. *J. anim. Ecol.* **30**, 1–8.

Southwood T. R. E. (1977) The stability of the trophic milieu, its influence on the evolution of behavior and of responsiveness to trophic signals. In: Labyrie V. (Ed) *'Comportement des insectes et milieu trophique.' Coll. Int. C.N.R.S.* **265**, 472–492.

Strong D. R. (1974) Rapid asymptotic species accumulation in phytophagous insect communities: the pests of cacao. *Science, N.Y.* **185**, 1064–1066.

Strong D. R. Jr. (1977*a*) Rolled-leaf hispine beetles (Chrysomelidae) and their Zingiberales host plants in middle America. *Biotropica* 9, 156–169.

Strong D. R. Jr. (1977*b*) Insect species richness: hispine beetles of *Heliconia latispatha. Ecology* **58**, 573–582.

Turner J. R G. (1976) Mullerian mimicry: classical 'beanbag' evolution and the role of ecological islands in adaptive race formation. In *'Population genetics and ecology,'* Ed. S. Karlin and E. Nevo, pp. 185–218. Academic Press, New York and London.

Note added in proof

Results of W. A. Haber's research on the Ithomiinae ('Evolutionary Ecology of Tropical Mimetic Butterflies (Lepidoptera: Ithomiinae)', PhD Dissertation, University of Minnesota, Minneapolis/St. Paul, 1978) were made available to us in May 1978 by the author. A few highlights are worth mention in the context of our paper:

1. At Monteverde, Costa Rica, 26% ($N = 5$) of the 19 ithomiines with host records were restricted to a single host. While this departs drastically from Haber's own data at other localities, and from data collected by Drummond or Gilbert (see above), the fraction using three or less of available host species at Monteverde (91%; $N = 17$) is consistent. Other localities may show a Monteverde-type pattern with further sampling.

2. Haber has examined Ithomiine/host-plant diversity along an elevational gradient as we suggest above (p. 92). As suspected, the increase in ithomiine species numbers from lowland to mid-elevation sites is closely paralleled by an increase in Solanaceae species richness.

3. Haber plotted numbers of ithomiine species against numbers of Solanaceae species for 12 Costa Rican localities. The resulting graph is virtually identical to that for *Heliconius* x Passifloraceae in our Fig. 6.4.

7 • Host-plant influences on insect diversity: the effects of space and time

J. H. LAWTON

Department of Biology, University of York, York Y01 5DD

This paper is concerned with predicting the numbers of kinds of insects on individual species of plants in temperate regions. In it I want to show that the communities of phytophagous insects which are found on particular species of plants are not random assemblages, chewing away independently of one another; rather they have a structure which can be unravelled by a consideration of space and time.

We will be concerned with space in three ways. First on a large scale, by examining the relationship between the number of species of insects associated with various species of plants, and the geographical ranges of these plants. Then on a local scale by looking at small clumps of plants (islands). Finally in a totally different way by considering the physical structure of the plants themselves, the 'architecture' which forms the insects' living-space. Time enters the story in several ways. What I have to say presupposes the existence of relatively stable assemblages of species over periods of time measured in hundreds, if not thousands of years; I do not propose to discuss the vexing question of whether ecological communities in general, and communities of plant-feeding insects in particular 'saturate' in evolutionary time (Gilbert, 1978; Southwood, 1973, 1977; Strong, 1974*a*, *b*; Whittaker, 1969). The main consideration of the effects of time will be with very short periods; that is with seasonal effects.

The paper is organised as follows. It starts by counting insects on plants that have different sized geographical ranges, and then asks how many of these species will occur at any one locality within that total range. Next, seasonal changes in insect diversity are considered, together with associated changes in plant architecture and chemistry. I then use the idea that plant architecture plays a key role in determining insect diversity to make some preliminary suggestions about the ways in which insects exploit plants.

The number of kinds of insects on different kinds of plants

Plants with large geographical ranges characteristically support more species of insects than rare plants. This is illustrated in Fig. 7.1 for British plants other than trees. The same phenomenon has been documented for trees both in this country (Southwood, 1961; Strong, 1974*a*) and elsewhere (Opler, 1974; Southwood, 1960*a*, 1973, 1977; Strong, 1974*b*) as well as in a variety of other plants (Strong & Levin, in press; Strong *et al*, 1977). If we view the number of species on a particular plant as a balance between colonisation and extinction, then geographically more widespread plants have a larger number of species associated with them because they are more likely to be found and colonised, given sufficient time (Janzen, 1968; Lawton & Schröder, 1977; Opler, 1974; Southwood, 1960*a*, 1977; Strong & Levin, in press). 'Experimentally' reducing or increasing the geographical area occupied by a plant leads respectively to extinctions in, and additions to its fauna (Southwood, 1961).

The four regression lines shown in Fig. 7.1 do not differ significantly from one another in slope, but they differ markedly in intercept. Southwood's data, subsequently reanalysed by Strong (*loc. cit*), for British trees show that they support even more kinds of insects than do the woody shrubs or perennial herbs illustrated in Fig. 7.1. Hence comparing plants of a similar sized geographical range, the richness of the associated insect fauna falls in the sequence trees > woody shrubs > perennial dicotyledonous herbs > weeds and other annuals < monocotyledons excluding grasses, for which food plant records tend to be unrealiable.

One hypothesis which would explain these results is that the number of species of insects associated with a particular kind of plant is a function not only of the geographical range of the plant, but also of its architecture, with trees to monocotyledons (many of which are little more than strips or cylinders of vegetation) forming a series of decreasing size and structural complexity in their above-ground parts. As both Lawton & Schröder (1977) and Strong & Levin (in press) point out, there are presumably more ways of making a living on a bush than there are on a bluebell, so that larger and more complex plants with more 'niches' (Southwood, this symposium, Chapter 2) support a greater variety of insects. Precise measurement of plant architecture – quantifying the variety of edible parts like leaves, bracts, flowers or fruits, the strata that bear them and their seasonal phenology – is not

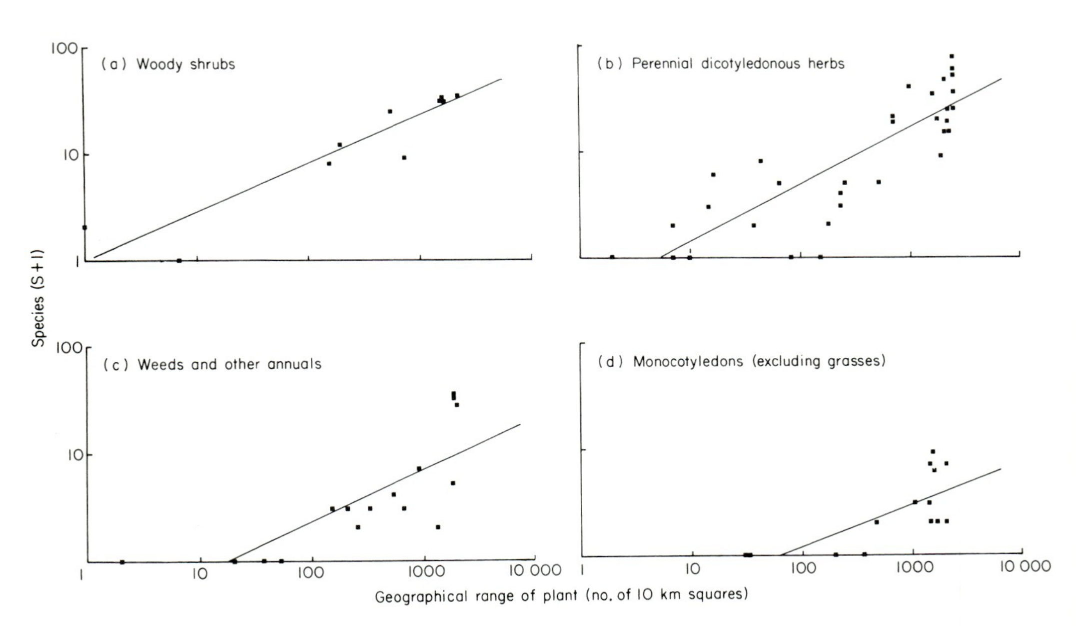

FIG 7.1. The insect species associated with various groups of British plants, as a function of the size of the plant's geographical range (measured in terms of the number of 10-km squares occupied by the plant in the *Atlas of the British Flora*, excluding Ireland; Perring & Walters (1962)). Data from Lawton & Schröder (1977, and in press).

essential for the present paper; however, an index like Foliage Height Diversity (MacArthur & MacArthur, 1961; Murdoch *et al*, 1972) would provide a suitable measure, although as originally defined it ignores differences in the type of plant tissue. The classical study of broom (*Sarothamnus*) by Waloff (1968) provides ample documentation of the richness of feeding, oviposition and overwintering sites that are used by insects on a bush of average structural complexity.

For completeness, notice that the number of species of insects associated with weeds and other annuals may be lower than that on perennial herbs for reasons other than just their architecture. In common with other, although by no means all, types of ecological community in which resources are unpredictable in time and space (see Abele, 1976; Brown, 1975; Karr, 1976; MacArthur, 1975), Schröder and I hypothesised that the pool of specialist species able to exploit weeds and other annuals would be reduced. This hypothesis is consistent with the data in Fig. 7.1, but needs further testing.

Local differences in diversity on one plant

Obviously, all the species of insects associated with a particular plant will not be found on every individual throughout the plant's geographical range; and the prospect of understanding enough about plant-insect interactions to be able to say which species from the total pool will occur where seems a very long way away. A full understanding of the problem requires approaches at a series of levels, geographical (Taylor, 1977; Ward, 1977), ecological (Atsatt & O'Dowd, 1976; Cromartie, 1974; Pimentel, 1961; Root, 1973) and even physiological and behavioural (Chew, 1975; Kogan, 1977; Wiklund, 1974, 1975, 1977) in scale. In this paper I want to trace some of the more obvious links in the chain.

Common bracken fern, *Pteridium aquilinum,* has between 18 and 20 species of insects, plus a mite, which regularly feed on its foliage in Britain, and at least another 20 insects which do so occasionally (Lawton, 1976). The plant is very widespread (it occurs in nearly 2400 '10 km' squares, excluding Ireland) and, for a plant with a growth-form like a perennial herb, 18–20 key-species is roughly what we would expect (Fig. 7.1b); certainly the fauna is neither very impoverished nor very rich (Lawton, 1976).

What proportion of the pool of key-species would we expect to find at any one locality? One of my students, Caroline Rigby, is investigating this problem on the North Yorkshire Moors. Within the study area are a series of well-defined patches (islands) of bracken, isolated by heather from other islands and from still larger 'continents'. Apparently one of the main things which determines the local diversity of the bracken fauna is the size of the patches (Fig. 7.2), but the distance of the island from the nearest large expanse of bracken (or 'continent') seems to have no effect, probably because source-areas are numerous and difficult to define. These preliminary results are very similar to those of others who have treated clumps of plants as islands (Cromartie, 1974; Davis, 1975; Tepedino & Stanton, 1976; Simberloff, 1976; Ward & Lakhani, 1977) drawing on classical island biogeography theory (MacArthur & Wilson, 1967). There is a whole paper devoted to the problem in this symposium (Simberloff, Chapter 9), so that here I need only refer briefly to the bracken results.

The loss of species with decreasing island size is not random (Fig. 7.3). Some species drop out sooner than others, and we will see later that one effect of this is that the overall balance of the community in terms of what the species do (rather than what they are) remains remarkably constant (Fig. 7.13). The ecological processes which determine losses and gains within islands are discussed by Simberloff (1976, and this symposium).

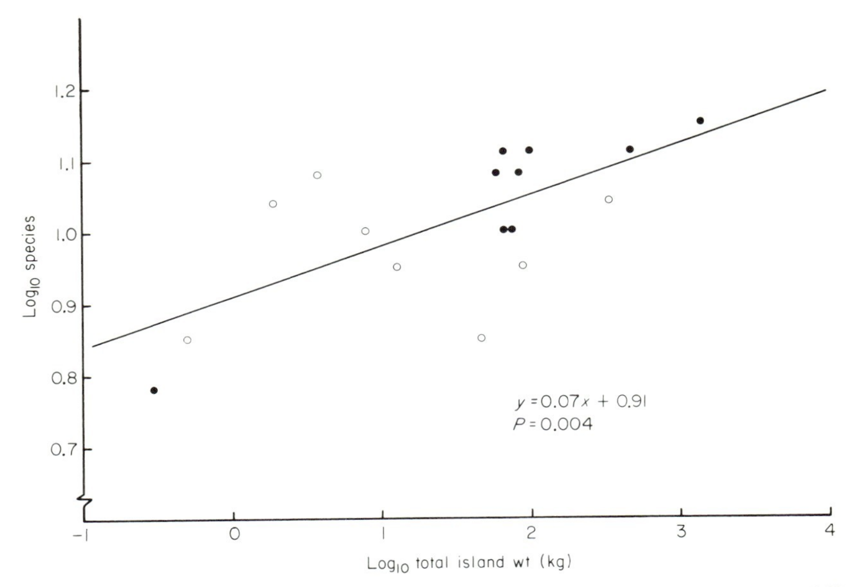

FIG. 7.2. The number of species of bracken-feeding insects on bracken 'islands' of different sizes (North Yorkshire Moors study area). Island size is expressed as weight, because this also incorporates differences in frond-density and frond-size, as well as the area of the island. ○, near islands; •, islands remote from large 'continents' of bracken. (C. Rigby, personal communication).

A typical bracken island does not contain all the species present in our study area until it is of the order of $1-2 \times 10^3$ m^2. There must then be a very marked change in the slope of the graph relating the number of species of insects to area, because extending our search over several thousand square kilometers adds no new species to the list.

The only data comparable to that on bracken is Ward's work on Juniper (*Juniperus communis*) (Ward & Lakhani, 1977). In the south of England, a Juniper patch must contain at least 3000 bushes to have a reasonable chance of holding all 15 of the common native species of insects present in the regional pool. Hence, just like bracken, large islands accumulate all the possible species in the area when they are still smaller by many orders of magnitude than the size of the geographical region under investigation. This encourages me to think that it is sensible to try and interpret broad geographical patterns in insect species diversity on plants by focussing on local ecological processes. Of course different species of insects will always be encountered in different parts of the geographical range of a widespread plant (Lawton, 1976; Southwood, 1961; Ward, 1977; Zwölfer, 1965). My point is that as one increases the area sampled, local patches appear to saturate with species rather quickly. On average one has then to move much larger (geographical) distances before any new species are encountered.

One major component of the 'colonisation and extinction' arguments used to explain results like those shown in Fig. 7.1 (e.g. Kuris & Blaustein, 1977; Opler, 1974; Strong & Levin, in press) is that widespread plants will be colonised by insects from a larger variety of habitats and climates than geographically restricted plants. But do widespread plants have more species *at any one locality*? Opler's oak-trees certainly do. We would expect this to be the case if the widespread species originally acquired their fauna from different parts of their geographical range, but then some or all the insect-species infiltrated other parts of their

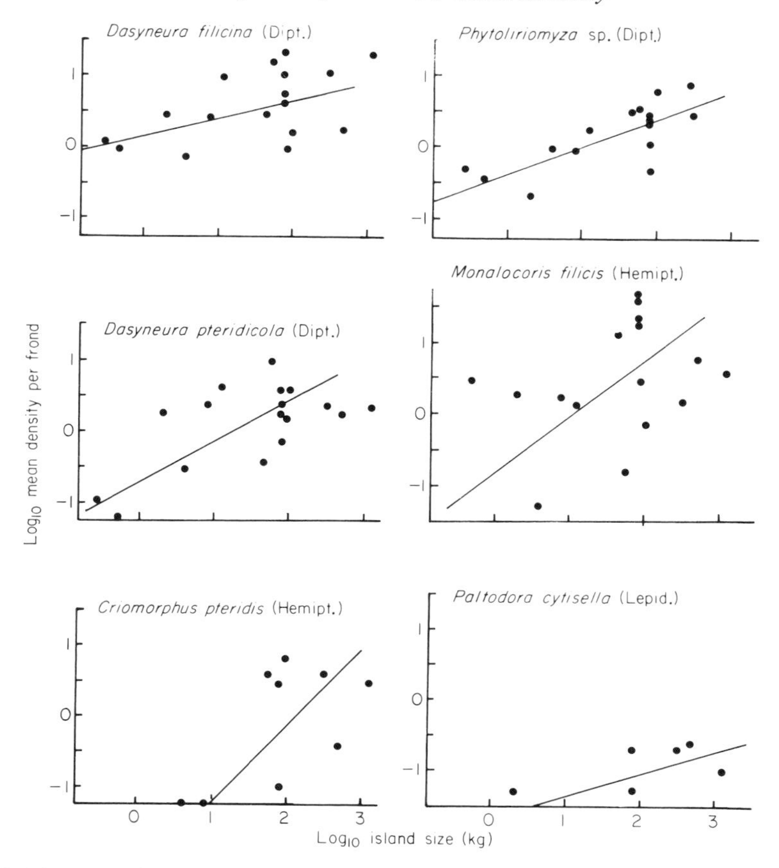

FIG. 7.3. Mean population densities of six representative species, on bracken islands of various sizes. Regression lines have been fitted by eye. Because of the log. scales, absences (zero populations) cannot be shown on these graphs but species which are scarce (e.g. *Paltodora*) or species which decline quickly (e.g. *Criomorphus*) are absent on small islands more often than abundant species which decline slowly with island size. Note how all species tend to be rarer *per frond* on smaller islands. (C. Rigby, personal communication).

hosts' total range thus raising the local diversity (Strong & Levin, in press). The distribution of eight species of *Erythroneura* leaf-hoppers (Cicadellidae, Homoptera) on American sycamore (*Platanus occidentalis*) can be interpreted in this way (McClure & Price, 1976) as can that of the two oak mirids in the genus *Phylus* discussed by Southwood (this symposium, Chapter 2).

With the exception of Opler's oak-trees we suspect, but we do not know, that geographically widespread species have a more diverse insect fauna associated with them at any one locality than structurally similar but geographically more restricted species. It is disturbing that our data should be so poor on so fundamental a point. It might even matter where we look to test the idea. Thus several coniferous trees have more insect-species near the presumed centre of their geographical range (and possibly evolution) in Sweden and Russia than they have in Britain (Southwood, 1960*b*, 1961, and this might well be typical of other plants (Goeden, 1971; Goeden & Ricker, 1974). We urgently need more data which provides us with species-lists from different parts of a plant's geographical range.

The progression of species within a season: the effects of plant architecture and plant chemistry

The total number of insects on a plant is one thing: the seasonal distribution of those species is quite a different problem. Very few studies have followed the seasonal progression of the fauna on one species of plant in detail. Fig. 7.4 summarises the normal seasonal progression in the number of phytophagous insects on a patch of bracken at Skipwith Common, in Yorkshire. There is a gradual increase in the number of species, reaching a peak in late July and early August, with very few being found on the plant throughout the whole season; most come and go as shown in Fig. 7.5. Other examples are given in Lawton (1976).

In my 1976 paper I suggested that this pattern of increasing species diversity with season was a direct consequence of changes in the chemical composition of the bracken frond. For convenience, let us call this the 'chemical hypothesis'. Bracken fronds in the spring and early summer contain at least two chemicals that would make exploitation by herbivores difficult. These are the cyanogenic glucoside prunasin, which releases free cyanide when the frond is crushed or eaten (complicated by a polymorphism in the plant; see Cooper-Driver, 1976; Cooper-Driver & Swain, 1976; Lawton, 1976), and thiaminase, which destroys vitamin B_1; all insects need this vitamin for normal development. Levels of both cyanide and thiaminase decrease as the season progresses (Fig. 7.6a & b).

Recently we have also monitored seasonal changes in three other defence mechanisms. Firstly, Clive Jones at York has shown that phytoecdysteroids (mimics of insect moulting hormones; see Carlisle & Ellis, 1968; Kaplanis *et al*, 1967) are present in Skipwith fronds in quantities that are too small by at least three or four orders of magnitude to have any effect on phytophagous insects, either on their development or as antifeedants. Levels rise slightly in fronds in the autumn, but too late and by too small an amount to affect the insects (Jones & Firn, 1978). Secondly, Jones (1977) has also isolated an antifeedant from bracken, of which a major fraction is the sesquiterpene pterosin F (see Cooper-Driver, 1976). Seasonal

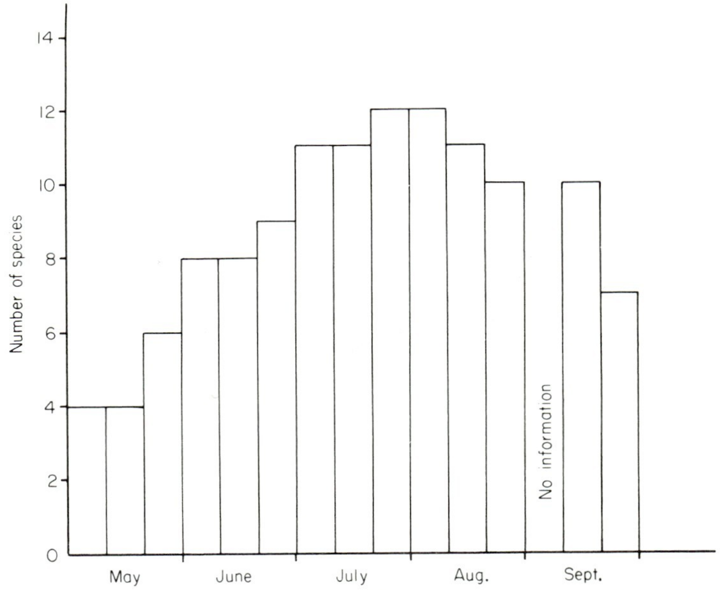

FIG. 7.4. Typical seasonal build-up in the herbivorous insects on bracken at a Skipwith (Yorkshire) study site, based on six years data (1972–1977).

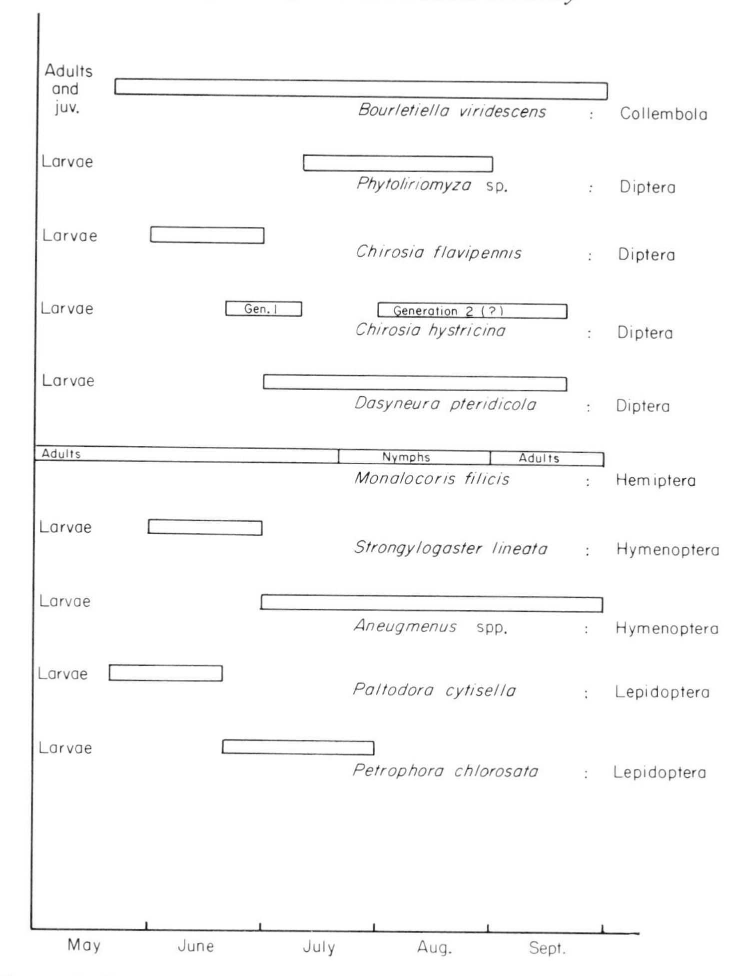

FIG. 7.5. Seasonal distribution of the feeding stages of 10 bracken insects at Skipwith, illustrating various life-history patterns.

changes in the levels of this compound follow closely those of cyanide and thiaminase (Fig. 7.6e). Finally, in my 1976 paper I suggested that the extra-floral nectaries of the plant also served a defensive role by attracting ants which in return removed foliage-feeding insects. There was nothing particularly novel about this suggestion (Bentley, 1976; 1977; Darwin, 1877; Hocking, 1975) and the hypothesis has now been confirmed. The detailed observations will be published elsewhere; Fig. 7.6d summarises the seasonal activity of ants visiting the nectaries, and hence the seasonal distribution of their defensive role. The pattern is again one of a spring high, followed by a decline during the summer and early autumn.

Confronted with this veritable armoury in the spring, the hypothesis that insect species diversity on the plant was determined by, and inversely related to, the level and number of kinds of defences deployed by the plant was obviously not unreasonable. But I now think that it was wrong.

Consider the data in Fig. 7.1. Here we invoked a hypothesis based on plant architecture to account for the differences in total insect diversity between types of plants. Why should

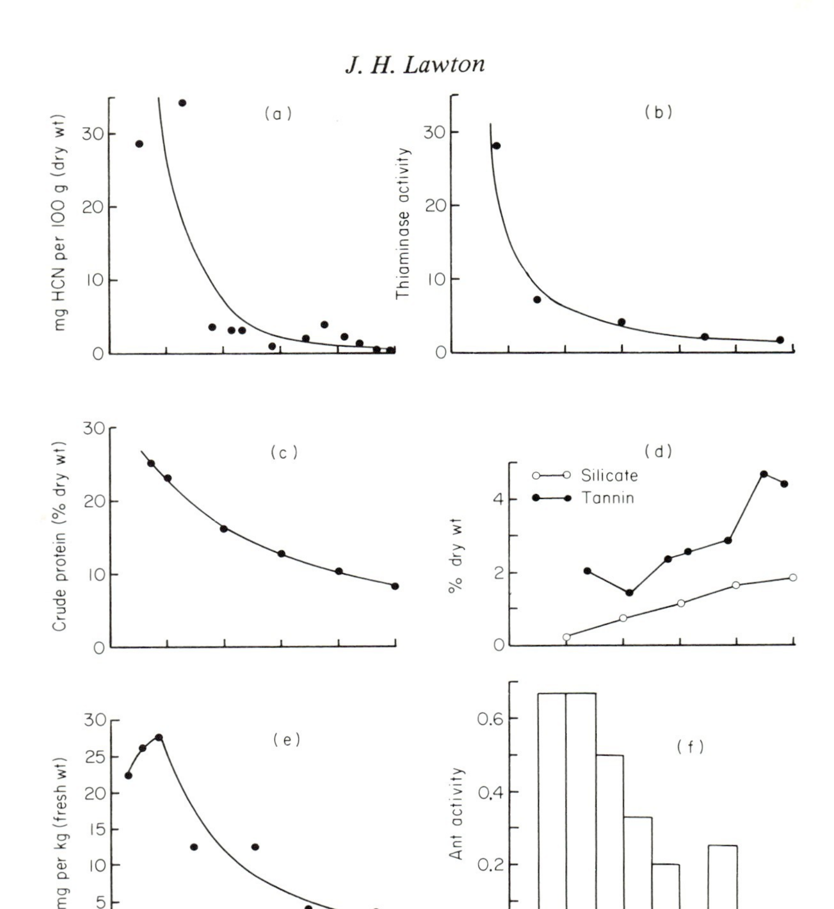

FIG. 7.6. Seasonal changes in the defences of bracken against herbivores. (a) Cyanide (Lawton, 1976) (b) thiaminase (Evans, 1976) (activity is expressed as μg thiamine destroyed min^{-1} g (dry wt)$^{-1}$ under standardised conditions); (c) protein [mean of several studies, from Lawton (1976)]; (d) tannin and silicate [both from Lawton (1976)]; (e) pterosin F, a sesquiterpene with antifeedant activity (Jones, 1977 and personal communication); (f) ant activity at the nectaries of the plant (see text); activity is expressed as the proportion of sampling days on which ants were seen visiting the nectaries.

we not also do this to account for the differences in diversity within a season on one plant? Except very early in the season all the insects exploit the blade of the plant or its rachis rather than the stalk (we are not concerned here with rhizome feeders). To a good approximation we can treat the blade as a triangle and follow seasonal changes in its surface area as shown in Fig. 7.7; as the blade expands, the bracken frond becomes architecturally more complex in the sense of this paper. Thus on a mature frond it is possible for different insects to feed on the basal pinnae or at the frond tip under very different microclimatic and nutritional conditions; for miners and gall-formers in particular to exploit different parts of the pinnules; for insects to utilise the developing sporangia, and so on. A young crozier presents none of these opportunities. Hence the area of the blade provides a crude index of the variety and quantity of resources that are available for insects to exploit, and insect

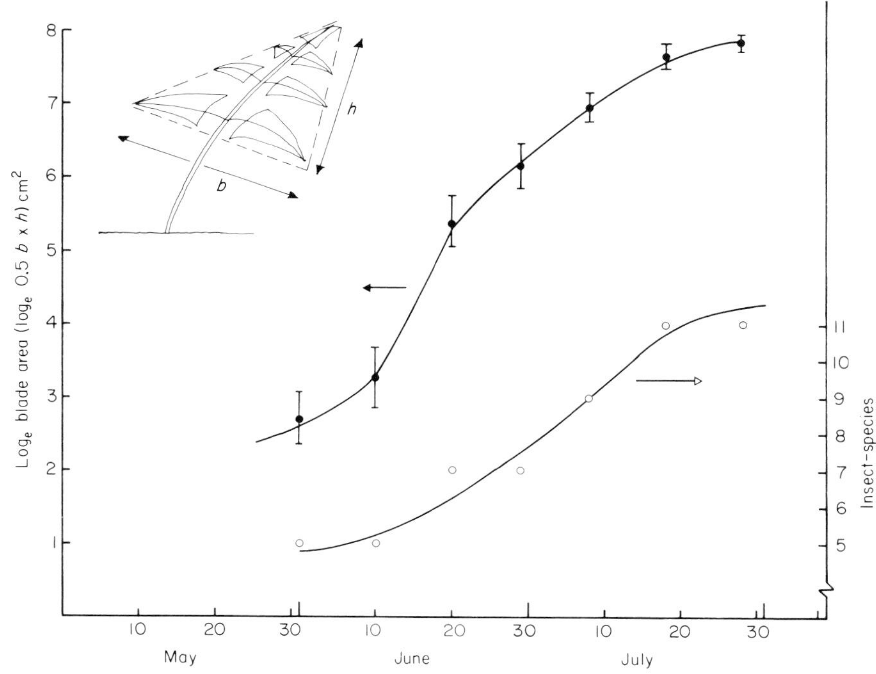

FIG. 7.7. Expansion of bracken fronds at Skipwith, 1977, together with the build up in insects on the plant in the same year: (vertical lines are 95% confidence intervals of $\log_e$ area). Area has been calculated by assuming that the frond is triangular (see diagrammatic sketch of frond).

species diversity follows rather closely the expansion of the frond (Fig. 7.7), with the number of species divided by the log. of the blade area remaining approximately constant throughout the season (Fig. 7.8).

Correlation, of course, does not imply causation. Unfortunately, it is very difficult to distinguish the architecture hypothesis from the chemical hypothesis using data from bracken alone, because nature will not allow us to vary architecture and plant chemistry

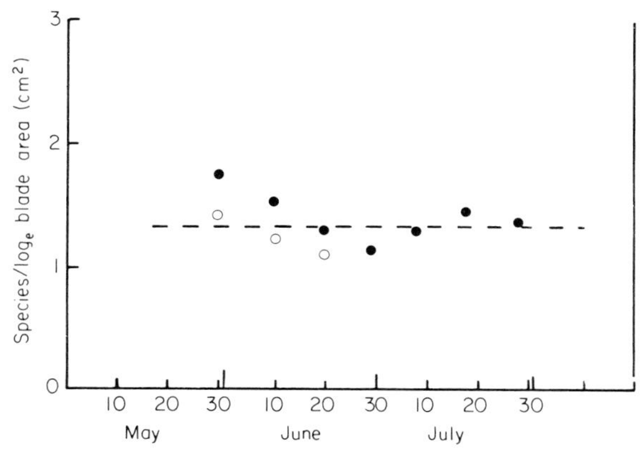

FIG. 7.8. The number of insect-species per unit area of frond, Skipwith 1977. •, all species; ○, excluding *Paltodora* caterpillers (Lepidoptera) which mine the stalk, and only rarely mine the blade of the plant.

independently. To do this we need to examine other evidence, both on seasonal changes in the number of species of insects as well as on the total number of species associated with other species of plants.

Which hypothesis?

A major advantage offered by the plant architecture hypothesis is one of economy, in that it allows us to explain both gross differences between types of plants in the diversity of insects which they support as well as the seasonal changes in species diversity within a plant, by one and the same mechanism. On this view, plant-chemistry has a profound effect on *which* species of insects attack each plant (most plant-feeding insects are monophagous or oligophagous), and a profound effect on the *season* at which each species is able to exploit its host plant (Lawton, 1976; McNeill, 1973; McNeill & Southwood, 1978), but only a small or in some cases perhaps no effect on the total number of insect-species which eventually evolve to exploit a plant. This is primarily determined by the geographical range of the plant and its architecture, as are the number of species capable of exploiting it at any one time during the growing season.

Feeny (1976) and Rhoades & Cates (1976) recognised two broad kinds of chemical-defences in plants which Feeny termed 'qualitative' and 'quantitative' and Rhoades and Cates 'toxic' and 'digestability reducing' defences respectively. Qualitative defences (e.g. cyanide or thiaminase) are effective at low concentrations against non-adapted insects but are apparently rather easily circumnavigated by specialists. The efficiency of quantitative defences (e.g. resin, or tannin) depends very much on their concentration (more is better)

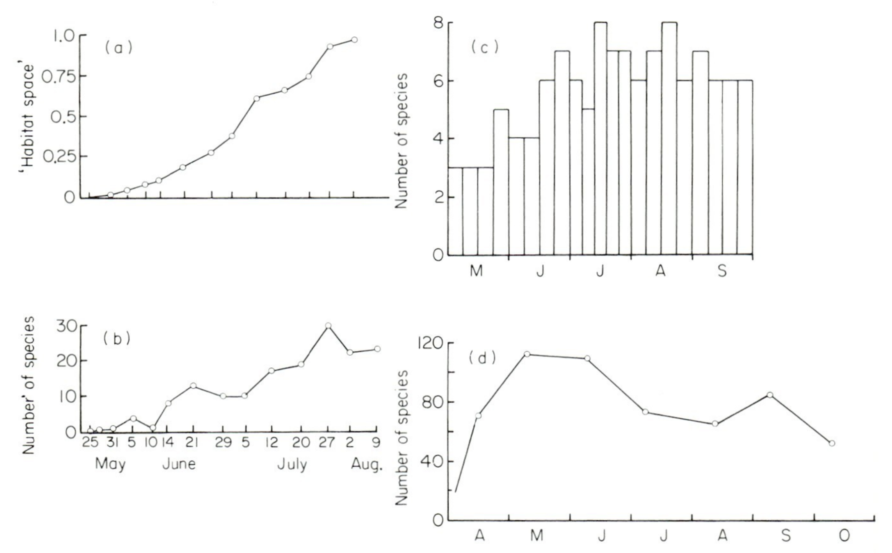

FIG. 7.9. Seasonal build-up in herbivorous insects on three plants, for comparison with the pattern seen in bracken (Fig. 7.4). (*a* & *b*) Soybeans (Price, 1976), showing the growth of the plant (*a*) (habitat space is an index of plant development composed of leaf number and leaf width), and (*b*) the parallel increase in insect diversity. (*c*) Nettle (Davis 1973), a perennial herb with a seasonal growth pattern like bracken. (*d*) Foliage-feeding Lepidoptera on oak (Feeny, 1970). Notice the gradual build-up in diversity in *b* and *c*, and the much more rapid increase in *d*.

and they are extremely difficult for any insect to overcome when they are present in large amounts. It is possible that the most easily detected effects of plant chemistry (on the number of species of insects associated with an established plane i.e. one that has not been recently introduced) will be those produced by quantitative defences, which effectively eliminate certain plant-parts from attack by insects. McNeill and Southwood (1978) document several examples of changes in feeding sites by insects during the course of a plant's growing season, as previous sites pass below some minimum nutritional threshold. Predicting which parts of a plant are actually utilisable as food and which parts are 'scaffolding' is not going to be easy, but is likely to develop out of the detailed study of quantitative defences of the type summarised by McNeill & Southwood (*loc. cit.*).

Several things follow from these general arguments. For example, any plant which starts off small and gets larger and more complex as the season progresses should show a build-up in the number of species of insects which feed on it (like bracken). But this pattern should not be influenced to any great extent by plant chemistry. Data from Price (1976) for cultivated soybeans show this phenomenon exactly (Fig. 7.9a, b). However, as with other annual crops (Cromartie, 1974; Pimentel, 1961) it is difficult to distinguish a build-up due to immigration as the insects find the crop from a build up of the type seen in bracken, when all the species are already present on-site. An even better example is therefore provided by nettle (*Urtica dioica*) (Davis, 1973). The seasonal growth pattern of this widespread perennial herb is very similar to bracken. So too is the number of species of insects which feed on it (27 specialists plus 19 others; Davis, 1973). The seasonal build-up in insect species on nettle follows very closely that seen on bracken (Fig. 7.9c; c.f. Fig. 7.4), but as far as is known (B. N. K. Davis, personal communication), nettle has nothing to compare with the battery of defences which bracken deploys in the spring (unintentional sampling suggests that it might sting more readily early in the season, but not by much!). On present evidence these results are consistent with the architecture hypothesis but not with the plant chemistry hypothesis (see also Denno, 1977).

In contrast to herbs which get very much bigger and more complex as the season progresses, trees and woody bushes, with most of their structure already in place, might be expected to fill up rapidly with species as soon as their leaves open. The foliage-feeding Lepidoptera on oak trees (*Quercus robur*) follow this pattern exactly (Fig. 7.9d) (Feeny, 1970), although because oak leaves are not chemically defended in the spring (Feeny, 1970), these results allow more than one interpretation. However, oak trees do appear to illustrate the modifying influence of quantitative defences fairly clearly. As tannin levels increase with season and protein levels fall (Feeny, 1970) so some of the oak foliage is no longer available and the diversity of the Lepidoptera larvae decreases (Fig. 7.9d). The seasonal distribution of the full oak fauna is obviously much more complex than this, and I do not know what would happen to Fig. 7.9d if other insect groups were included as well. However, it may not be coincidence that the number of species on bracken (Fig. 7.4) also falls slightly after the mid-summer peak in response to similar changes in the levels of quantitative defences. In bracken, tannin and silicate levels rise (Fig. 7.6d) and protein levels fall (Fig. 7.6c), exactly as in the oak, so that the palatability of the foliage progressively decreases during the growing season. Fig. 7.10 illustrates, in an idealised manner, the proposed interaction between plant architecture and quantitative defences during the growing season in a hypothetical tree and herb.

In passing, we should note that the tannins in bracken may be less effective quantitative defences than those of oak trees. Only dicotyledons have hydrolysable tannins; the condensed tannins of more primitive plants are about five times less effective than

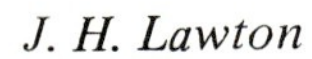

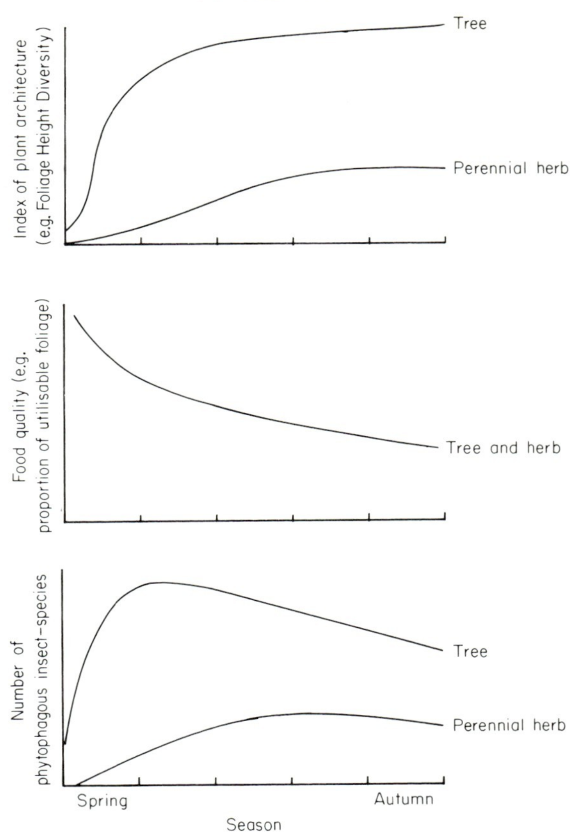

FIG. 7.10. Hypothetical seasonal changes in insect diversity on a typical deciduous tree and perennial herb in response to seasonal changes in plant architecture and chemistry. Species diversity increases as the leaves appear and as the plant grows, but decreases in response to seasonal changes in plant 'quantitative defences' (see text). Scales are arbitrary, but an appropriate measure of plant architecture would be Foliage Height Diversity (e.g. Murdoch *et al*, 1972), and for the proportion of utilisable foliage, the protein: tannin ratio (Feeny, 1970).

hydrolysable tannins as digestibility reducing compounds (Swain, 1977). This could explain why sawfly caterpillars are able to feed on bracken much more successfully in late summer than are their ecological analogues (Lepidoptera) on oak (Feeny, 1976; Lawton, 1976).

Other data on the seasonal development of the insect communities on woody bushes and trees are scarce. There appear to be roughly as many species of phytophagous insects on broom (*Sarothamnus scoparius*) in March and April as there are in July and August (Waloff, 1968) with a temporary and rather small rise in diversity in mid-season as the seed-pods develop. This is exactly what we would expect, but obviously more data are needed to test Fig. 7.10 adequately.

Finally, consider the total number of insects associated with a particular plant, rather than seasonal changes in the diversity of the insects on one plant. Different species of plants differ enormously in their chemistry, so that, if plant chemistry played a major role in determining the total number of species of insects on particular types and species of plants, the patterns displayed in Fig. 7.1 should be nowhere near as clear as they are. (Of course this is negative evidence, but it is none the less useful for that.) Exactly this point has previously been made by Southwood (1960*b*) for British trees. However, this does not rule out the possibility that differences in plant chemistry might contribute to some of the scatter round these regression lines. Just as quantitative defences appear to have a slight, but detectable, effect on changes in the number of species of insect attacking bracken and oak trees within a season, so chemical defences may also have a slight but again detectable effect on the total species-pool; indeed the former is likely to contribute to the latter. Thus several native British trees (e.g. holly, *Ilex*, and yew, *Taxus*) have smaller numbers of insects associated with them than would be expected from their abundance, and Southwood (1961, 1973, 1977) has already suggested that this might be a result of peculiarities in their chemistry.

Though attractive, this idea is difficult to test rigorously. Southwood (1960*b*, 1973) has suggested that the 'degree of predilection' for changing host plants by insects will depend on the physical and biochemical similarities of the plants involved. Hence 'exchange' of herbivores (on an evolutionary time scale) between closely related plants should be easier than 'exchange' between unrelated plants; and taxonomically isolated, and therefore in the main chemically distinct, plants might be expected to have fewer species associated with them than plants with many relatives in the same geographical region. Southwood (1977) presents some evidence for this in the case of British trees. However taxonomic isolation (the number of species of plants in each genus) explains none of the residual scatter round the regression-lines shown in Fig. 7.1 (Lawton & Schröder, 1977), except in the case of the monocotyledons where there is a small but nevertheless significant effect.

A major barrier to demonstrating an effect of taxonomic isolation in this way, or to a direct and clear demonstration of the effects of plant chemistry, is that their influence seems slight and likely to be obscured both by inadequacies in the data as well as by other processes which also need to be considered. For example, small differences in architecture within each plant category (see also Southwood, 1960*a*, again for trees) and the plant's abundance within its geographical range are also likely to influence the scatter round the regression lines shown in Fig. 7.1, as is the length of time a plant has been a member of the flora (Southwood, 1961; 1973; 1977), and it is not difficult to think of still other factors. This is merely another way of saying that the effects of a plant's geographical range and its architecture are much more important in determining the number of species of insects which eventually evolve to exploit it than are its chemistry, or other details of its biology.

What you are and what you do

How does plant architecture influence insect diversity? One possibility is that phytophagous insects compete for a limited food supply in the classical manner; the number of species on a plant being determined by niche overlap and niche width along resource axes which are the various parts of the plant, its leaves, stems, fruits and so on (MacArthur, 1972; May, this symposium, Chapter 12; Pianka, 1976; Southwood, this symposium, Chapter 2).

Very few people, I think, would cling to the view that because 'the world is green' herbivores cannot be food limited (Hairston *et al*, 1960). We have already seen that not all plant material may actually be edible, and a net surplus at the end of the year does not imply that critical food shortages have not occurred within the year (Sinclair, 1975). There is also a growing literature demonstrating competition for food amongst phytophagous insects which does not, at the same time, involve complete defoliation of the host (Gruys, 1970; Hill, 1976; McNeill, 1973; McNeill & Southwood, 1978).

However, in many phytophagous insects there really is no evidence of food-shortage or interspecific competition for food most of the time; Rathcke (1976) provides an excellent summary of the literature. A major factor limiting the diversity of phytophagous insects on a particular species of plant may therefore be competition for 'enemy-free space' (Gilbert & Singer, 1975; Lawton & Schröder, 1976 and in press; Zwölfer, 1975). Considerable attention has been paid to the effects of monophagous insect parasitoids and predators on the population dynamics of plant-feeding insects (Beddington *et al*, 1976; Hassell, 1976; May, this symposium, Chapter 12). Interestingly, the number of polyphagous enemies attacking one species of insect is usually at least equal to, and often much greater than, the number of strictly monophagous species (Askew & Shaw, 1974; Miller & Renault, 1976; Pschorn-Walcher & Zinnert, 1971; Rathcke & Price, 1976; Richards, 1940; Schröder, 1974; Shaw & Askew, 1976). Not only that, polyphagous parasitoids often attack a variety of hosts doing similar things to a particular part of the plant; that is they tend to be plant-specific, and location specific but not insect-specific in their attacks (Askew, 1971; 1975; Askew & Shaw, 1974; DeBach, 1964; Richards, 1940: Shaw & Askew, 1976; Vinson, 1975, 1976). Avoiding generalist predators and parasitoids that are already in residence is presumably a major problem for any insect invading a new host plant. Indeed, attempts to introduce phytophagous insects for the biological control of weeds often fail because of the impact of these polyphagous enemies (Goeden & Louda, 1976).

The effects of polyphagous enemies on invasability is complex, and I do not want to develop it further here. I raise it simply to establish that plant architecture may influence species diversity in ways other than through conventional competition for food. Larger and more complex plants permit greater niche diversification *both* for classical competitor avoidance *and* the avoidance of your neighbours' predators and parasitoids.

These considerations are useful in one other way. They serve to remind us that the number of species which attacks a plant is a very crude thing to measure. Perhaps we should be increasingly concerned not just with numbers of species but with what those species are, and even more important, what they do to the plant.

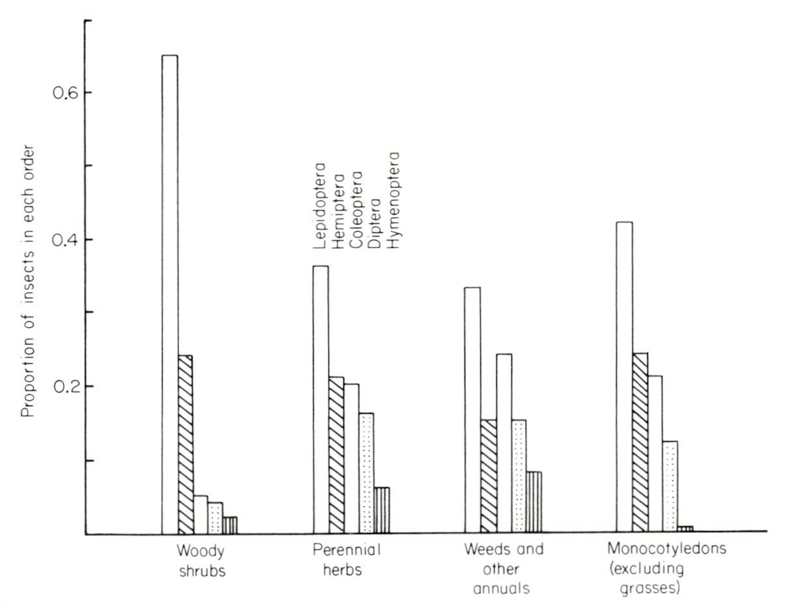

FIG. 7.11. Taxonomic composition of the insect faunas on the four groups of plants illustrated in Fig. 7.1. (Data from Lawton & Schröder, in press).

Figure 7.11 summarises the types of insects (by order) which go to make up the total species' complements of the plants in Fig. 7.1 (Lawton & Schröder, in press). A simple pair-wise comparison using χ^2 shows that the proportions of insects in each taxonomic group do not differ significantly between the kinds of plants, except for woody shrubs, which have proportionately more species of Lepidoptera than the other kinds of plants. Note that Southwood's (1961) data on British trees present a picture very similar to that for woody bushes and shrubs; Lepidoptera on trees are approximately four times more important (0.67 of the records) than the Hemiptera and Coleoptera, which, as on most of the other plant-types, are recorded in approximately equal proportions (0.16 and 0.17 of the records for trees respectively).

These results give us an average picture of the sorts of insects typical plants of different types have feeding on them. Individual species may differ markedly from the norm (Lawton & Schröder, in press). Thus plants in the genus *Sonchus* (sow thistles) appear to have proportionately more Diptera than expected. A further example is provided in Fig. 7.12a which summarises the taxonomic composition of the key species of insect on bracken in Britain. The most striking things about this community are the lack of any Coleoptera and the importance of sawflies; indeed the whole balance is quite different to the average picture conveyed by Fig. 7.11. The full faunal list for bracken reinforces this impression (Lawton, 1976).

In the light of the regular patterns in the total number of species of insects found on a plant (Fig. 7.1), differences in the *kinds* of insects involved on particular plants are intriguing. To what extent, for example are the major differences between plant types, or the residual scatter round the regression lines in Fig. 7.1 due not only to the sorts of processes already discussed, but also to the particular success or failure of different groups of insects on different types or species of plants? Some of the increase in the number of species on woody bushes, for example, can be accounted for by their greater numbers of Lepidoptera, but other generalisations are less easy to make.

However, the fact that clear patterns are discernible in Fig. 7.1 suggests that either we have been very lucky, or that what the species are taxonomically may be less important than what they do to the plant. Thus Lepidoptera can mine, gall or feed externally (chew): so can Coleoptera. What is more, adults and larvae, or even different larval stages, often do different things (see Southwood, this volume). If competition for food and enemy free space are important then we might expect greater interaction within guilds, irrespective of taxonomy, than between similar taxa which exploit the plant in different ways. I would therefore predict greater constancy in the ways in which insects exploit particular kinds or species of plants, than in the taxonomic composition of those insects.

If this idea has any substance, it should be easiest to test on a widespread plant with a well-established and diverse fauna; for example, bracken. The insects on the above-ground parts of bracken exploit it in four main ways: chewers, feeding externally on the pinnae; suckers, all Hemiptera; miners burrowing into the pinnae and rachis; and gall-formers. The proportions (in the British Isles as a whole) within each guild are shown in Fig. 7.12b. Now consider our islands on the North Yorkshire Moors. We have already seen that as the size of a bracken patch falls, species drop out. But the sequence in which this happens is very far from random in terms of what they do the plant. In fact, the proportions in the various guilds stay approximately constant (Fig. 7.13).

Recently, Alan Kirk (Kirk, 1977, and personal communication) has carried out preliminary surveys on bracken insects in Papua and New Guinea. It would be difficult to imagine a more strikingly different place to look at the plant, and because of the lack of

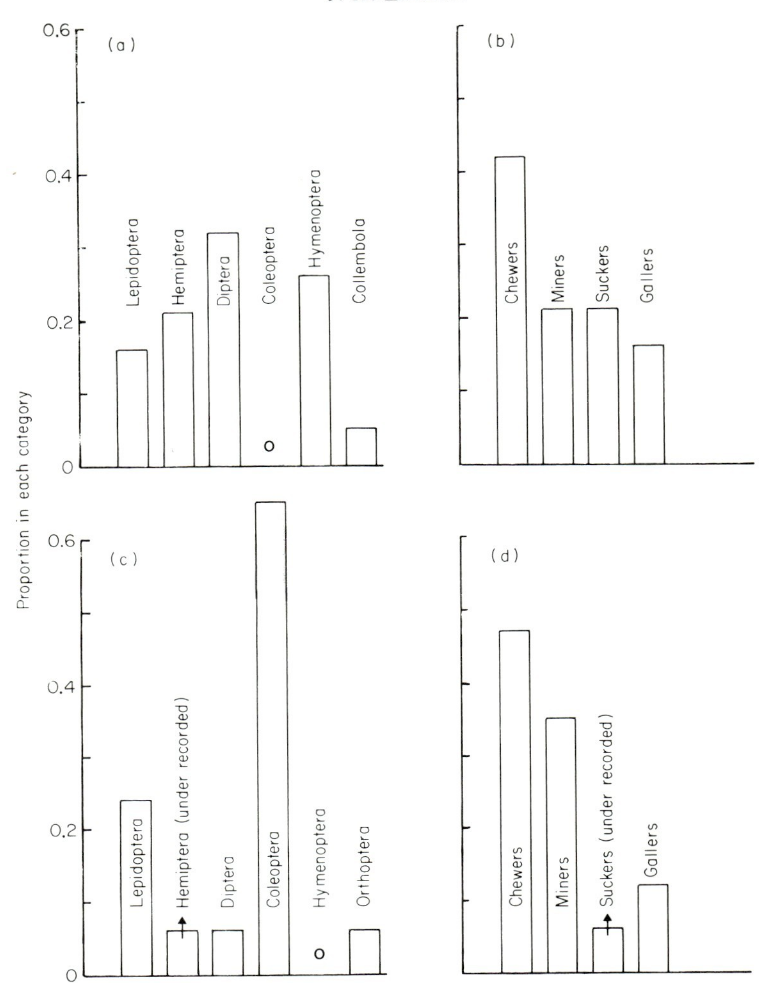

FIG. 7.12. Bracken insects by taxonomy and guild-structure. (a) Taxonomic composition of 19 key-species (excluding rhizome-feeders) in Britain; the Hymenoptera are all sawflies (Tenthredinidae). (b) distribution of the same species by feeding-type; (c) taxonomic composition and (d) feeding-type of 17 species from bracken in Papua and New Guinea (Kirk, 1977 and personal communication). Some of the New Guinea Coleoptera mine both the stalk and the rhizome; half the species which do this have arbitrarily been excluded from the analysis.

seasonality all developmental stages, from croziers to mature pinnae, are present at one and the same time. However the structure of the individual plant, its basic architecture, remains the same. Taxonomically, the bracken fauna in Papua and New Guinea is totally different to that found in Britain (Fig. 7.12c). There are no sawflies (although sawflies as a group occur on other plants), and Coleoptera dominate the system. However, what the insects do to the plant is not too dissimilar (Fig. 7.12d). Given the major differences in climate, habitats and natural enemies I would not expect the guilds to be identical, and there are some important differences which Fig. 7.12 hides. Far more of the miners do so in the rachis and stalk, for example, in Papua and New Guinea compared with Britain. But the convergence is sufficiently similar to make me wonder whether it is entirely due to coincidence.

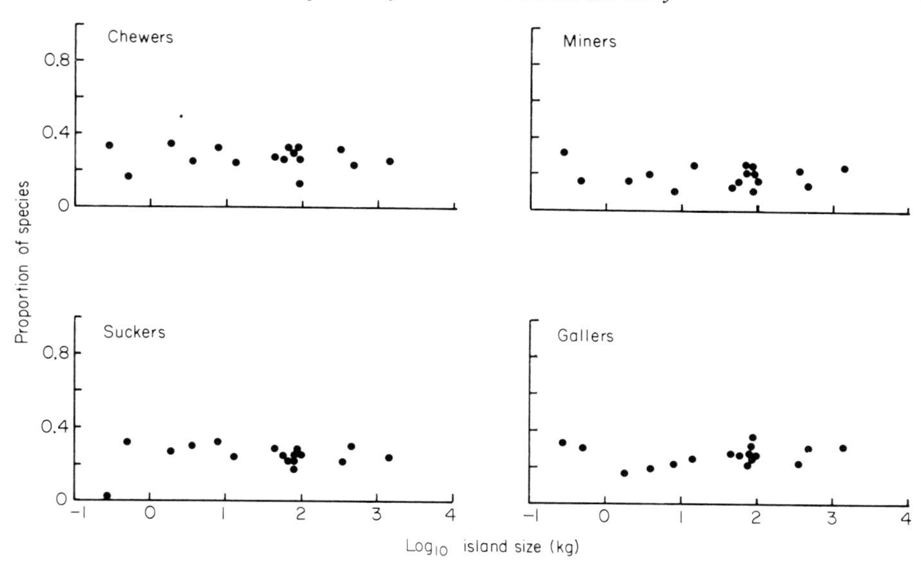

FIG. 7.13. Proportion of feeding-types on bracken islands of different sizes. Although the number of species changes on different sized islands (Figs. 7.2 and 7.3) the structure of the community stays constant. Not all the 19 key-species shown in Fig. 7.12 occur on these islands. There is also a gall-forming mite which is rare nationally and which has not been included in the key-species. Hence the structure of these island communities is not exactly the same as is shown in Fig. 7.12. (C. Rigby, personal communication).

I think that a consideration of what insects do to plants, rather than just counting the numbers of species, may well reveal unsuspected patterns in the structure of phytophagous insect communities, but we cannot test this idea without more detailed information.

Concluding remarks

Viewed in the wider context of the ecology of other groups of organisms the ideas outlined in this paper are not novel. Habitat complexity influences species diversity in a variety of other taxa, including aquatic invertebrates (Harmon, 1972; Macan & Kitching, 1972), Psocoptera (Turner & Broadhead, 1974), small mammals (Rosenzweig & Winakur, 1969) and birds (MacArthur & MacArthur, 1961; Recher, 1971). Arising directly from MacArthur's work on bird species-diversity Murdoch *et al* (1972) attempted to disentangle the effects of plant species-diversity and vegetation complexity on the diversity of plant-sucking Homoptera. They found that either plant species-diversity or vegetation-complexity (foliage height diversity) were almost equally good predictors of insect-diversity. Since different species of Homoptera tend to occur on different kinds of plants, a correlation between plant-diversity and insect-diversity is to be expected. Equally, the present work suggests that more complex plants tend to have more insects on them anyway. Hence a correlation between vegetation complexity and species diversity would also be expected. The relative contributions of each are impossible to sort out because plant species-diversity and foliage height-diversity are themselves highly correlated.

I want to finish with a caveat. By piecing together a series of hypotheses, and not too much data, we can trace arguments which run from the seasonal changes in the structure of an insect community on the one hand, and small-scale changes in community structure with plant patch-size on the other, to the much broader and more general patterns displayed in Fig. 7.1. But our view of the world is only as good as our data, so that it is disturbing to find that Strong & Levin (in press) in an independent analysis similar to that shown in Fig. 7.1 come to different conclusions on some (but not all) key points. For example, they find no difference like that seen in Fig. 7.1b and 7.1d between monocotyledons and perennial dicotyledons. This can partly be explained by small, but important, differences in the methods of analysis; in particular they have pooled all the insect-records for plant genera, and have analysed these rather than use the data for the individual species of plants. However, the differences between us might also reflect weaknesses in our data, or theirs, or both. In other words, as at several other points in this paper, we need more and better data. But then in ecology we often do.

Acknowledgements

Professor T. R. E. Southwood, F.R.S., Professor M. H. Williamson and Dr. M. P. Hassell made extensive comments on an early draft of this paper, and greatly improved it. I have benefited enormously from discussions and correspondence about insects and plants, both of them and with Drs J. M. Anderson, A. A. Kirk, S. McNeill, D. Schröder, and D. R. Strong. Miss L. C. Rigby and Mr C. G. Jones kindly gave me permission to quote from their unpublished thesis work. I am most grateful to them, and to those other friends and colleagues listed above who also gave me generous access to unpublished manuscripts.

References

Abele L. G. (1976) Comparative species richness in fluctuating and constant environments: coral-associated decapod crustaceans. *Science* **192**, 461–3.

Askew R. R. (1971) *Parasitic Insects*. xvii + 316 pp. Heinemann, London.

Askew R. R. (1975) The organisation of chalcid-dominated parasitoid communities centred upon endophytic hosts. In *Evolutionary Strategies of Parasitic Insects and Mites*, Ed. P. W. Price, pp. 130–53. Plenum, New York & London.

Askew R. R. & Shaw M. R. (1974) An account of the Chalcidoidea (Hymenoptera) parasitising leaf-mining insects of deciduous trees in Britain. *Biol. J. Linn. Soc.* **6**, 289–335.

Atsatt P. R. & O'Dowd D. J. (1976) Plant defence guilds. *Science* **193**, 24–9.

Beddington J. R., Free C. A. & Lawton J. H. (1976). Concepts of stability and resilience in predator-prey models. *J. anim. Ecol.* **45**, 791–816.

Bentley B. L. (1976) Plants bearing extrafloral nectaries and the associated ant community: interhabitat differences in the reduction of herbivore damage. *Ecology* **57**, 815–20.

Bentley B. L. (1977) The protective function of ants visiting the extrafloral nectaries of *Bixa orellana* (Bixaceae). J. *Ecol.* **65**, 27–38.

Brown J. H. (1975) Geographical ecology of desert rodents. In *Ecology and Evolution of Communities*, Eds. M. L. Cody & J. M. Diamond, pp. 315–41. Belknap Press, Harvard University, Cambridge, Mass.

Carlisle D. B. & Ellis P. E. (1968) Bracken and locust ecdysones: their effects on moulting in the desert locust. *Science* **159**, 1472–4.

Chew F. S. (1975) Coevolution of pierid butterflies and their cruciferous foodplants I. The relative quality of available resources. *Oecologia* **20**, 117–29.

Cooper-Driver G. (1976) Chemotaxonomy and phytochemical ecology of bracken. *Bot. J. Linn. Soc.* **73**, 35–46.

Cooper-Driver G. & Swain T, (1976) Cyanogenic polymorphism in bracken in relation to herbivore predation. *Nature* **260**, 604.

Cromartie W. J. Jr. (1974) The effects of stand size and vegetational background on the colonization of cruciferous plants by herbivorous insects. Unpublished Pd.D. thesis, Cornell University.

Darwin F. (1877) On the nectar glands of the common brake fern. *J. Linn. Soc. (Bot.)* **15**, 398–409.

Davis B. N. K. (1973) The Hemiptera and Coleoptera of stinging nettle (*Urtica dioica* L.) in East Anglia. *J. appl. Ecol.* **10**, 213–37.

Davis B. N. K. (1975) The colonization of isolated patches of nettles (*Urtica dioica* L.) by insects. *J. appl. Ecol.* **12**, 1–14.

DeBach P. (1964) *Biological Control of Insect Pests and Weeds*, xxiv + 844 pp. Chapman & Hall, London.

Denno R. F. (1977) Comparison of the assemblages of sap-feeding insects (Homoptera-Hemiptera) inhabiting two structurally different salt marsh grasses in the genus *Spartina*. *Env. Ent.* **6**, 359–72.

Evans W. C. (1976) Bracken thiaminase-mediated neurotoxic syndromes. *Bot. J. Linn. Soc.* **73**, 113–31.

Feeny P. (1970) Seasonal changes in oak leaf tannins and nutrients as a cause of spring feeding by winter moth caterpillars. *Ecology* **51**, 565–81.

Feeny P. (1976) Plant apparency and chemical defence. In *Biochemical Interaction between Plants and Insects*. Eds. J. W. Wallace & R. L. Mansell. *Recent Adv. Phytochem.* **10**, 1–40.

Gilbert L. (1978) Development of theory in the analysis of insect-plant interactions. In *Analysis of Ecological Systems*, Eds. D. Horn, R. Mitchell and G. R. Stairs. Ohio State University Press, Colombus, (in press).

Gilbert L. E & Singer M. C. (1975) Butterfly ecology. *A. Rev. Ecol. Syst.* **6**, 365–97.

Goeden R. D. (1971) Insect ecology of silverleaf nightshade. *Weed Science* **19**, 45–51.

Goeden R. D. & Louda S. M. (1976) Biotic interference with insects imported for weed control. *A. Rev. Ent.* **21**, 325–42.

Goeden R. D. & Ricker D. W. (1974) The phytophagous insect fauna of the ragweed, *Ambrosia acanthicarpa*, in Southern California. *Envir. Ent.* **3**, 827–34.

Gruys P. (1970) Growth of *Bupalus piniarius* (Lepidoptera: Geometridae) in relation to larval population density. *Verh. Rijks Inst. Nat.* **1**, 1–127. PUDOC, Wageningen.

Hairston N. G., Smith F. E. & Slobodkin L. B. (1960) Community structure, population control, and competition. *Am. Nat.* **84**, 421–5.

Harmon W. N. (1972) Benthic substrates: their effect on fresh water molluska *Ecology* **53**, 271–7.

Hassell M. P. (1976) Arthropod predator-prey systems. In *Theoretical Ecology*, Ed. R. M. May, pp. 71–93. Blackwell Scientific Publications, Oxford.

Hill M. G. (1976) The population and feeding ecology of five species of leafhoppers (Homoptera) on *Holcus mollis* L. Unpublished D.Phil. thesis, University of London.

Hocking B. (1975) Ant-plant mutualism: evolution and energy. In *Coevolution of Animals and Plants*, Eds. L. E. Gilbert & P. H. Raven. pp. 78–90. University Texas Press, Austin & London.

Janzen D. H. (1968) Host plants as islands in evolutionary and contemporary time. *Am. Nat.* **102**, 592–5.

Jones C. G. (1977) Chemical content and insect resistance in bracken fern (*Pteridium aquilinum* (L.) Kuhn). Unpublished D.Phil. thesis, University of York.

Jones C. G. & Firn R. D. (1978) The role of phytoecdysteroids in bracken fern (*Pteridium aquilinum* (L.) Kuhn) as defences against phytophagous insect attack. *J. Chem. Ecol.* (in press).

Kaplanis, J. N., Thompson M. J., Robins W. E. & Bryce B. M. (1967) Insect hormones: alpha ecdysone and 20-hydroxyecdysone in bracken fern. *Science* **157**, 1436–8.

Karr J. R. (1976) Seasonality, resource availability, and community diversity in tropical bird communities. *Am. Nat.* **110**, 973–94.

Kirk A. A. (1977) The insect fauna of the weed *Pteridium aquilinum* (L.) Kuhn (Polypodiaceae) in Papua New Guinea: a potential source of biological control agents. *J. Aust. Ent. Soc.* **16**, 403–9.

Kogan M. (1977) The role of chemical factors in insect/plant relationships. *Proc. XVth Int. Cong. Ent., Washington D.C. 1976*, 211–7.

Kuris A. M. & Blaustein A. R. (1977) Ecotoparasitic mites on rodents: application of the island biogeography theory? *Science* **195**, 596–8.

Lawton J. H. (1976) The structure of the arthropod community on bracken. *Bot. J. Linn. Soc.* **73**, 187–216.

Lawton J. H. & Schröder D. (1977) Effects of plant type, size of geographical range and taxonomic isolation on number of insect species associated with British plants. *Nature* **265**, 137–40.

Lawton J. H. & Schröder D. (1979) Some observations on the structure of phytophagous insect communities: the implications for biological control. *Proc. 4th Int. Symp. Biol. Control Weeds, Gainesville, Florida, 1976*, (in press).

Macan T. T. & Kitching A. (1972) Some experiments with artificial substrata. *Verh. Int. Ver. Limnol.* **18**, 213–20.

MacArthur J. W. (1975) Environmental fluctuations and species diversity. In *Ecology and Evolution of Communities*, Ed. M. L. Cody & J. M. Diamond pp. 74–80. Belknap Press, Harvard University, Cambridge, Mass.

MacArthur R. H. (1972) *Geographical Ecology*. Harper & Row, New York.

MacArthur R. H. & MacArthur J. W. (1961) On bird species diversity. *Ecology* **42**, 594–8.

MacArthur R. H. & Wilson E. O. (1967) *The Theory of Island Biogeography*. Princeton U.P., Princeton N.J.

McClure M. S. & Price P. W. (1976) Ecotope characteristics of coexisting *Erythroneura* leafhoppers (Homoptera: Cicadellidae) on sycamore. *Ecology* **47**, 928–40.

McNeill S. (1973) The dynamics of a population of *Leptoterna dolobrata* (Heteroptera: Miridae) in relation to its food resources. *J. anim. Ecol.* **42**, 495–507.

McNeill S. & Southwood T. R. E. (1978) The role of nitrogen in the development of insect/plant relationships. In *Biochemical Aspects of Insect/Plant Interactions*, Eds. J. B. Harborne & H. F. van Emden. Academic Press, London and New York.

Miller C. A. & Renault T. R. (1976) Incidence of parasitoids attacking endemic spruce budworm (Lepidoptera: Tortricidae) populations in New Brunswick. *Can. Ent.* **108**, 1045–52.

Murdoch W. W., Evans F. C. & Peterson C. H. (1972) Diversity and pattern in plants and insects. *Ecology* **53**, 819–28.

Opler P. A. (1974) Oaks as evolutionary islands for leaf-mining insects. *Am. Sci.* **62**, 67–73.

Perring F. J. & Walters S. M. (Eds) (1962) *Atlas of the British Flora.* BSBI, Nelson, London and Edinburgh.

Pianka E. R. (1976) Competition and niche theory. In *Theoretical Ecology*, Ed. R. M. May, pp. 114–141. Blackwell Scientific Publications, Oxford.

Pimentel D. (1961) The influence of plant spatial patterns on insect populations. *Ann. Ent. Soc. Am.* **54**, 61–9.

Price P. W. (1976) Colonization of crops by arthropods: non-equilibrium communities in Soybean fields. *Envir. Ent.* **5**, 605–11.

Pschorn-Walcher H. & Zinnert K. D. (1971) Investigations on the ecology and natural control of the larch sawfly (*Pristiphora erichsonii* Htg., Hymenoptera: Tenthredinidae) in central Europe Part II: Natural enemies. *Tech. Bull. Commonw. Inst. Biol. Control* **14**, 1–50.

Rathcke B. J. (1976) Competition and coexistence within a guild of herbivorous insects. *Ecology* **57**, 76–87.

Rathcke B. J. & Price P. W. (1976) Anomolous diversity of tropical ichneumonid parasitoids: a predation hypothesis. *Am. Nat.* **110**, 889–93.

Recher H. F. (1971) Bird species diversity: a review of the relation between species number and environment. In *Quantifying Ecology*, Ed. H. A. Nix. *Proc. Ecol. Soc. Australia* **6**, 135–52.

Rhoades D. F. & Cates R. G. (1976) Toward a general theory of plant antiherbivore chemistry. In *Biochemical Interactions between Plants and Insects,* Ed. J. W. Wallace & R. L. Mansell, *Rec. Adv. Phytochem*. **10**, 168–213.

Richards O. W. (1940) The biology of the small white butterfly (*Pieris rapae*) with special reference to the factors controlling its abundance. *J. anim. Ecol.* **9**, 243–88.

Root R. B. (1973) Organization of a plant-arthropod association in simple and diverse habitats: the fauna of collards (*Brassica oleracea*) *Ecol. Monogr.* **43**, 95–124.

Rosenzweig M. L. & Winakur J. (1969) Population ecology of desert rodent communities: habitats and environmental complexity. *Ecology* **50**, 558–72.

Schröder D. (1974) A study of the interactions between the internal larval parasites of *Rhyaconia buoliana* (Lepidoptera: Olethreutidae). *Entomophaga* **19**, 145–71.

Shaw M. R. & Askew R. R. (1976) Ichneumonoidea (Hymenoptera) parasitic upon leaf-mining insects of the orders Lepidoptera, Hymenoptera and Coleoptera. *Ecol. Ent.* **1**, 127–33.

Simberloff D. (1976) Species turnover and equilibrium island biogeography. *Science* **194**, 572–8.

Sinclair A. R. E. (1975) The resource limitation of trophic levels in tropical grassland ecosystems. *J. anim. Ecol.* **44**, 497–520.

Southwood T. R. E. (1960*a*) The abundance of the Hawaiian trees and the number of their associated insect species. *Proc. Hawaii. ent. Soc.* **17**, 299–303.

Southwood T. R. E. (1960*b*) The evolution of the insect-host tree relationship – a new approach. *Proc. XIth Int. Cong. Ent., Vienna* 1960: 651–4.

Southwood T. R. E. (1961) The number of species of insect associated with various trees. *J. anim. Ecol.* **30**, 1–8.

Southwood T. R. E. (1973) The insect/plant relationship – an evolutionary perspective. In *Insect/Plant Relationships Symp. R. Ent. Soc. Lond.*, Ed. H. F. Van Emden **6**, 3–30.

Southwood T. R. E. (1977) The stability of the trophic milieu, its influence on the evolution of behaviour and of responsiveness to trophic signals. *Coll. Int. Cent. natn. Res. Scient.* **265**. *Comportement des Insectes et Milieu Trophique.* pp. 471–93.

Strong D. R. Jr. (1974*a*) Nonasymptotic species richness models and the insects of British trees. *Proc. natn. Acad. Sci. USA* **71**, 2766–9.

Strong D. R. Jr. (1974*b*) Rapid asymptotic species accumulation in phytophagous insect communities: the pests of Cacao. *Science* **185**, 1064–6.

Strong D. R. & Levin D. A. (1980) Species richness of plant parasites and growth form of their hosts, (in press).

Strong D. R., McCoy, E. D. & Rey, J. R. (1977) Time and the number of herbivore species: the pests of sugarcane. *Ecology* **58**, 167–75.

Swain T. (1977) The effect of plant secondary products on insect plant coevolution. *Proc. XVth Int. Cong. Ent., Washington D.C. 1976:* 249–56.

Taylor L. R. (1977) Migration and the spatial dynamics of an aphid, *Myzus persicae. J. anim. Ecol.* **46**, 411–23.

Tepedino V. J. & Stanton N. L. (1976) Cushion plants as islands. *Oecologia* **25**, 243–56.

Turner B. D. & Broadhead E. (1974) The diversity and distribution of psocid populations on *Mangifera indica* L. in Jamaica and their relationship to altitude and micro-epiphyte diversity. *J. anim. Ecol.* **43**, 173–190.

Vinson S. B. (1975) Biochemical coevolution between parasitoids and their hosts. In *Evolutionary Strategies of Parasitic Insects and Mites*, Ed. P. W. Price pp. 14–48. Plenum, New York & London.

Vinson S. B. (1976) Host selection by insect parasitoids. *A. Rev. Ent.* **21**, 109–133.

Waloff N. (1968) Studies on the insect fauna on Scotch broom *Sarothamnus scoparius* (L.) Wimmer. *Adv. ecol. Res.* **5**, 87–208.

Ward L. K. (1977) The conservation of juniper: the associated fauna with special reference to southern England. *J. appl. Ecol.* **14**, 81–120.

Ward L. K. & Lakhani K. H. (1977) The conservation of juniper: the fauna of food plant island sites in southern England. *J. appl. Ecol.* 14, 121–135.

Whittaker R. H. (1969) Evolution of diversity in plant communities. *Diversity and Stability in Ecological Systems. Brookhaven Symp. Biol.* **22**, 178–95.

Wiklund C. (1974) Oviposition preferences in *Papilio machaon* in relation to the host plants of the larvae. *Ent. exp. appl.* **17**, 189–98.

Wiklund C. (1975) The evolutionary relationship between adult oviposition references and larval host plant range in *Papilio machaon* L. *Oecologia* **18**, 185–97.

Wiklund C. (1977) Oviposition, feeding and spatial separation of breeding and foraging habitats in a population of *Leptidea sinapis* (Lepidoptera). *Oikos* **28**, 56–68.

Zwölfer H. (1965) Preliminary list of phytophagous insects attacking wild Cynareae (Compositae) in Europe. *Tech. Bull. Commonw. Inst. biol. Control* **6**, 81–153.

Zwölfer H. (1975) Speciation and niche diversification in phytophagous insects. *Verh. dt. zool. Ges.* **67**, 394–401 (in German, English summary).

8 • Urbanisation and the diversity of insects

B. N. K. DAVIS

Institute of Terrestrial Ecology, Monks Wood Experimental Station, Abbots Ripton, Huntingdon, Cambs

This paper considers some of the short-term changes in insect faunas resulting from urban development. Urbanisation in fact is as old as civilisation but it is only within the last two centuries and particularly within the last 50 years that its effects have come to be felt as major intrusions upon the countryside. In North America, places like Chicago grew up from a village on the prairie to become a 'city of the plains' in less than half a century, and so impinged on habitats little altered by man. In Britain, however, the countryside has been subject to agricultural and forestry practices for many centuries before giving way to bricks and tarmac. Already, therefore, the fauna and flora have been much altered from their primeval state by an increase in woodland margin, culture-steppe and ruderal species adapted to the patchwork landscape of enclosure farming.

Urbanisation causes a further reduction in the size and number of semi-natural habitats but provides a variety of new ones. The degree of disturbance increases markedly and, in the past at least, air pollution has become a limiting factor for plant growth. The combined effects on the fauna and flora are twofold: the extinction of the less tolerant and more specialist species and an increase in eurytopic and synanthropic ones. These effects are relatively well documented in the case of London and I shall draw most of my examples from here in an initial review. This is followed by a brief study of geographical and historical factors as determinants of faunal diversity. In the absence of quantitative data in most instances, I shall mainly consider species richness as a component of diversity, N_0 in Hill's (1973) terminology.

Review

Fitter (1945) has given a vivid account of the growth of London, and Kent (1975) has provided a clear picture of the gradual loss of woodland, meadow, marsh and heathland plants in the former county of Middlesex. In the early nineteenth century there was open country beyond Marylebone but by about 1900 it had retreated to Notting Hill, Kilburn and Hampstead, and in the next 30 years to Brentford, Ealing, Hendon and Friern Barnet with urban fingers reaching out as far as Harrow and Enfield. Today one can travel through built up land as far as Watford some 30 km from the city of London. In the past 100 years, 78 species of native and naturalised plants known to Trimen and Dyer (1869) have become extinct and many more have declined. Recent examples of urban pressures and the erosion of scientific interest are given in a general review of this subject (Davis, 1976; see Tables 4 and 5).

Changes in the invertebrate fauna have not been catalogued in nearly such detail but the disappearance of rural habitats has undoubtedly been accompanied by the loss of many insects directly or indirectly associated with them. A seventeenth century collection of Coleoptera from the London area described by Hammond (1975) provides useful historical data but few surprises. Of the 106 species identified, one is scarce in Britain and eight others scarce or infrequent in the London area today. One of the best accounts of the loss or deterioration of habitats for Lepidoptera is given by Adkin (1934) for the period since about 1860. He describes changes occurring in the fauna caused by housing developments at Lewisham and Blackheath, by the construction of a munitions factory and gas works on Greenwich marshes and by increased public pressure at West Wickham Wood and Shirley Heath. Changes in the fauna of Hampstead Heath provide good case histories. Guichard and Yarrow (1948) gave details of 216 species of aculeate Hymenoptera recorded from the Heath since 1832 and considered that more than 150 species had occurred there during the present century. Forty-nine species had apparently not been recorded since 1900 though Satchell (1965) estimated that probably only 30 had really disappeared in this period. The species remaining are clearly well adapted to this area and therefore able to tolerate considerable modifications and disturbance of the habitat. More recently, Carter (1967) has reviewed the changes occurring in the Lepidoptera of Hampstead Heath in four decades up to 1950–8. He noted the absence of 90 species recorded by O'Brien-Ellison (Hampstead Scientific Society, 1913) and the presence of 35 species not previously recorded. The effects of urbanisation upon the Lepidoptera have also been examined by Taylor *et al* (in press) using collections of moths made in the Rothamsted Insect Survey. He showed that light traps situated in five suburban and urban areas yielded very impoverished faunas compared with traps in mature woodland sites whilst agricultural habitats were intermediate.

Although the larger parks make important contributions to biological diversity, it is the waste land and above all the suburban gardens which provide the characteristic facies of town faunas. Industrial yards, railway lines and derelict land have become invaded by weedy species many of which, like *Chamaenerion angustifolium, Melilotus officinalis* and *Reseda lutea*, were relatively rare in Middlesex a century ago (Kent, 1975). In addition, there are now over 100 adventive alien species which have become well established since then. *Buddleja davidii* which is so attractive to butterflies and *Cardaria draba* which supports the monophagous weevil *Ceuthorhynchus turbatus* are just two which have shown rapid spread on waste ground since their introduction into Britain about 1896 and 1866 respectively.

An example of the rapid colonisation of urban waste ground by plants and invertebrates was seen in the bombed sites of central and south east London. Parmenter (1953) recorded over 200 species of Diptera at Cripplegate including several previously unknown in Britain. Oxford ragwort *Senecio squalidus*, the dominant plant of the area and itself a widespread colonist only since the advent of the railways, was principal host to the leaf-mining agromyzid *Phytomyza atricornis* and trypetid *Spilographa zoe*. Likewise, eight other Compositae, Groundsel *Senecio vulgaris*, Mugwort *Artemisia vulgaris*, Creeping and Spear thistles *Cirsium arvense* and *C. vulgare*, Coltsfoot *Tussilago farfara*, Sowthistle *Sonchus oleraceus*, Nipplewort *Lapsana communis* and Mayweed *Matricaria* sp. were mined by six other agromyzids and mined or galled by eleven trypetids (now Tephritidae). Records of Lepidoptera, Symphyta and Araneida were discussed by Owen (1951, 1954), Currie (1952) and Le Gross (1949) respectively.

Quarries, gravel pits and other mineral workings share some features in common with urban waste land and may be considered an adjunct of urbanisation. Their scattered rural location, however, and special soil conditions set them in a class apart and their invertebrate faunas are correspondingly distinctive (Davis & Jones, in press).

Gardens possess a fine-grained pattern of structural diversity produced by areas of mown or unmown grass, herbaceous borders, shrubberies and scattered fruit or ornamental trees, cultivated vegetable plots and compost heaps together with a variety of artifical structures such as gravel or paving-slab paths, rockeries, walls, wooden fences, cellars and miniature ponds providing habitats for a wide range of synanthropic species. Gardens are also very rich botanically. They often boast an enormous range of exotic species as well as many close relatives of our native flora which attract and support many insects.

Laburnum anagryoides is the main host of the agromyzid *Phytomyza cytisi* and likewise Monkshood, Larkspur and Columbine, *Aconitum* spp., *Delphinium* spp. and *Aquilegia* spp., are the main hosts for *Phytomyza aconiti* and *P. aquilegiae* (Spencer, 1972). *Amauromyza flavifrons* attacks garden *Dianthus* and *Saponaria* and the more polyphagous *Liriomyza strigata* occurs on Canterbury bells *Campanula medium*, Red Vallerian *Centranthus ruber* and Hollyhock *Althaea rosea* (Parmenter, 1953). Amongst the Lepidoptera there are several which now occur on garden conifers, notably *Eupithecia intricata, E. pusillata, E. phoeniciata* and *Lithophane leautieri*, the latter two exhibiting marked extensions in range in the last 20 years (Heath, 1974). A list of 22 other conifer feeding moths taken in light traps from two suburban gardens in South-east London is given by Allen (1965). A comprehensive study of phytophagous garden insects has yet to be made in this country but Pappa (1976) has compiled a list of 167 species attacking 100 ornamental plant species in Hamburg – 3 Heteroptera, 69 Homoptera, 14 Hymenoptera, 12 Coleoptera, 48 Lepidoptera and 29 Diptera.

A large number of predatory and parasitic insects also occur in gardens though it is less easy to determine their degree of dependence on them. Owen and Owen (1975) collected 455 species (6445 individuals) of Ichneumonidae in two years from a suburban garden in Leicester. Since they were all caught in a Malaise trap 2.6 m^2 sited over an herbaceous border, the spatial concentration of species created by the attraction of the plants is clearly very high. Seventy-four species of Syrphidae and 15 species of butterflies were also collected and several Syrphinae and three Pieridae were seen to breed in the garden. However, most individuals and species of these three groups may merely have visited the plants for nectar or honey-dew before flying on. Many bees and wasps have adapted to urban habitats. Guichard and Yarrow (1948) list 72 species of Aculeata recorded from 6 gardens around Hampstead Heath whilst Haeseler (1972) describes the use of buildings and pruned shrubs as nesting sites.

A. A. Allen's records of more than 700 species of Coleoptera from a garden in South-east London provide some of the most complete data for insects in a suburban area. He has very kindly provided me with notes supplementing the published list of Carabidae (Allen, 1951), and, although the data are not quantitative they are sufficiently detailed to provide a unique record of change spanning 46 years from 1927 until 1973. In 1926 the site was a damp meadow in the north-east corner of Kent about 9 km from the centre of London whereas today it is about equidistant from the centre and the limits of urban development.

Sixty-four species of Carabidae were taken in this garden altogether including two first British records. Table 8.1 provides an analysis of records of 30 species showing marked trends in abundance; species which occurred very sporadically or which remained fairly common or uncommon throughout the period are omitted. (Two were taken only at a mercury vapour light, never in the garden itself.) Four species disappeared or declined after the first few years. On the other hand, 18 species increased in numbers or frequency in the last two decades though only four of these, *Harpalus rufipes*, *Leistus fulvibarbis*, *Metabletus foveatus* and *Platyderus ruficollis* were ever common. Another eight species showed more variable patterns of occurrence with a maximum or minimum at some time during the whole period.

Table 8.1. Species of Carabidae from a suburban London garden exhibiting change in status between 1927 and 1973. Dates indicate first or last date of capture (e.g. 1950– and –1950) or approximate period of change (e.g. –1949–) (data from Allen, 1951, and personal communication)

Becoming more common		Becoming less common	
Agonum dorsale	1950s–	*Amara familiaris*	–1950–
Amara anthobia	1949–	*Carabus nemoralis*	–1931
A. aulica	1950–	*Clivina fossor*	–1930s
A. bifrons	1950–	*Pterostichus melanarius*	–1927
A. lunicollis	1950s–	Increasing then decreasing again	
Asaphidion flavipes	–1949–	*Badister bipustulatus*	Max. 1950s
Bembidion lunulatum	1960s–	*Bembidion tetracolum*	Max. 1949
Bradycellus harpalinus	1950–	*Carabus violaceus*	Max. 1950s–60s
Calathus piceus	–1960s–	*Harpalus tardus*	Max. 1950s
Dromius meridionalis	1950–	Decreasing then increasing again	
Harpalus latus	1960s	*Amara convexior*	Min. 1940s–60s
H. rufibarbis	1950–	*A. plebeja*	Min. 1930s–60s
H. rufipes	–1950s–	*Calathus fuscipes*	Min. 1940s–60s
Leistus fulvibarbis	1949–	*Pterostichus madidus*	Min. 1940s–50s
Metabletus foveatus	1950s–		
Notiophilus rufipes	1949–		
Platyderus ruficollis	–1949–		
Trechus quadristriatus	1959–		

Many occurrences must have been overlooked and one cannot, of course, be certain that changes for any particular species were the result of gradual urban development in the district; population changes in some species are to be expected for a variety of reasons over such a period of time (Hammond, 1974). However, the overall diversity (species richness) of the Carabidae fauna in this garden has, if anything, increased slightly over the past 40 years which implies more than simple inertia to considerable environmental perturbations. Orians (1975) discussed various aspects of stability and diversity which in this instance must reflect a long history of adaptation to disturbance.

Different degrees of tolerance to urban conditions are certainly likely. In order to examine this question I made a comparative survey of the ground fauna of gardens in Greater London ranging from the city centre through 'mid-urbia' to the present urban fringe.

Effects of age and density of urban development on invertebrate faunas

Methods

Fifteen gardens were chosen in the north-west quadrant of London, divided roughly into an inner, middle and outer series. The middle and outer series of sites contained some well-established as well as newer developments whilst the inner group were necessarily all old. They ranged in size from about 75 m^2 at site 10 (Kensington) to about 1700 m^2 at site 11 (Totteridge) but 10 of them were between 200–400 m^2. The Chelsea Physic Garden (site 12) is about 1.4 ha but sampling was confined to one corner to avoid undue bias. Fig. 8.1 shows the positions of the sites in relation to three main phases of urban expansion but the picture is simplified by omitting many islands of early development which were engulfed by later urban spread. The sites are numbered from west to east and south to north according to their grid references. Sampling was by means of eight pitfall traps (73 mm diameter) partly filled with ethylene glycol and left down at each site for four weeks between 17/18 May and 14/15 June 1977. This technique was chosen, despite its known limitations for quantitative sampling, in order to concentrate on elements of the fauna likely to be residing in the sample areas and therefore adapted to the prevailing conditions.

Three main habitat factors were considered which were thought likely to influence the fauna (Table 8.2):

Table 8.2. Location of fifteen London gardens with measurements of their position, age and % open space within ¼, 1 and 2 km radius

Site no.	Locality	Grid ref.	Dist. from St. Paul's (km)	Dist. from urban edge (km)	Approx. age (yrs)	% open space within radius		
						¼ km	1 km	2 km
1	Denham	023877	30	0.2	22	78	84.5	93.0
2	Hillingdon	074847	25	1.8	15	67	76	74.0
3	Watford	097978	27	0.5	45	72	71	69.5
4	Pinner	119911	22	4	47	46.5	79	69.0
5	Harrow	154875	18	8	177	74.5	78.5	74.0
6	Elstree	187959	20	0.2	16	56.5	83.5	87.5
7	Chiswick	206772	12.5	10	62	66	72.5	49.0
8	Dollis Hill	228863	11	8	52	63.5	50.5	49.0
9	Putney	248747	10	16	82	50.5	53	55.5
10	Kensington	249791	7.5	14	142	38	37	32.5
11	Totteridge	253941	14.5	0.2	72	78	83	72.5
12	Chelsea	277777	5.5	15	304	20	42.5	30.0
13	Highgate	277877	8	6	112	74.5	82	69.0
14	Friern Barnet	277918	11.5	3	44	64	68	62.5
15	St Pancras	300824	4	12	177	19	21.5	25.0

1. The degree of urbanisation as measured by the distance from the city centre or nearest boundary of urban development. The nearest boundary was a variable and somewhat arbitrary point but distance from open country might directly affect the degree of penetration of species into urban habitats. Proximity to the city centre (St Paul's Cathedral) was a more precise and not exactly complementary measurement. This could affect species indirectly through their ability to survive in small units of habitat and their tolerance of pollution. (The absence or rarity of many common spiders from central London was ascribed by Bristowe (1939) to the heavy soot deposits there.)

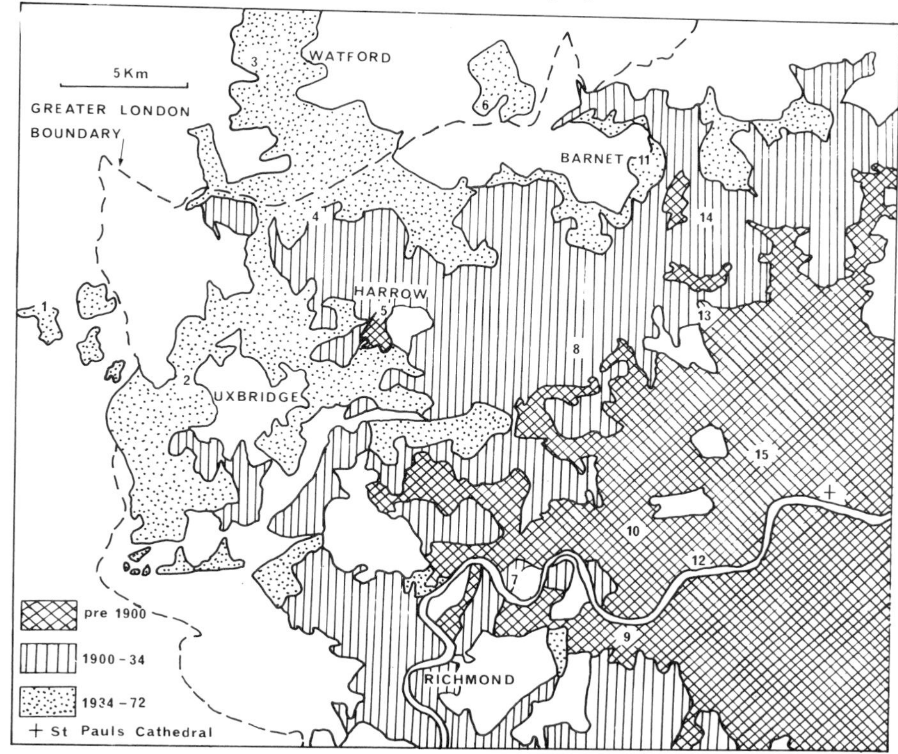

Fig. 8.1. Map of North-west London showing the position of the 15 sampling sites in relation to three periods of urban development. Blank areas represent undeveloped land and large parks. Based upon a map of the Growth of Greater London produced by the London County Council in 1939 and updated from the Ordnance Survey Map. Crown Copyright Reserved.

2. The age of a site, determined from the date of development, could have a direct relationship with the extinction rates of non-tolerant species and the spread of more synanthropic ones. Allen's records suggest that this may be a relatively rapid process but longer term effects might be experienced through cumulative pollution or the development of mature garden habitats.

3. The amount of open space around each site – gardens, parks, railway and river banks etc, but not roads or open water – could again affect the fauna directly. Since the most appropriate unit of area was unknown, estimates were made of the percentage open space within circles of ¼, ½, 1, 1½ and 2 km radius using a random point technique with 1:25000 scale maps. These measurements were rather crude, partly because of the small scale of the maps (larger scale maps are generally too out of date to be reliable) and partly because they took no account of the degree to which gardens were connected or isolated from one another by roads and buildings. The values for successively larger circles alter according to the proximity and size of parks and other larger open areas. The values for ½km and 1½ km radius are omitted from Table 8.2 though used in the following analysis.

Results

Thirty-one species of Carabidae were obtained altogether with seven species of Isopoda, eight Diplopoda, five Chilopoda and two Opiliones. (Spiders, the other principal group of

Table 8.3. Numbers of arthropod species and individuals in pitfall traps from London gardens and indices of diversity for the whole fauna

Site no.	Locality	Carabidae Species	Carabidae Individuals	All arthropods* Species	All arthropods* Individuals	Simpson's Index
1	Denham	10	57	21	340	.711
2	Hillingdon	9	53	21	216	.818
3	Watford	6	21	15	255	.714
4	Pinner	7	72	17	387	.747
5	Harrow	12	205	21	459	.803
6	Elstree	12	63	24	133	.894
7	Chiswick	9	92	18	256	.859
8	Dollis Hill	1	4	7	518	.584
9	Putney	5	65	10	359	.683
10	Kensington	5	45	12	685	.642
11	Totteridge	6	38	17	252	.845
12	Chelsea	4	46	10	78	.818
13	Highgate	7	58	14	213	.809
14	Friern Barnet	8	33	17	455	.680
15	St Pancras	5	105	10	864	.426

*Carabidae, Isopoda, Diplopoda, Chilopoda and Opiliones.

ground fauna, have yet to be determined.) Table 8.3 gives for each site the numbers of species and individuals of Carabidae alone and of all five groups combined. Details for each species will be published elsewhere. The total species richness of sites was about twice the respective values for Carabidae alone whilst total catches of *individuals* were often much greater than for Carabidae alone owing to the widespread occurrence of four common woodlice and one harvestman. Simpson's index of diversity

$$1 - \frac{\Sigma N_j(N_j - 1)}{N(N-1)}$$

where N_j = number of individuals of the jth species and $N = \Sigma N_j$ was calculated for the whole fauna in order to provide a measure of the community structure at each site although strictly this assumes equal catchability of species which, with pitfall trapping, does not hold. This index is sensitive to the abundance of the more plentiful species in a sample and is therefore, as Hill (1973) has shown, at the opposite end of the spectrum of possible diversity indices from species richness.

Analysis

The assessment of the relationship between the species richness and the various habitat factors discussed above was based on the standard multiple regression technique (Ward & Lakhani, 1977). The total species count and the index of diversity were separately treated as the dependent variables, with the eight sets of site measurements as independent variables. The analyses involving 'age of development' were then repeated using the logarithmic transformation of age.

For the analyses using the species count as the dependent variable, all the habitat variables except the untransformed age variable were found to be individually significant

Table 8.4. Regressions of species richness Y_1 and Simpson's index of diversity Y_2 (Table 8.2) against distance, age and open space variables (Table 8.3)

Independent variables		Regression Sum of Squares Y_1	Y_2
X_1	Distance to St Paul's	172.1 **	0.0247
X_2	Distance to urban edge	170.0 **	0.0274
X_3	Age	79.4	0.0041
X_4	Log age	131.6 *	0.0153
X_5	% open space $\frac{1}{4}$ km	102.0 *	0.0431
X_6	% open space $\frac{1}{2}$ km	150.1 **	0.0732 *
X_7	% open space 1 km	217.8 ***	0.1174 *
X_8	% open space $1\frac{1}{2}$ km	174.0 **	0.0727 *
X_9	% open space 2 km	213.0 ***	0.0575 *
Independent variable additional to X_7		Additional contribution to regression sum of squares	
	X_1/X_7	15:7	0.0128
	X_2/X_7	6.8	0.0239
	X_3/X_7	0.1	0.0288 *
	X_4/X_7	11.2	0.0099
Total sum of squares		353.6	0.2105

Significance levels * = 5%, ** = 1%, *** = 0.1%

Best regression models:

$\hat{Y}_1 = 2.528 + 0.200\ X_7$ $(R^2 = 0.62; P < 0.001)$

$\hat{Y}_2 = 0.2582 + 0.0063\ X_7 + 0.0007\ X_3$ $(R^2 = 0.69, P < 0.001)$

(Table 8.4). The highest regression sum of squares was obtained with % open space within 1 km radius. This variable accounted for 61.6% of the observed variation in species counts ($F_{1,13} = 20.9$, $P < 0.001$). The % open space within smaller or larger radius was less strongly correlated with species count indicating that the optimum scale for this factor had been bracketed. When taken in pairs, % open space (1 km) and the distance to St Paul's produced the highest regression sum of squares, but the *additional* contribution of the distance variable was not significant ($P \sim 0.2$). The addition of a third variable in the regression model did not increase the regression sums of squares appreciably. Surprisingly, age rather than log age contributed more when combined with distance or open space measurements. Fig. 8.2 a–c illustrate the simple linear relationships with distance (St Paul's), log age and % open space (1 km). Very similar pictures are obtained if numbers of carabid species are plotted in place of total species as shown in Fig. 8.2d. The 95% confidence limits for the predicted Y, shown by the broken lines, are quite wide. This is not surprising since the variation in species richness is likely to be due to more than the one habitat factor in each case.

The analyses using Simpson's index of diversity as the dependent variable gave a similar result in that the % open space (1 km) was still the best single predictor of diversity. However, X_3 the age of the garden was the best additional variable and the inclusion of this variable in the regression model significantly improved the prediction model ($P < 0.05$). The combination of three variables gave no further significant improvement. Unlike the earlier analyses using species richness, measurements of distance to the city centre, urban fringe and log age were not significantly correlated with this index of diversity.

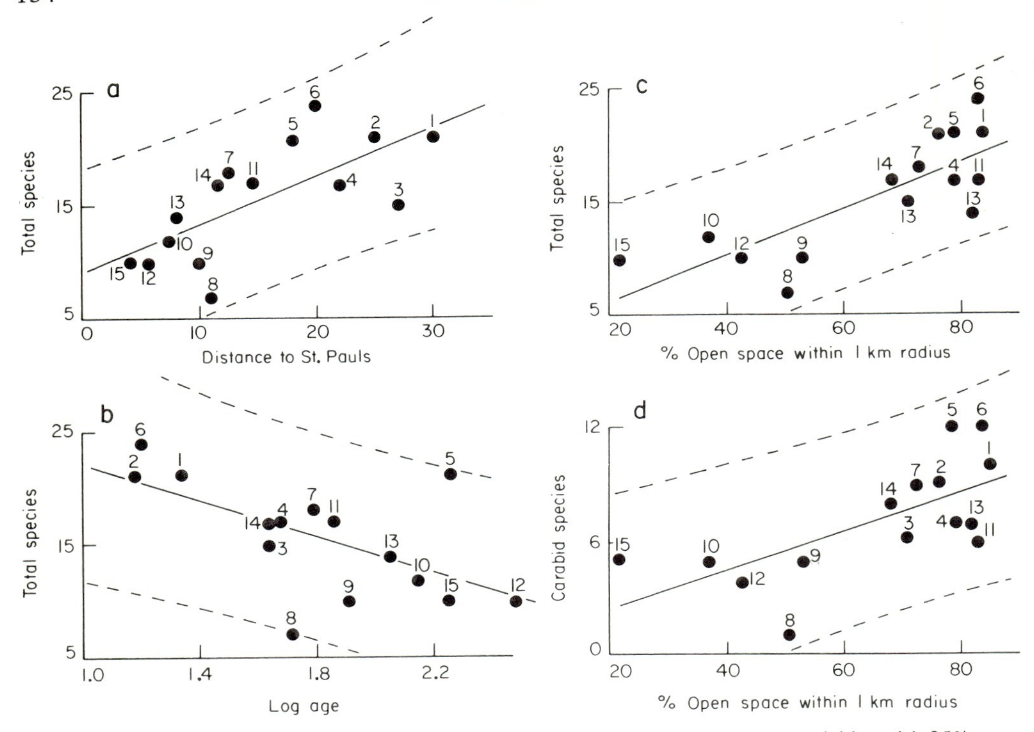

Fig. 8.2. Regression relationships between species richness and three habitat variables with 95% confidence limits for the predicted variable.
a. Total species. $\hat{Y} = 9.099 + 0.431\ X_1; P < 0.01$
b. Total species. $\hat{Y} = 29.955 - 7.930\ X_4; P < 0.05$
c. Total species. $\hat{Y} = 2.528 + 0.200\ X_7; P < 0.001$
d. Carabidae species. $\hat{Y} = 0.414 + 0.102\ X_7; P < 0.01$

These analyses produced logical results in that one could expect open space to have the most direct effect upon the fauna. That it should be so clearly related to faunistic diversity is perhaps surprising considering that the quality or variety of habitat was not assessed. Attempts were made to score various habitat features within the individual gardens but this was abandoned as impracticable. It would be interesting to see whether the diversity of the ground fauna collected within the larger parks could be adequately predicted from the regression models produced by these garden samples (Table 8.4). Correlation with any of these factors cannot of itself be taken to imply causation and I am indebted to Mr Vane-Wright for the suggestion that air temperature might be a more direct measure of urban influence than distance from an arbitrary city centre.

Fig. 8.2 shows that site 8 (Dollis Hill) was anomalous in having so few species (only one species of Carabidae). No good explanation can be advanced but some very local factor would seem to be operating. It was known that a hedgehog visited the garden regularly but such a selective and effective reduction of Coleoptera seems unlikely. The predicted number of species for this site is 13 which suggests that six or seven species of Carabidae would be more typical of the area. The effect of rejecting the Dollis Hill site as an outlier was examined. The best single predictor of species richness was still % open space (1 km) ($R^2 = 0.63$). However, the additional distance or age variables X_1, to X_4 also reached significance with X_7 and X_4 jointly providing the best prediction model:

$$\hat{Y} = 15.692 + 0.128\, X_7 - 4.390\, X_4 \qquad (R^2 = 0.72)$$

Site 8 is not so anomalous in the regressions with Simpson's index and its exclusion from these analyses has only marginal effects.

Site 5 (Harrow) on the other hand had more species than would be predicted from its log age or distance from the urban edge. This site was unusual in having a small wood immediately adjacent to the more formal part of the garden – a feature taken into account in the measurements of open space. Site 6 also had more species than predicted from any single factor but this may be accounted for by its recent development (low age) in combination with low density housing and closeness to undeveloped countryside.

Discussion

For this group of invertebrates at least, the losses due to increasing urbanisation were not balanced by gains in synanthropic ones. Four Carabidae appeared to be most tolerant of dense urban conditions namely *Notiophilus rufipes*, which occurred in all sites except Elstree the most recently developed garden, *Nebria brevicollis*, in all except Dollis Hill and Friern Barnet, *Notiophilus biguttatus* and *Asaphidion flavipes* in 11 and 9 sites respectively including Putney, Chelsea and St. Pancras. *Pterostichus madidus* was one of the most common and widespread species but absent from five of the innermost sites.

A more general appraisal of urban Carabidae can be gained by comparison with reports by Henderson (1945, 1946), Richards (1964) and Kevan (1945) for individual gardens as well as Allen's (1951) supplemented list. Nield (1974) has more recently sampled the Coleoptera from gardens in Manchester and has kindly made data available to me. Together these surveys list 80 species of Carabidae of which 75 are from the London area – 79 including the Plukenet collection mentioned earlier. Twenty-nine species common to three or more of these surveys are given in Table 8.5. Many of them are considered to be generally eurytopic species widespread in cultivated or disturbed habitats (Hammond, 1974, Table 4). Morrison (1974) collected 28 species of Carabidae from his more rural Perthshire garden

Table 8.5. Carabidae occurring in three or more lists of Coleoptera from urban gardens in Britain. A = Allen, R = Richards, D = Davis, H = Henderson, K = Kevan, N = Nield. Additional records: M = Morrison, P = Plukenet, T = Topp.

Species	Records	Species	Records
Abax parallelepipedus	ADHN(M)	*Carabus violaceus*	ADHN(M)
Agonum dorsale	ARDHN(T)	*Dromius linearis*	AHN
Amara aenea	ARDN(P)	*Harpalus affinis*	ARD(PT)
A. aulica	ARHN	*H. rufibarbis*	ADH(T)
A. bifrons	AHKN(T)	*H. rufipes*	ADN(T)
A. familiaris	ADK(MP)	*Leistus fulvibarbis*	ADN(M)
A. plebeja	ADKN	*Metabletus foveatus*	ADH(P)
A. similata	AHN	*Nebria brevicollis*	ARDHKN(MPT)
Asaphidion flavipes	ARDN	*Notiophilus biguttatus*	ARDHKN(MPT)
Badister bipustulatus	ADH	*N. rufipes*	ADH
Bembidion lampros	ADHKN(M)	*N. substriatus*	ADN(M)
B. tetracolum	ARDKN(MPT)	*Platyderus ruficollis*	ARD
Bradycellus harpalinus	ARDN(M)	*Pterostichus madidus*	ARDHKN(MP)
B. verbasci	ARHN(T)	*Trechus quadristriatus*	ADKN(PT)
Calathus fuscipes	ARHKN(M)		

whilst Topp (1972) recorded 31 species from the botanic gardens in Kiel, North Germany. These and the Plukenet records are shown in brackets in Table 8.5. Nield (1974) found that woods, woodland edges and town edge fields had many more Coleoptera species than town centre gardens whereas town edge gardens were intermediate. Relatively speaking, however, the Carabidae became more important as one progressed from rural to urban habitats i.e. they represented 47% of total species from woodland sites, 63% from woodland edge sites, 52% from fields, 67% from allotments, 73% from town edge gardens and 80–92% from town centre gardens.

On a very local level it appears that parks and suburban gardens can sometimes produce surprisingly rich insect faunas: Bradley and Mere (1964, 1966) collected 367 species of Lepidoptera from Buckingham Palace gardens over two years, Owen and Owen (1975) nearly a third and a quarter respectively of the British Syrphidae and Ichneumonidae from their garden in two years and Allen (1951–1964) one fifth of the British Coleoptera (735 species) from his garden over 37 years. A rich bird fauna as well as a high breeding density was similarly noted by Simms (1962) in a suburban habitat at Dollis Hill. Much of this faunistic diversity is a consequence of the plant and habitat diversity mentioned earlier. However, the variations between suburban gardens are relatively small compared with rural habitats and so the increase in species expected with increasing sample size is likely to be small.

Fig. 8.3 illustrates a simple model representing suburban gardens and a variety of rural habitats which they might have replaced between say Hampstead and Harrow. The insect fauna in each garden is, on average, richer in species and has a higher index of diversity than their rural counterparts though the sample sizes are the same in all cases. Taken together,

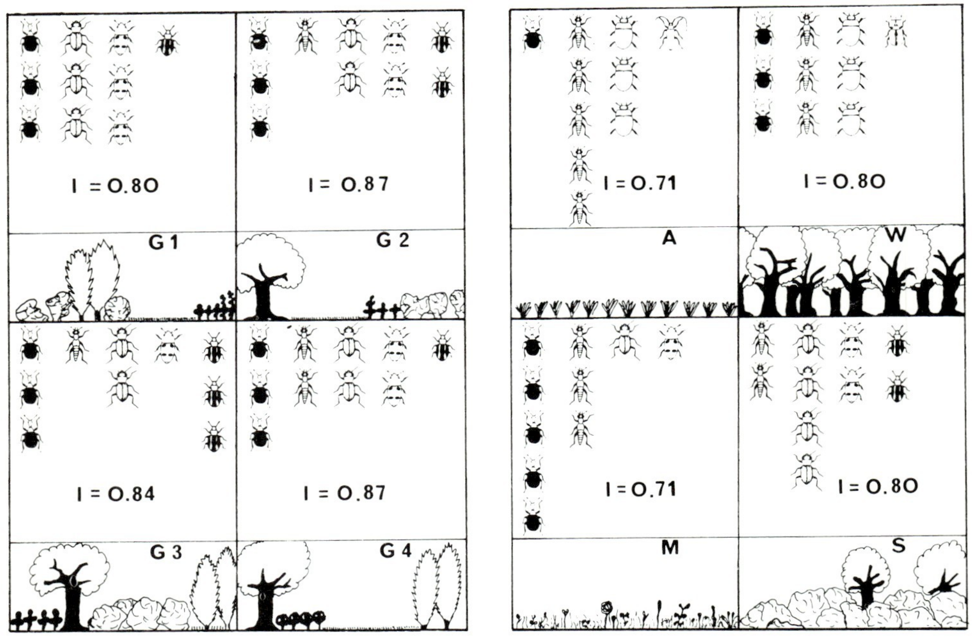

Fig. 8.3. Model representing insect diversity in gardens compared with diversity in the same unit areas before urbanisation, represented here by: agricultural land A, woodland W, meadow M and scrub S.

however, the rural habitats are more varied than the gardens and this is reflected in the more diverse fauna – eight species as against five with Simpson's index 0.82 as against 0.80.

Measures of diversity alone take no account of the biology of individual species and may disguise important community differences – compare garden 1 with woodland in the model in which only a quarter of the species are the same. Hammond (1974) has pointed out that the gain in culture-favoured species is a poor substitute for the loss of species with special habitat requirements such as old woodland. This does not, of course, lessen the scientific interest in successful species. Since urban expansion is still continuing in many places there is ample scope for studying the changes in the fauna that this produces not only in diversity but in the shifting balance of adaptive behaviour among species.

Acknowledgements

It is impossible to thank all those who helped me to find suitable gardens and offered facilities with this study but I am especially grateful to Mrs J. Cummings, Mrs G. Dain, Mr R. M. Hamilton, Mr and Mrs P. King, Mr and Mrs. T. J. Marshall, Mr. G. Myers, Mr A. P. Paterson, Mrs M. Rey, Mr G. H. Roberts, Mr and Mrs K. G. V. Smith, Mr and Mrs A. Sturdy, Mr G. Treasure, Mr R. A. Wallace, Mr M. R. Weaver, Mrs S. White and Mr J. S. Wickramaratna. I should also like to thank Mr P. E. Jones and Mr H. Outhuis for their assistance and Mr K. H. Lakhani for valuable suggestions and statistical analysis of the data. The beetles in Fig. 8.3 were adapted from originals provided by Dr R. C. Welch.

References

Adkin, R. (1934) Some lost suburban hunting grounds. *Proc. S. Lond. ent. nat. Hist. Soc.* **1934**, 125–31.

Allen A. A. (1951–64) The Coleoptera of a suburban garden. *Entomologist's Rec. J. Var.* **63**, 61–4, 187–90, 256–9; **64**, 61–3; 92–3; **65**, 225–31; **68**, 215–22; **71**, 16–20, 39–44; **76**, 237–42, 261–4.

Allen A. A. (1965) A comparison of the conifer-feeding Lepidoptera of two suburban localities. *Entomologist's Rec. J. Var.* **77**, 68–70.

Bradley J. D. & Mere R. M. (1964) Lepidoptera. In *Natural history of the garden of Buckingham Palace*, Ed. D. McClintock. *Proc. S. Lond. ent. nat. Hist. Soc.* pt II. **1963**, 55–74.

Bradley J. D. & Mere R. (1966) Natural history of the garden of Buckingham Palace. Further records and observations 1964–5. Lepidoptera. *Proc. S. Lond. ent. nat. Hist. Soc.* **1966**. 15–17.

Bristowe W. S. (1939) *The comity of spiders*. Ray Society, London.

Carter C. I. (1967) A change in the Lepidoptera of Hampstead after four decades. *Proc. S. Lond. ent. nat. Hist. Soc.* **1967**, 123–5.

Currie P. W. E. (1952) The sawflies of the Cripplegate bombed sites. *Lond. Nat.* **31**, 89–91.

Davis B. N. K. (1976) Wildlife, urbanisation and industry. *Biol. Conserv.* **10**, 249–91.

Davis B. N. K. & Jones P. E. (1978) The ground arthropods of some chalk and limestone quarries in England. *J. Biogeogr.* **5**, 157–170.

Fitter R. S. R. (1945) *London's natural history*. Collins, London.

Guichard K. M. & Yarrow J. H. H. (1948) The Hymenoptera Aculeata of Hampstead Heath and the surrounding district, 1832–1947. *Lond. Nat.* **27**, 81–111.

Haeseler V. (1972) Anthropogene Biotope (Kahlschlag, Kiesgrube, Stadtgärten) als Refugien für Insekten, untersucht am beispiel der Hymenoptera Aculeata. *Zool Jb.* **99**, 133–212.

Hammond P. M. (1974) Changes in the British coleopterous fauna. In *The changing flora and fauna of Britain*, Ed. D. L. Hawksworth. Academic Press, London.

Hammond P. M. (1975) Seventeenth century British Coleoptera from the collection of Leonard Plukenet (1642–1706). *Entomologist's Gaz.* **26**, 261–8.

Hampstead Scientific Society (1913) *Hampstead Heath: its geology and natural history*. Fisher Unwin, London.

Heath, J. (1974) A century of change in the Lepidoptera. In *The changing flora and fauna of Britain*, Ed. D. L. Hawksworth, pp. 275–92. Academic Press, London.

Henderson J. L. (1945) The beetles of a suburban London garden in Surrey. *Entomologist's mon. Mag.* 81, 63–5.

Henderson J. L. (1946) More beetles of a suburban London garden in Surrey. *Entomologist's mon. Mag.* 82, 38.

Hill M. O. (1973) Diversity and evenness: a unifying notation and its consequences. *Ecology* **54**, 428–32. 81, 112–3.

Kent D. H. (1975) *The historical flora of Middlesex.* Ray Society, London.

Kevan D. K. (1945) The coleoptera of an Edinburgh garden. *Entomologist's mon. Mag.* **81**, 112–3.

Le Gros A. E. (1949) Notes on the spiders of the bombed sites. *Lond. Nat.* 28, 37–9.

Morrison B. (1974) Observations on ground beetles (Col. Carabidae) in a Perthshire garden. *Proc. Trans. Br. ent. nat. Hist. Soc.* **6**, 97–103.

Nield C. E. (1974) An investigation into the Coleoptera of a town and its environs, with particular reference to the families Carabidae and Staphylinidae. Ph. D. Thesis, Manchester University.

Orians G. H. (1975) Diversity, stability and maturity in natural ecosystems. In *Unifying concepts in ecology*, Ed. W. H. van Dobben & R. H. Lowe-McConnel. Junk, The Hague.

Owen D. F. (1951) Bombed site Lepidoptera. *Entomologist* **84**, 265–72.

Owen D. F. (1954) A further analysis of the insect records from the London bombed sites. *Entomologist's Gaz.* **5**, 51–60.

Owen J. & Owen D. F. (1975) Suburban gardens: England's most important nature reserve? *Environ. Conserv.* **2**, 53–9.

Pappa B. (1976) Zierpflanzenschädlinge in und um Hamburg. *Ent. Mitt. zool. Mus. Hamb.* **5**, 25–47.

Parmenter L. (1953) City bombed sites survey. The flies of the Cripplegate bombed site, City of London. *Lond. Nat.* **33**, 89–100.

Richards O. W. (1964) Coleoptera (Beetles). In: Natural history of the garden of Buckingham Palace, ed. by D. McClintock. *Proc. S. Lond. ent. nat. Hist. Soc.* pt. II. **1963**, 87–92.

Satchell J. E. (1965) Extinctions and invasions – some case histories and conclusions. *Symp. Monks Wood Expt Stn No.* **1**, pp. 19–28. Abbots Ripton.

Spencer K. A. (1972) Diptera, Agromyzidae. *Handbk Ident. Br. Insects* **10** (5g): 136 pp.

Simms E. (1962) A study of suburban bird life at Dollis Hill. *Br. Birds* **55**, 1–36.

Taylor L. R., French R. A. and Woiwod I. P. (1978) The Rothamsted Insect Survey and the urbanization of land in Great Britain. In *Perspectives in urban entomology*, Ed. G. W. Frankie and C. S. Koehler. Schmid-McCormick, Berwyn, PA.

Topp W. (1972) Die Besiedlung eines Stadtparks durch Käfer. *Pedobiologia* **12**, 336–46.

Trimen H. & Dyer W. T. T. (1869) *Flora of Middlesex*. Robert Hardwicke, London.

Ward L. K. & Lakhani K. H. (1977) The conservation of Juniper: the fauna of food-plant island sites in southern England. *J. appl. Ecol.* **14**, 121–135.

9 • Colonisation of islands by insects: immigration, extinction, and diversity

DANIEL S. SIMBERLOFF

Department of Biological Science, Florida State University
Tallahassee, Florida 32306

Preston (1962) and MacArthur and Wilson (1963) independently suggested a new model for the size of island biotas, namely, that these biotas are dynamic equilibria determined by a balance between immigration of species new to an island and extinction of species already there. This model was an inevitable deductive outgrowth of the intersection of two active ecological foci: (1) dynamics of single-species population growth and pairwise interactions initially studied by Nicholson on *Lucilia*, Park on *Tribolium*, and others (especially Andrewartha & Birch) in the 1940s and 1950s, and (2) the distributions of population sizes in collections from nature, most notably for birds by Preston and MacArthur in the 1950s but also exhaustively described for insects by Williams (1964). Quickly applied in the study of a variety of more or less insular habitats (reviewed by Simberloff, 1974), this model struck a responsive chord by focussing with straightforward mathematics on a familiar statistic (species number) and painting a simple, clear picture of a nature which was dynamic but still readily understood by virtue of its division into small subunits.

Although this has not always been recognised, the equilibrium theory of island biogeography is implicitly reductionist, and so runs counter to the holistic concept of nature which dominated ecology for the first half of this century (Simberloff, 1979). The island-wide immigration and extinction rates which drive the equilibrium are primarily viewed as the sum of species-specific propagule invasion rates and extinction probabilities (Simberloff, 1969). These, in turn, are determined by population characteristics such as birth, death, and dispersal rates. The latter, especially death rate, may be influenced by species interactions, but the primary approach is through population parameters. This model views not only species number but also species composition as simply a consequence of population phenomena. It derives expressions for such traits as 'what constitutes a good colonising species' and 'what is a species' expected residence time on an island' directly from population traits (MacArthur & Wilson, 1967; Richter-Dyn & Goel, 1972); Crowell (1973) has performed the only direct experimental test of this aspect of the model. This view of community size and composition is strikingly different from that of Levins (1968; see also May, 1973), which sees both traits as resulting from the mathematical characteristics of a matrix of pairwise species-interaction coefficients, and differs equally from that of holistic ecosystem models (e.g., Walters, 1971) with component difference or differential equation systems dominated by the effects of interaction.

A second aspect of the equilibrium model which sets it apart from other ecological traditions, particularly the holistic one, is its explicitly stochastic nature. Number and composition of island species are constantly changing, so of course cannot be deterministic entities. The best possible prediction, given all conceivable information on all species in the species pool, would be a confidence interval about some expected species number (Simberloff, 1969), and a statement that a given set of species is likely or unlikely to persist for some specified period (Wilson, 1969; Simberloff & Wilson, 1970). Tiwari & Hobbie (1976) observe that ecosystem models have traditionally been deterministic and propose a general method of developing stochastic versions. The community matrix (Levins, 1968) is consummately deterministic, and in a deterministic situation the eigenvalue signs alone suffice to predict stability. Thus May's observation (1973) that stochastic fluctuation of the physical environment would make the magnitudes of the real parts of the community matrix eigenvalues also important is the first, and still unpursued, tentative move to stochasticise this branch of ecological holism.

Finally, the utility of the equilibrium model in understanding, predicting, and possibly manipulating the dynamic structure of nature, and particularly the insights which it can provide into the forces governing evolution, rest on the extent to which populations in nature are truly insular (Simberloff, 1976*c*). Lynch and Johnson (1974) have debunked the model, suggesting that many of the avian data interpreted as evidence for immigration and extinction in fact represent transient movement and not true population processes; Smith (1975) observes that if the birds alighting on and leaving a tree are monitored, one can compute immigration and extinction rates, but that these are unlikely to be ecologically informative. The equilibrium model appears to support the notion that classical group selection, operating by survival or extinction of entire populations, is an important force in evolution. But if observed immigration and extinction rates are artifacts of transient intrapopulational movement, then it is likely that gene flow would preclude most group selection. In short, we must know how much local population recruitment arises from breeding within the population and how much results from invasion. Such data are exceedingly scarce for birds, still scarcer for insects.

An experimental system

In recent community ecology there has been a singular dearth of direct experiment designed to falsify hypotheses (MacFadyen, 1975; Simberloff, 1979). With this in mind, E.O. Wilson and I set out in 1966 to find a system of islands on which unequivocal experiments could be performed to test the equilibrium hypothesis and several subsidiary hypotheses. We required islands which were sufficiently numerous to provide replication, physically simple enough to manipulate, yet contained sufficiently large communities that the expected stochastic temporal variation in species number would not obscure an equilibrium should one exist. Further, we required a system with high enough dispersal rates and fast enough population growth that events (including possible attainment of an equilibrium) would occur quickly, and one in which dispersal rates themselves could be estimated to verify that the system was, in fact, insular. Finally, we had to be able to census non-destructively, for presence or absence, entire island communities, not just subsets like single taxa, and to be certain that no two species of observed colonists were confused with each other.

The small red mangrove (*Rhizophora mangle*) islands of Florida Bay admirably satisfied these criteria (Fig. 9.1). There are many thousands, and except for the configurations in

Fig. 9.1. Red mangrove island E3 (Rattlesnake Lumps, Florida Keys).

which they are embedded the islands are replicates of one another, with one plant species creating a quite homogeneous matrix. Physically simple, the islands consist of trees growing directly from a subtidal substrate with no supratidal ground; yet each island contains a sizable subset of several trophic types of insects, spiders, and other terrestrial arthropods from a species pool of some 4000 in the Florida Keys, of which perhaps 500 commonly colonise mangrove swamps. Their simplicity enabled us to census the islands with assurance for presence or absence of species by exhaustive examination of all microhabitats, hollow twigs, dead bark, leaves, etc., and the nearness of the islands and nature of their inhabitants encouraged us to think that events happen quickly. Finally, the initial censuses of nine small islands showed strong correlation between species number and both distance and area, suggesting that these islands were miniatures of the many large-scale biogeographic theaters discussed in the literature. The islands and their fauna are more thoroughly described by Wilson and Simberloff (1969) and Simberloff (1976*b*).

A direct test of the equilibrium hypothesis

The first experiment on this system consisted of removing all animals on six of these islands by tent fumigation with methyl bromide, just as one would fumigate a house (Wilson & Simberloff, 1969). Two control islands were censused but untreated, and the treated islands were censused regularly to document the course of recolonisation. Observations are recorded and interpreted by Wilson and Simberloff (1969), Simberloff and Wilson (1969, 1970), and Simberloff (1969, 1976*a*, 1976*c*, 1978*a*) and may be summarised as follows.

EQUILIBRIUM AND ITS ATTAINMENT

The islands are clearly equilibrium entities. By 300 days after defaunation, species number on all islands but the most distant (E1) had achieved the original level and appeared to be

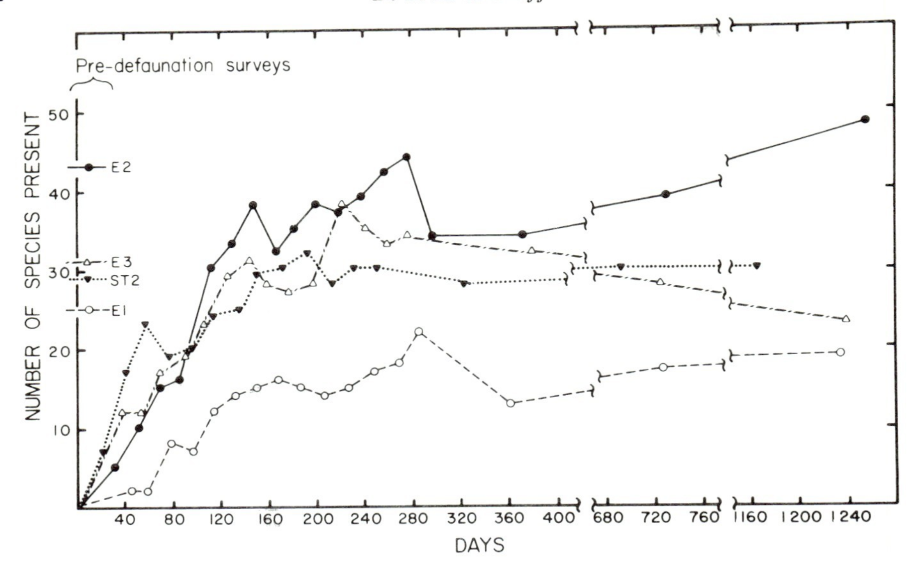

Fig. 9.2. Colonisation curves for islands near Sugarloaf Key. Pre-defaunation species numbers on ordinate.

oscillating irregularly about that level (Fig. 9.2). By three years, the E1 community was within one species of its original size. Species numbers on the two control islands remained approximately constant for a year after the defaunation, while an island used as a control for experiments to be described below had almost unchanged species number from two years after fumigation (1969) through the present.

Except for island E1, the colonisation curves of species number on the defaunated islands all rose to a point slightly above the original level, then fell and oscillated. Except for these overshoots, the shapes of the curves are all consistent with a deductive model (Simberloff, 1969) which assumes fixed, independent probabilities of invasion and non-interactive extinction per unit time. The curves in no way prove the model, which has wide 95% confidence limits about the 'expected' colonisation curves, and a rigorous test of the model would require precise characterisation of the distributions of dispersal and non-interactive extinction rates for all 500 species in the pool. The characteristic overshoot we have tentatively suggested is a consequence of reduced species interactions during early colonisation (Simberloff & Wilson, 1969). For at least six months after defaunation, population sizes of most species on all islands were low compared to those on the control islands. Ephemeral population 'explosions' of a thrips, a few psocopterans, and a weevil constitute the major exceptions. In particular, the several species of ants which are the clear ecological dominants on these islands, and which are involved in well over half of all observed interspecific actions (especially predation), were all rather slow colonists, many not arriving until well after the pre-defaunation species number had been achieved, and of course not producing large colonies for several months thereafter. Our hypothesis, then, is that low population sizes early in colonisation so reduced the probability of interactions – predation, parasitism, and competition – that more species could temporarily coexist than on an untreated island. As population sizes increased such interactions occurred with sufficient frequency that extinction rates were raised slightly, while immigration rates remained unchanged, producing a small drop in the equilibrium species number. E1, the most distant island at 533 m from a source area, had such low immigration rates that it never contained many populations with few individuals each. Rather, a few of the species which managed to colonise early, particularly a thrips (*Pseudothrips inequalis*) and a weevil (*Pseudoacalles sablensis*), achieved enormous population sizes before most other species arrived. This may have eliminated the overshoot.

TURNOVER AND THE NATURE OF THE EQUILIBRIUM

The equilibrium was indeed a dynamic one, generated by a balance between high immigration and extinction rates. For a year after defaunation (until well after equilibrium had been achieved on all islands but E1), censuses were taken approximately every 18 days, but even at this frequency true immigration and extinction rates would be underestimated since species appeared and disappeared between censuses. By computer simulation of colonisation from a species pool with hypothetical distributions of invasion and species-extinction rates, I estimated that about as much turnover was being missed as was being observed (Simberloff, 1969), suggesting that for islands like E3 or ST2 (*ca* 160 m from source areas) actual turnover rates at equilibrium were between a half and one species per day. Gilroy (1975), analogising the island colonisation process to one of molecular adsorption, deduced a more optimistic estimate that only about a tenth, rather than half, of all turnover is unobserved. Whichever, it is clear that there is an enormous amount of extinction on these islands and one would wish to know whether it represents real population build-up and extinction or just chance temporary presence or absence of mobile populations from small fragments of their ranges. The results of this experiment have been widely cited (Levins, 1970; Boorman & Levitt, 1973; Levin & Kilmer, 1974, Wilson, 1975) as evidence that group selection, acting through elimination of entire populations, is sufficiently frequent to be an important evolutionary force. But if Smith's birds-in-trees model accurately depicts its dynamics, this system can hardly be construed as evidence for group selection in nature.

Direct observation of dispersal and extinction is scant, but implies that a number of extinctions simply consisted of death or emigration of a chance individual or pair which probably did not breed on an island. For example, the termite *Neotermes castaneus* disappeared from island E2 between years 1 and 2, but I had originally observed only a founding pair, not a colony, and such pairs must be very unlikely to found a colony successfully. Similarly many times single ant queens, particularly *Crematogaster ashmeadi* and *Xenomyrmex floridanus*, were observed on these and other islands, but no colonies were found subsequently. I suspect that most of these never even produced workers, let alone reproductives, and in those instances where the species was otherwise unrepresented on the island this would technically constitute extinction (Simberloff, 1976*c*); however, this is not very interesting evolutionarily. Another class of recorded extinction which probably ought to be viewed simply as 'noise' consists of species which cannot possibly survive and breed on *Rhizophora mangle*: phytophagous insects host-specific for other plants, ground-nesting ants, parasites of absent hosts, etc. Perhaps as many as half of all observed extinctions fall into these two categories: obligate transients, and 'valid' mangrove colonists which never underwent initial population increase.

The remaining turnover, then, still constitutes a substantial number of extinctions which could be either real population extinction or transient movement. For organisms which require extraordinary events (e.g., phoresy) to reach islands, disappearance of an ongoing population cannot constitute transient movement, and the great delays before such organisms as pseudoscorpions, centipedes, terrestrial isopods, and millipedes colonised the defaunated islands confirm our intuition that they disperse poorly. But many species, even wingless ones like spiders, appear capable of inter-island movement over the relatively short distances in this system either under their own power or anemochorously; spider ballooning, for example, is a well-known phenomenon. However, even highly dispersive animals need not frequently cross water. Island E2 provides evidence that these islands are not just passively sampling large, active, widely dispersed demes, for even though it is just 2 m from the large, mangrove-fringed Snipe Keys and was rapidly colonised after defaunation, most of the insects from the adjacent source, even strongly flying ones, were never recorded there. Three larger islands used in an experiment described below provide similar evidence that these island animals are not just fragments of larger populations, and that overwater dispersal in this system is not commonplace. To judge the amount of inter-island movement, the colonisation record from the six islands for one year after defaunation was used to estimate, for each species, the average number of days between propagule invasions (Simberloff, 1976*c*). Since each of the islands was originally sterile, the number of days until the first appearance of the species on each island is presumably one sample from the distribution of intervals between invasions. The mean of these six values serves to estimate the mean interval, conservatively since some of the species never colonised some of the islands during the year of 18-day periodic censuses; in these instances, 270 days was arbitrarily assumed to be the time to first invasion, when the real time was certainly longer. For 116 species present during at least one of the predefaunation (year 1, year 2, and year 3) censuses, respectively, the mean interinvasion interval is 215 days, with a few species having very low intervals (*ca* 90 days) and many more than half having intervals greater than the mean. Almost all species breed all the year-round in the Keys, and though generation times are not rigorously established it is clear that the interinvasion interval greatly exceeds generation length for most species. For Hemiptera, Homoptera, and Psocoptera, for example, generation lengths are probably about one month, and interinvasion intervals about half a year. So dispersal in this system is not so frequent as to preclude turnover of populations isolated for several generations. One would wish to know the mean persistence times of these isolated populations, and though such data are not available since equilibrium was established (*ca* 1 yr), colonisation episodes *before* equilibration, which if anything should have been longer, averaged about 50 days (with high variance), much less than the inter-invasion intervals. Apparently, then, there is much turnover which cannot simply be transient movement in widely ranging populations, and could contribute substantially to group selection. From a sample of 51 extinctions on four islands over two years, presumably representative of extinctions generally, at least 12 (or 1.5 extinctions/island/yr) can almost certainly be viewed as elimination of long-standing, isolated, breeding populations (Simberloff, 1976*c*). This estimate involved ultraconservative criteria at all steps, and the true number of 'legitimate' extinctions is likely to be several times higher.

Finally, Wilson (1969) and Simberloff & Wilson (1970) have proposed that as time passes, chance alone dictates that occasionally a particular set of species well adapted to one another and/or the physical environment will find itself on an island; when this happens, species-extinction rates will be depressed, invasion rates will be unchanged, so island-wide turnover rates should decrease and species number should increase. This slight increase has been termed the 'assortative equilibrium.' There appeared to be slightly more species on the islands three years after the first reattainment of pre-defaunation levels, while species composition seemed not to be changing as rapidly as it was initially (Simberloff, 1976*c*).

Both observations support a very tentative suggestion that the defaunated islands are undergoing assortative equilibration.

SPECIES COMPOSITION DURING COLONISATION

Heatwole and Levins (1972) suggested that recolonisation data from the defaunated mangrove islands indicate a dynamic equilibrium phenomenon even more exciting and fundamental than approximate constancy of species numbers. In brief, they claimed that the trophic structure of mangrove island communities is an equilibrium characteristic independent of their size, that the recolonisation data show initially great deviations from some common equilibrium trophic structure and progressively closer approach through time to that equilibrium, and that the trophic structure equilibrium is determined by interspecific interactions among colonising species. They subsequently extended this conclusion to the fauna of a small Puerto Rican sandcay (Heatwole & Levins, 1973). This is a very important claim since, if correct, it would constitute one of the strongest pieces of evidence that communities of animals in nature are deterministically organised and structured, and not simply chance collections, from among many possible, of species with similar physical requirements. Trophic structure has traditionally been seen as the major organising principle in ecology (Smith, 1975), and determinism has dominated interpretations of ecological community dynamics (Simberloff, 1979); both are associated with the predominantly holistic view which ecologists have taken of nature, and contrary observations on the importance of individual populations' dynamics and responses to the physical environment have not been widely incorporated into ecologists' worldview (Simberloff, 1979).

Heatwole and Levins characterise trophic structure by the distribution of colonists into eight broad trophic categories (herbivores, detritivores, wood-borers, etc.) and compute deviation of an island from 'equilibrium' trophic structure as the sum of squares of observed proportions of species in each category minus the 'equilibrium' proportion in that category. Even when statistical improprieties in the original construction of 'equilibrium' trophic proportions are rectified, this method turns out not to be sufficiently powerful to demonstrate a trophic structure equilibration through time (Simberloff, 1976*a*). The main reason is that the deviation statistic is a strongly monotonically decreasing function of number of species even for 'communities' randomly drawn from the species pool; since the mangrove island communities increased in species number as colonisation proceeded, the decrease in the deviation statistic with time is an artifact.

An even more fundamental difficulty with this method is that it rests on a model of community trophic pattern in which calories are paramount and species' identities secondary. We are all familiar with the food chain model of energy flow in which producers are lumped together in one box, herbivores in another, and so forth, with calories flowing unidirectionally from box to box (e.g., Odum, 1971). Sophistication of the food chain model to the food web model has not fundamentally changed the conception of species in one trophic class as energetically interchangeable, since the food web depictions stress rich interconnections from level to level, each species eaten by many species and eating many species (e.g., Phillipson, 1966). This concept of species' interchangeability has as a consequence that species in the same trophic level compete, which, in turn, implies that a limited number of them can coexist in a community. But Root (1973) indicates that insects on plants are more commonly organised into 'component communities' of parasites and hyperparasites restricted to one or a few hosts, so that insect food webs are as richly constrained 'vertically' (species' identities) as they are 'horizontally' (general trophic classes). Much evidence from the entomological literature confirms this view (Simberloff, 1976*a*). Furthermore, entomologists have been far less impressed with the ecological importance of competition (Ehrlich & Birch, 1967; Rathcke, 1976) than have vertebrate ecologists; I suggest that this is partly a function of insects' narrow diets.

Species-interactions and patterns of energy flow may constrain colonisation, but these interactions and patterns must first be documented, preferably by experiment, and this is why Heatwole and Levins' imaginative attempt could not have succeeded. Without detailed autecological observations, there is little indication that the species compositions during successive censuses on each island are other than stochastic draws from a pool of species with differing dispersal and persistence probabilities. I constructed a species pool of all colonists recorded on 15 small mangrove islands, each species weighted proportionally to the number of islands on which it was found. Then for each defaunated island near Sugarloaf Key, I determined by computer simulation the number of species expected in common between the original, pre-defaunation community and each subsequent one if these were just randomly drawn from the species pool; this expected number of shared species is a function of the numbers of species in the pair of communities examined. From the species lists I found how many species were actually shared between the pre-defaunation community and subsequent faunas, for comparison to the expected overlap (Simberloff, 1978*a*). Fig. 9.3, for island E3, exemplifies the results: observed and expected numbers of shared species are usually close, often not statistically significantly different, though observed usually exceeds expected. This small excess may be interpreted as that component of the colonisation

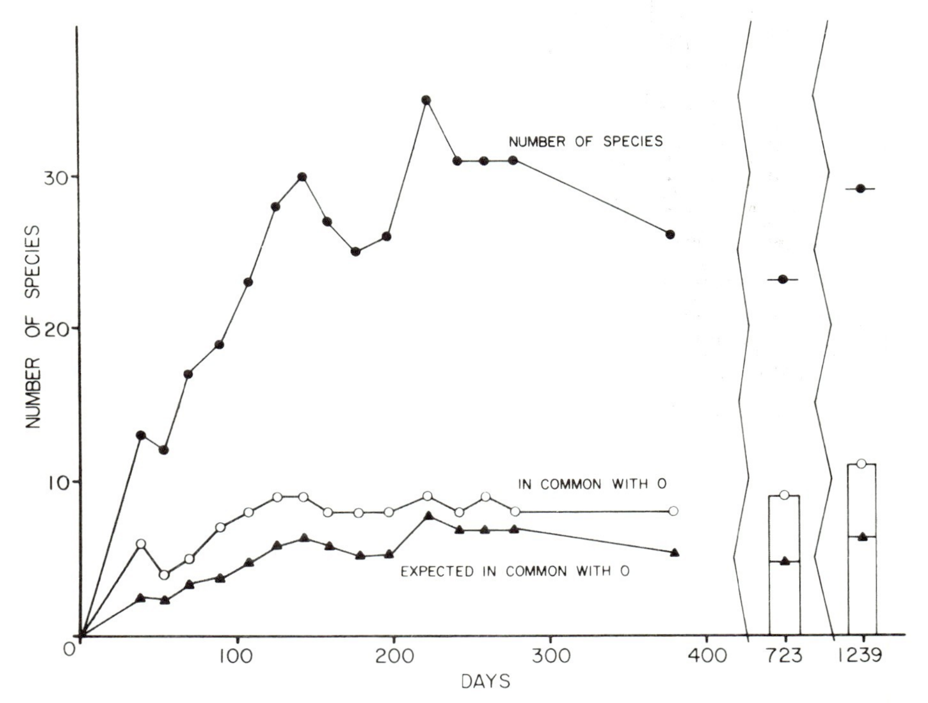

Fig. 9.3. Observed numbers of species present during colonisation of island E3, plus numbers of species shared with pre-defaunation set, and numbers of species expected to be shared on random draw hypothesis.

process which cannot simply be ascribed to the stochastic aspects of individual species' characteristics, since it represents the only compositional similarity beyond that which one would find between randomly drawn communities, given that certain species are particularly adept colonists.

Of course the species sets are not likely to be random, but knowing only the names and general habits of the members does not permit us to say more. Even if the species sets *were* random, it would be interesting to know why certain species colonise more frequently. For the mangrove arthropods it appears that aerial transport far exceeds swimming or rafting (Simberloff & Wilson, 1970), and those groups 'over-represented' among early colonists on the mangrove islands (spiders, lepidopterans, psocopterans) are all either strong fliers or readily transported by the wind. Ants were among the last colonists, though within this family there was a rather rigid order of colonisation which seemed well correlated with frequency of the respective species on undisturbed islands. These observations are rationalised by the ants' well-known poor dispersal ability and admit the hypothesis that, so long as attention is restricted to arboreal ants, frequency of occurrence on islands depends mostly on vagility.

Discussion of possible causes of extinction and importance of species interactions is deferred until the last section.

A direct experiment on the area effect

It has long been known that larger islands (or quadrats) have more species (Simberloff, 1974) though precise statistical generalised description of the relationship remains obscure (Connor & McCoy, 1978). Three non-exclusive bases suggest themselves. First, large islands tend to have more habitats than small ones; since each habitat has its own set of species, large islands have more species. Second, particularly for non-oceanic islands the species-area effect might only reflect the larger samples of individuals on larger islands (just as bigger nets catch larger plankton samples), and larger numbers of individuals would be expected to represent larger numbers of species (Simberloff, 1978*b*). In this view, the island 'communities' are not integral communities but just samples from a larger community of transient populations like Smith's birds-in-trees. Finally, the dynamic equilibrium model implies a species-area effect because the bigger islands' larger populations would be less likely to be extinguished by whatever forces cause extinction. Consequently, island equilibrium turnover rates would be lowered and equilibrium species numbers raised (Fig. 9.4). If larger islands also have higher immigration rates (c.f. Whitehead & Jones, 1969), the species equilibrium would be still higher, but the equilibrium turnover rate would be increased.

For two reasons shapes of the species-area curves from many taxa on many archipelagoes do not permit a choice among these three hypotheses. Firstly, different sets of data best fit different models (Connor & McCoy, 1978) and in any event, depending on assumptions about species-individuals distributions in nature, one probably could predict some observed model (say, a linear log-log plot) from any of the three causal hypotheses. Secondly, correlations can never demonstrate causality as convincingly as experiments, and since two of the proposed causal agents – area and habitat diversity – are usually highly intercorrelated, the species-area question is particularly unlikely to be resolved by correlation.

With this in mind, I designed an experiment on mangrove island entomofaunas to elucidate species-area relationships in this system (Simberloff, 1976*b*). Censuses of the fumigated islands and others generally indicated more species on larger islands, but since

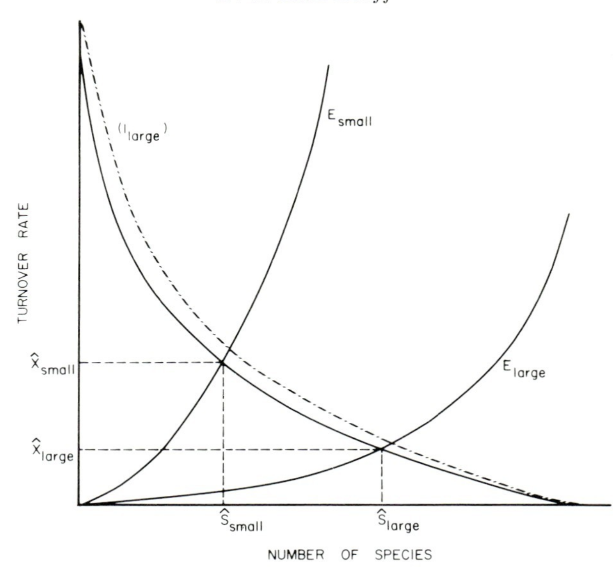

Fig. 9.4. Predicted effects of area change, acting through changed immigration and extinction rates, on species number and turnover rate at equilibrium species number.

islands may differ from one another in a host of ways, some not even recognised, in addition to area, it became clear that the best procedure would be to modify island areas and to observe the effects on their faunas. Because these islands are virtual monocultures of *Rhizophora mangle* and are structurally homogeneous, habitat diversity ought not to be a factor, and any observed faunal change, beyond that anticipated from the dynamic equilibrium aspects which had already been demonstrated for this system, could be ascribed to area change. Habitat diversity may have both subtle and surprisingly large effects on species number, however. On island E6, a control for the defaunation, a fourth of all species at the end of the experiment were recorded *only* in a few small bird's nests. Similarly, bracket fungi and birds' nests in the experiment to be described housed a few species found nowhere else on these islands, and a few species (primarily beetles and collembolans) were found only in intertidal mud and algae, not on the islands proper. All these species were omitted from species counts.

Nine islands, up to five times the size of the defaunated ones, were exhaustively censused, after which the areas of eight of them were reduced by sawing off the component trees and pulling up the remaining stumps until the removed areas were totally submerged by the daily high tides. Removed sections were towed by barge to distant deep water and sunk. This procedure presumably produced a series of islands which were out of equilibrium (oversaturated, as opposed to undersaturated as in the defaunation experiment). After seven months for re-equilibration, all experimental islands plus the control were recensused; then further sections of four of them were removed. A year later, final censuses were taken on these four islands, the control island, and an island which served as a control for the period allotted for re-equilibration since its area was reduced the first time only. Results (Fig. 9.5) clearly indicated that area affects species number independently of habitat diversity: the fauna on the control island (IN1) remained the same size, then increased slightly, the time control island (MUD2) fauna decreased by 17 species after area reduction, then by only one species by the final census; the remaining censuses, each representing the effects of an area reduction, showed 12 decreases and no increase. If area *per se* did not affect species number, the probability of such an extreme result would be only $(½)^{12} = 0.0002$.

Not only the habitat diversity hypothesis, but also the different sized sample hypothesis, can be ruled out as an explanation for this observed effect of area. For three of the islands, censusing was first restricted to the section which was to remain after the experimental treatment. The censuses of these three islands minus the fractions to be removed produced 116%, 116%, and 112%, respectively, of the numbers of species found seven months later, when the same territories comprised whole islands rather than parts of islands. If the islands were not really insular, but were just collecting samples proportionally to their sizes from a constantly active pool of propagules, there should have been no difference between species numbers for these territories. Put another way, this directly confirms MacArthur & Wilson's hypothesis (1967) that the exponent in an allometric species-area relationship is lower for mainland quadrats than for an archipelago. Their explanation, that species which cannot maintain themselves in a small area may be transients if the small area is embedded in a larger one supporting breeding populations, is the likely reason for my result.

The turnover rate data are equivocal. If the decreased area simply increased extinction rates, turnover rates should have increased; if immigration rates were decreased, no prediction is possible (Fig. 9.4). Three rates rose and two fell following decrease in area, while the control island (IN1) rate changed more than any other.

An examination of the data to determine which species contributed most to the species-area effect confirms that turnover is stochastic. Some species (e.g., the ant *Camponotus abdominalis floridanus* and the scorpion *Centruroides keysi*) are found much more often on large islands, but there are exceptions to all such rules, and it is clear that any species is more likely to disappear from a smaller island. Particularly good colonists are not taxonomically restricted, and a few (e.g., the oniscid isopod *Rhyscotus* sp., the polyxenid millipede *Lophoproctinus bartschi*, and the spider *Ariadna arthuri*) are frequently present

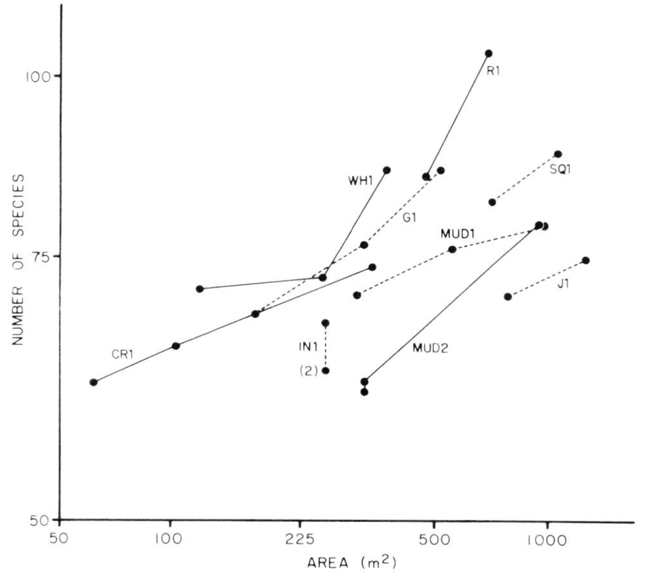

Fig. 9.5. Changes in species number with area on several modified mangrove islands. Control island is IN1 and partial control is MUD2.

and rarely extinguished in spite of poor dispersal. Other well-adapted colonists include four ants, two crickets, two roaches, two weevils, a termite, and a moth.

Current experiments and observations

One island-manipulation experiment is nearing completion; after censusing, five mangrove islands of various sizes were turned into archipelagoes by cutting and digging channels 1 m wide through them (Fig. 9.6). Two of the archipelagoes consist of two islands, while there is one each of three, four, and five islands. A channel this wide is apparently a major barrier to invasion by many species, and not only those which do not fly; the composition of island E 2, discussed above, perhaps provides the best evidence for this. The motivation for this experiment was to examine the degree of similarity between islands in the same archipelago and the relative sizes of the communities on the archipelago as compared with the original single island. As a null hypothesis one might predict fewer species on the archipelago, reasoning that the component islands would have mostly the same species, while the single large island would have all these plus several others by virtue of its larger size. This size might allow sufficiently large populations to lower extinction rates of several species to the point where they would be commonly observed colonists. An alternative hypothesis is that some species might exist on different subislands of an archipelago but could not coexist on a single island, also that extinctions on a subisland would be quickly compensated for by immigration from the other subislands, and that the archipelago as a whole would maintain

Fig. 9.6. Mangrove island SN1 (near Snipe Keys, Florida Keys) turned into archipelago.

more species than the original single island. Two of the archipelagoes have been censused (Simberloff & Abele, 1976). The larger contains more species than the original island from which it was constructed, the smaller contains fewer. Compositions will not be analysed until the complete data become available.

Another current project is to determine the causes of population decline, and the consequent turnover that makes the equilibrium dynamic. Interspecific competition does not appear to be an important force in these communities, but it would be very difficult to detect competition simply from an examination of the distributions of different species on a few islands (Simberloff, 1978*a*). Experimental introductions and removals would be far more convincing, and appear to be feasible in this simple system, but this work has just begun. From distributional data alone, I have inferred competition between the ants *Crematogaster ashmeadi* and *Camponotus abdominalis floridanus* and between the centipedes *Orphnaeus brasilianus* and *Cryptops* sp. (Simberloff, 1976*b*); in both instances active replacement confirmed a suspicion of competition engendered by the observation of mutually exclusive distributions. Until further data are available, however, I tentatively propose that the vagaries of the physical environment acting on these exposed tiny islands cause more extinction than any other single force. Frequent observation of drowned individuals after hurricanes and even ordinary thunderstorms supports this proposal, as do the small sizes of the overshoots on the colonisation curve.

Ingenious experiments by Onuf (Onuf *et al*, 1977) indicate that herbivore competition may play a role. Onuf examined a mangrove island which serves as a rookery for many marine birds, and so has an enormous input of nitrogen and phosphorus from guano. Compared to a control island, the rookery island mangrove production was increased by a factor of 1.4 because of this enrichment. The standing crop was apparently unchanged because of greatly increased herbivory by several folivorous caterpillars and a scolytid beetle. Leaves on the rookery island contained more nitrogen, and at least the mangrove skipper, *Phocides pigmalion*, sustained faster growth on the enriched island, suggesting that intra- and interspecific competition for food may have been limiting the population sizes of herbivores on the control island, but several aspects of this problem require further study. First, Onuf worked with very small sample sizes, and the control island is much nearer the mainland than is the rookery. This, in turn, suggests that greater rates of predation and parasitism may occur on the control. Second, probability of a caterpillar's successfully pupating did not differ between control and rookery, but sources of mortality were not systematically examined. Finally, preliminary work by Earl McCoy, Linda King, and myself indicates that a large fraction of mangrove folivores are killed by wasp parasitoids, and that this fraction may be lower on distant islands. James Beever and I are currently examining herbivory and parasitism in much larger samples, as well as attempting systematic augmentation and exclusion of several of the caterpillars.

Finally, I have gradually gathered data on similar *Rhizophora mangle* islands and swamps in many parts of the New World in an attempt to determine from a different direction what forces, if any, determine the composition of the mangrove insect community. Much remains to be done, but an intriguing observation so far is that small mangrove islands everywhere except Baja California have quite similar communities both in size and in trophic composition, with many genera but few species shared between locations. Tropical islands (off Panama and Jamaica) support a few more species, but the fringing swamps there contain more mangroves and epiphytes, and many more insect species. Before one assumes that the similar trophic composition, in spite of few shared species, provides evidence for a deterministic community organization, it should be noted that all sites examined so far have basically similar New World source faunas from which to draw. Evidence for deterministic structure will require data from similar islands in other parts of the world. The Baja California site is exciting, since the fauna is clearly depauperate, and the site comprises much less mangrove area than the other regions. This is consistent with Strong's observations (1974*a*, *b*, Strong *et al*, 1977) that area occupied by a plant, and its density over that area, are the major determinants of the size of a host plant's local insect community. Interestingly, in spite of its severe faunal depauperisation, there is only one strikingly different pattern of resource use between Baja California and other New World sites. A thrips, *Scirtothrips citri*, was a common insect on leaves of all ages and sizes, while its ecological analog elsewhere, *Pseudothrips inequalis*, is normally rare and restricted to very young leaves. However, on the depauperate and distant island E1, *Pseudothrips* was one of only two species observed to undergo ecological release, attacking older leaves and increasing its population size greatly.

References

Boorman S. A. & Levitt P. R. (1973) Group selection on the boundary of a stable population. *Theor. Pop. Biol.* **4**, 85–128.

Connor E. F. & McCoy E. D. (1978) The statistics and biology of the species-area relationship. *Am. Nat.* (in press).

Crowell K. L. (1973) Experimental zoogeography: Introductions of mice to small islands. *Am. Nat.* **107**, 535–58.

Ehrlich P. R. & Birch L. C. (1967) The 'balance of nature' and 'population control.' *Am. Nat.* **101**, 97–107.

Gilroy D. (1975) The determination of the rate constants of island colonization. *Ecology* **56**, 915–23.

Heatwole H. & Levins R. (1972) Trophic structure stability and faunal change during recolonization. *Ecology* **53**, 531–34.

Heatwole H. & Levins R. (1973) Biogeography of the Puerto Rican bank: Species – turnover on a small cay, Cayo Ahogado. *Ecology* **54**, 1042–55.

Levin B. R. & Kilmer W. L. (1974) Interdemic selection and the evolution of altruism. *Ecology* **28**, 527–45.

Levins R. (1968) *Evolution in Changing Environments*. Princeton Univ. Press, Princeton, N.J.

Levins R. (1970) Extinction. In *Some Mathematical Questions in Biology*, Ed. M. Gerstenhaber. Amer. Math. Soc., Providence, R.I.

Lynch J. F. & Johnson N. K. (1974) Turnover and equilibria in insular avifaunas, with special reference to the California Channel Islands. *Condor*. **76**, 370–84.

MacArthur R. H. & Wilson E. O. (1963) An equilibrium theory of insular zoogeography. *Evolution, Lancaster, Pa.* **17**: 373–87.

MacArthur R. H. & Wilson E. O. (1967) *The Theory of Island Biogeography*. Princeton University Press, Princeton, N.J.

MacFadyen A. (1975) Some thoughts on the behavior of ecologists. *J. anim. Ecol.* **44**, 351–364.

May R. M. (1973) *Stability and Complexity in Model Ecosystems*. Princeton Univ. Press, Princeton, N.J.

Odum E. P. (1971) *Fundamentals of Ecology*. W. B. Saunders, Philadelphia.

Onuf C. P, Teal, J. M & Valiela, S. I. (1977) Interactions of nutrients, plant growth, and herbivory in a mangrove ecosystem. *Ecology* **58**, 514–26.

Phillipson J. (1966) *Ecological Energetics*. Edward Arnold, London.

Preston F. W. (1962) The canonical distribution of commonness and rarity. *Ecology* **43**, 185–215, 410–32.

Rathcke B. J. (1976) Competition and coexistence within a guild of herbivorous insects. *Ecology* **57**, 76–87.

Richter-Dyn N. & Goel N. S. (1972) On the extinction of a colonizing species. *Theor. Pop. Biol.* **3**, 406–33.

Root R. B. (1973) Organization of a plant-arthropod association in simple and diverse habitats: The fauna of collards (*Brassica oleracea*). *Ecol. Monogr.* **43**, 95–124.

Simberloff D. S. (1969) Experimental zoogeography of islands: A model for insular colonization. *Ecology* **50**, 246–314.

Simberloff D. S. (1974) Equilibrium theory of island biogeography and ecology. *A. Rev. Ecol. Syst.* **5**, 161–182.

Simberloff D. S. (1976*a*) Trophic structure determination and equilibrium in an arthropod community. *Ecology* **57**, 395–98.

Simberloff D. S. (1976*b*) Experimental zoogeography of islands: Effects of island size. *Ecology* **57**, 629–48.

Simberloff D. S. (1976*c*) Species turnover and equilibrium island biogeography. *Science* **194**, 572–78.

Simberloff D. S. (1978*a*) Using island biogeographic distributions to determine if colonization is stochastic. *Am. Nat.*, in press.

Simberloff D. S. (1978*b*) The use of rarefaction and related methods in ecology. In *Quantitative and Statistical Analyses of Biological Data in Water Pollution Assessment*, Ed. J. Cairns, K. L. Dickson & R. J. Livingston, ASTM, Philadelphia. (in press).

Simberloff D. S. (1979) A succession of paradigms in ecology: Essentialism to materialism and probabilism. *Synthese* (in press).

Simberloff D. S. & Abele L. G. (1976) Island biogeography theory and conservation practice. *Science* **191**, 285–86.

Simberloff D. S. & Wilson E. O. (1969) Experimental zoogeography of islands: The colonization of empty islands. *Ecology* **50**, 278–96.

Simberloff D. S. & Wilson E. O. (1970) Experimental zoogeography of islands: A two-year record of colonization. *Ecology* **51**, 934–37.

Smith F. E. (1975) Ecosystems and evolution. *Bull. Ecol. Soc. Amer.* **56**, 2.

Strong D. R. (1974*a*) Nonasymptotic species richness models and the insects of British trees. *Proc. natn. Acad. Sci. U.S.A.* **71**, 2766–69.

Strong D. R. (1974*b*) Rapid asymptotic species accumulation in phytophagous insect communities: The pests of cacao. *Science* **185**, 1064–66.

Strong D. R., McCoy E. D. & Rey J. R. (1977) Time and the number of herbivore species: The pests of sugarcane. *Ecology* **58**, 167–75.

Tiwari J. L. & Hobbie J. E. (1976) Random differential equations as models of ecosystems: Monte Carlo simulation approach. *Math. Biosci.* **28**, 25–44.

Walters C. J. (1971) Systems ecology: The systems approach and mathematical models in ecology. In *Fundamentals of Ecology*, Ed. E. P. Opum. W. B. Saunders, Philadelphia.

Whitehead D. R. & Jones C. E. (1969) Small islands and the equilibrium theory of insular biogeography. *Evolution, Lancaster, Pa.* **23**, 171–79.

Williams C. B. (1964) *Patterns in the Balance of Nature and Related Problems in Quantitative Ecology*. Academic Press, New York.

Wilson E. O. (1969) The species equilibrium. In *Diversity and Stability in Ecological Systems*, Eds. S. H. Woodwell & H. H. Smith. U.S. Dept. Commerce. Springfield, Va.

Wilson E. O. (1975) *Sociobiology*. Harvard Univ. Press, Cambridge, Mass.

Wilson E. O. & Simberloff D. S. (1969) Experimental zoogeography of islands: Defaunation and monitoring techniques. *Ecology* **50**, 267–78.

10 • Evolution and diversity under bark

W. D. HAMILTON

Museum of Zoology, University of Michigan, Ann Arbor

A dying tree opens a wide variety of habitats for colonisation by insects. The basic anatomy of the tree remains: there are heartwood, sapwood, phloem, cambiums and bark. The various layers differentiate further by size, regimes of dryness and temperature, height above ground and in other ways. Underground the tree supplies rotting roots and rootlets and in the soil surface a litter of fallen branches and twigs; below ground, on its surface, and in the tree's standing or fallen hulk come patchy invasions and the fruit bodies of the major fungi.

Such a complex of habitats would be expected to have a rich fauna of insects and this is the case. The richness most peculiar to it, however, and that on which I will concentrate this account, is not the most obvious kind. Compared to certain other broadly defined habitats the dead-tree complex is not outstanding in its total muster of species. A more remarkable diversity lies in the host of genera, families and higher taxa, which, themselves tending to be few in species or even monotypic, are to be found only in this complex (e.g., Stubbs, 1972), or which, if found elsewhere, extend only into other kinds of decaying plant litter.

In trying to understand this 'phytonecrophily' and perhaps specific 'arbornecrophily' of so many of the odd and 'primitive' taxa of entomology, it is necessary to bring into view another peculiar and under-commented richness of the dead-tree arthropod fauna. Besides the many cases of deep phyletic *divergence*, illuminating common ancestry with outside groups, dead trees are rich in examples of functional *convergence*. Examples of such convergence will also provide ongoing threads in the theme of this essay and interweave with the problem of phylogeny. Four groups are outlined in the following paragraphs. Jointly they refer to certain similarities in *breeding structure*. Effects of breeding structure will form the essay's tentative thread of explanation.

1. Dead trees, and especially their habitats that lie immediately beneath the bark, are very productive of examples of *wing polymorphism* (list in Appendix, Tables 6.2 and 6.3). Such polymorphisms seem to be almost never of the type that is simply switched by alternative genotypes; they are switched instead by environmental cues to habitat change (personal observations; Taylor, 1975; Hood, 1940; Bournier, 1961). Hence they tend to index the occurrence of colonies that inbreed to multiply* and then disperse. Convergent cases of paedogenesis, combined or not with thelytoky (*Plastosciara, Micromalthus, Heteropeza,*

**Ptinella errabunda* is at once polymorphic for wings and wholly thelytokous (Taylor, 1975). Similar exceptions from sexuality are known in Psocoptera (New, 1971*a*, *b*) and of course, within summer clones, are ubiquitous in aphids.

etc.), have rather similar significance and are an extension of this adaptive pattern. More trivially, convergences to phoresy and to caducescence of wings in adults, both characteristic in the habitat, are linked to the same pattern.

2. The occurrence and probable origin in this habitat of about four of the six or so inventions of *male haploidy* by insects (Hymenoptera, Thysanoptera, *Micromalthus*, certain Scolytidae; excluded are iceryine coccids and Aleyrodidae) is likely also to reflect a structure of local breeding. The situtation for male-haploid Acarina may well turn out to be similar, while certain parahaplo-diploid groups (in Sciaridae and Cecidomyiidae) may be arrested steps in a similar direction (see Borgia, 1978).

3. At least two of the origins of advanced insect *social life* (termites and ants)* have their place in this type of habitat (Malyshev, 1969). Here again special breeding structure and/or male haploidy are implicated (Hamilton, 1972, 1974). Cases of subsocial life in the dead tree complex are legion and without doubt many still await discovery. One example which seems novel enough to deserve mention here, even though somewhat marginal to our habitat and concerning a non-insect, is that presented by a certain social group-hunting pseudoscorpion which Dr. L. A. O. Campos pointed out to me in Brazil. By cooperatively securing legs or antennae, bands of these pseudoscorpions are able to capture ants (their principal prey) many times their own size and far larger than would otherwise be possible. Although these atemnids are found most commonly under bark flakes of living *Eucalyptus* (where they waylay mainly *Atta*), I had also noticed similar bands in the habitat more typical for the group, that is, under the main slabs of bark of dead trees. Here ants of a similar size, e.g., *Camponotus*, were common. A seemingly very similar gregarious and myrmecophilous species was described by Turk (1953) from Argentina.

4. Finally, as a less basic convergence, although one which is no less striking, we may note the very numerous developments of certain types of *sex dimorphism*. Frequently the cases are connected with subsociality and reflect particular roles of the male (including as usual, with regard to offspring care itself, a frequent lightness or lack of role). In and around dead trees and recently evolved out of them are found a great number of the wingless, or dwarf, or outsize, or bizarrely-armed insect males that are known (Hamilton, in 1978).

Following the hint from these four groups of examples, that habitats of the dead-tree complex tend to force certain kinds of breeding structure, it will be argued that, besides calling forth specifically the convergences mentioned, such breeding structure is favourable to rapid evolution in a most general way: several interconnected reasons for this will be given. As to what may have actually been achieved by such postulated rapid evolution in dead trees, the strongest claim – and, as I judge from preliminary reactions from entomologists, the most difficult of acceptance – will be that many of the major insect groups have diverged there. More precisely this claim is that the phyletic divergences which now give systematists their separations at generic, family and ordinal levels were initiated with disproportionate frequency among insects living in dead trunks and branches. Such a claim seems to be almost new: so far as I know only one author in a very brief note has argued for a special importance of dead trees in insect evolution (Mamaev, 1971). A weaker and doubtless more acceptable version of the claim will be simply the same with the proposed crucial habitat complex widened to refer to vegetation detritus generally, not just

*Patterns of both solitary and social aculeate wasps are also conceivably derivable from a biofacies of roughly bethylid type in rotten wood (Malyshev, 1969). In social bees wing polymorphism in a hypothetical ancestor of this type could have left an impress which is now renascent in the juvenoid-controlled caste differences between queen and worker (Campos *et al*, 1975).

to the hulks of woody plants. This version is much less novel (e.g., Ghilarov, 1956; Southwood, 1973; Hinton, 1977); but even if the phenomenon is not more special than this it is still perhaps useful to look over a theory of Ghilarov's (or Mamaev's) type with the focus on breeding structure instead of on the more physical factors that have usually been emphasised.

Breeding structure

The argument will refer primarily to the *cavernous* quality of the insect living spaces which dead trees offer. Besides this an important auxiliary factor will be the uneven and *scattered* distribution of dead trees. This second factor, however, is less peculiar to dead trees since living trees tend to show it as well. Moreover, its consequences for evolution are already appreciated.

Sewall Wright has long claimed that species are best able to make rapid evolutionary advance if their populations are divided into many small quasi-isolated demes. Obviously rotting trees tend to impose subdivision. Tree bodies are large enough and often die in a sufficiently piecemeal way to provide food for several generations: yet in the end each is eaten up and dispersal enforced. Returning to the cavernous quality already mentioned, dead-tree insects tend to live *in* dead trees and at least for those whose colonies can run several generations, without being forced to emerge, this affects the rigour of their isolation. Of course, in theory, colonies on the outside of live trees could perennate much longer still but in practice their isolation is much less certain; predators can make such insects fly or jump off and winds can mix them.

Support for Wright's thesis has come in recent years from demonstrations that evolutionary rates, as measured by morphology and chromosome rearrangements, can differ very markedly from group to group, and from recent surveys of such differences (Wilson *et al*, 1975; Bush *et al*, 1977) which strongly implicate an effect from breeding structures. This is seen to work partly on the lines of Wright's prediction and partly on the basis of a founder effect through local isolation which aids the spread of chromosome rearrangements in a process that amounts to a kind of sympatric speciation (Bush, 1975; Wilson *et al*, 1975). In so far as such speciation goes on in dead trees it may multiply species numbers less than it would elsewhere since the new karyotype may be more apt to take over the whole habitat from the old (see below); but if karyotype and morphology change are correlated in insects as they are in mammals the process is still important from the present point of view and one of the remarkable convergences of dead-tree fauna – that to male haploidy (also see below) – seems to attest this.

Wright's plasticity through drift and Bush's 'instant speciation' by no means exhaust the evolutionary facilitations that can arise from local isolation. Some others are briefly indicated in the following summary:

1. Evolutionary change facilitated through drift plus recombination (Wright, 1932 and later papers).
2. Social evolution promoted through kinship. Local kinship groups are an inseparable concomitant of genetic drift (Hamilton, 1964, and refs. in E. O. Wilson, 1975).
3. Karyotype plasticity (Wright, 1940; A. C. Wilson *et al*, 1975). Via male haploidy (Hamilton, 1974; Borgia, 1978) this synergises (2).
4. Subdivision resists wide-scale disaster from drive (Hamilton, 1967; Wright, 1969). This point connects closely with (1), (2) and (3) but emphasises deme extinction.
5. Interspecies reciprocation and symbiosis. The Trivers argument for individuals can be

extended to relations of multi-generation local stocks (Trivers, 1971; Hamilton, 1972; D. S. Wilson, 1976, and in prep.).

6. Polymorphism promoted by disruptive or alternating selection for sedentary breeders and dispersers (Hamilton & Taylor, unpublished). Such polymorphism has synergised (2) in ants and termites.

Diversity in dead trees

Compared to living oak trees (Southwood, 1961) dead oaks do not have an impressively long species list of insects (Larkin & Elbourn, 1964; Fager, 1968). Inclusion of Acarina would reduce but not eliminate this contrast. Extension of the survey to other tree species on the other hand would accentuate the difference: considering the fauna of a whole British wood, and still more for that of a tropical forest, the part associated with dead trees will prove far less species-diverse than that associated with live vegetation.

This relative poverty in species extends, of course, to other detritus habitats and the reason for it certainly has to do with the chemical and other special defenses of living green plants (Levin 1976; Swain, 1977; Gilbert, this symposium). It is well known that the stage of scolytid attack through the bark to the phloem is normally about the last where host specificity is at all marked. Subsequently arriving insects (perhaps no worse than most humans faced with debris of a tree) are poor botanists: in Britain, for example, they very roughly separate into degraders of conifers or of hard woods, but in general they tend to ignore species and go for convergently similar conditions of bark and wood and state of invasion by fungi. Thus faunal lists for dead tree overlap very widely (Elton, 1966).

If important evolutionary advance has an equal chance of occurring in all species (as is suggested by Mayr, 1963) the diversity of species on living plants would be expected to give these groups a great advantage with respect to initiating major new groups. Since direct phytophages and also parasitoids contribute disproportionately to the immense total species list of the Insecta, this should apply particularly to them. Yet it can be said with near certainty that the main forks of the tree of insect evolution have not occurred in phytophages or parasitoids. Even for the predominantly phytophagous groups Lepidoptera and Hemiptera there is plenty of room for doubt that the common ancestor was phytophagous (McKay, 1970). For the Coleoptera such an ancestor is hardly considered (Crowson, 1974). The scattered distribution of predation in generally phytophagous groups suggests that a transition from phytophagy to predation is fairly easy. It probably occurs most often via cannibalism in circumstances of severe competition (Fox, 1975). However, possible cases of transition to detritophagy or mycetophagy seem much less common. This topic will be reconsidered below. As one immediate example, the recently preferred phylogeny of the Thysanoptera which I used to view as providing the most probable case of a fairly important reversion from green plants to fungi (Stannard, 1968) has been thrown into doubt by Mound and O'Neil (1974): the group which they highlight (Merothripidae) raises a new possible image of the ancestral thysanopteran.

As might be expected the evidence for important reversions from parasitoidal existence is also weak. The most that can be claimed is that the ancestry of the aculeate Hymenoptera probably includes forms with at least ectoparasitoid habits (Malyshev, 1969). Even this reversion can be partly turned to advantage for the present thesis in that the distribution of primitivity in Hymenoptera suggests that it occurred among parasitoids, very possibly polymorphic for wings, living in dead trees, rather than in parasitoids whose hosts were in more aerial habitats (Malyshev, 1969; Brothers, 1975).

If reversions from phytophagy and parasitism are few and special, the weight of species numbers associated with living vegetation and the smallness of the numbers associated with detritus can largely be dismissed from our argument.

Primitivity in dead trees

PARADOX

Much more than in species numbers and superficial diversity, a serious, almost paradoxical, difficulty concerning the role claimed for dead trees arises when we consider how it is proper to interpret the present day taxonomic distribution of 'primitivity'. Dead-tree insects are especially apt to be categorised as primitive by taxonomists (see Table 10.1). It has to be admitted that the most obvious interpretation of this is not at all that dead trees are the sites of radical innovations but rather that they are quiet backwaters serving as a kind of refuge for insect forms that have failed and vanished in the course of faster evolution going on elsewhere. The primitivity of dead-tree insects has indeed been referred to in such terms (e.g. Mamaev, 1975), and no doubt there are examples, particularly in those insects which have adapted to the difficult and nutritionally poor diets in the tree, to which the concept of a refuge and genetic stagnation really applies. But often the claims of systematists are confusing: they present the idea of a refuge and at the same time suggest that the habit of mycetophagy, for example, is more primitive in a group than that of feeding on the living parts of plants. This applies to Mamaev's discussion of Cecidomyiidae. The first impression on pulling slabs of bark from a dead tree trunk and seeing the dense and varied community, very rich in carnivores, that is present at a certain stage of decay of phloem is certainly not that of a quiet refuge (see also Beebe, 1923 Chapter 7). Yet, at the same time, to mention one group in particular, the beetle species in this subcortical community often present to the viewer a kind of synopsis of the whole classsification of Coleoptera. Further, combining both the fossil and modern evidence on this group, Crowson (1974) has concluded that this is indeed the most probable habitat for the ancestral beetle.

Such a paradox may be partly resolved in the following way. Suppose a detritophagous stock evolves a phytophagous branch. Suppose that the lines diverge and attain, say, subfamily distinction, but all the time the phytophagous line speciates far more rapidly. A taxonomist trying to classify the whole group will have to heed and use far more characters in the phytophagous branch. The characters are needed to define numerous species and genera. Only a few characters will be needed in the detritophagous branch. Further, possibly it is just the characters that are most radical and pregnant with possibility of major evolutionary change which, in the few species of this detritophagous group, the taxonomist will neglect to emphasise because these characters are at once superfluous for definition and out of line with what are considered 'useful' kinds of character in the rest of his task. In other words, the primitivity of dead-tree insects could be partly an illusion arising out of the need to define species and create keys. For example, a horny quality of the adult forewings (e.g. *Issus*, *Merope*), or paedogenesis (e.g. Heteropezini), might be relatively overlooked as an apomorphous character of a dead-tree group because other characters had already sufficed to define it. Systematists have actually produced a name for such characters, autapomorphous, yet still seem to neglect them (Ashlock, 1976).

Consider *Plastosciara perniciosa*. This is a parallel to *Heteropeza* in that, besides being a dead-tree insect in the wild (Steffan, 1975), it is also a pest of protected cultivation (Hussey *et al*, 1969) and shows a kind of neoteny. Specifically, this species has a wingless and rather worm-like morph (genuinely adult and represented in both sexes) as an alternative to the normal alate morph (Steffan, 1975). The place of the species in sciarine taxonomy is established, however, mainly on characters of the normal adult and Steffan's discovery and the strange form of the claustral adults is unlikely to affect this. If the alate form were abandoned due to success with another method of dispersal (e.g. endoparasitic like a *Deladenus* nematode, Bedding, 1967) the worm-like form could conceivably become the ancestor of a future 'order' of vermiform soil-dwelling insects, even with potentiality to become a new class.

DIRECTION OF TRANSITION: INTO OR OUT OF DEAD TREES?

The example of *P. perniciosa* was chosen because the special evolutionary potential of neoteny is widely recognised, but, apart from this, the argument of the preceding section would apply equally well if the hypothesised original stock had been non-detritophage. In general whether a stock has emerged from dead trees (or simply from detritus) or has gone into it has to be settled from the particular evidence of the case including the hints as to adaptation that appear in the systematist's reconstruction of the original type of a group.

Thus the fact that, so far as I know, the only place where *Heteropeza pygmaea* can be reliably found in the wild is under dead bark, where, in Britain, it is an abundant and characteristic pioneer in the decay of phloem, strongly suggests that occurrence of this species in mushroom houses (Hussey *et al*, 1969) is secondary. The case of *P. perniciosa* is actually less certain because although known from rotting logs (Steffan, 1975) I do not know that this habitat is most typical in its native land, wherever this may be. Similarly, another sciarine, *Bradysia paupera*, may give a good illustration of a tentative switch of oviposition and larval feeding out of the usual habitat and on to green leaves (Hussey *et al*, 1969); but this is more likely to be a switch out of soil than out of dead trees. Nevertheless, dead trees are indeed rich in sciarine genera (Tuomikoski, 1957). A more direct transfer to phytophagy from a subcortical habitat is likely in the mite *Rhizoglyphus echinopus* (Fager, 1968; Obreen, 1967), and such transfer is virtually certain in the case of the bug *Aradus cinnamoneus* (Usinger & Matsuda, 1959; Turćek, 1964). At generic or higher levels Crowson (1974) instances other cases in Coleoptera.

Brief mention in this text of three other examples of probable emergence from dead trees must suffice.

Firstly, consider the beetle family Rhipiphoridae; this may be treated as a possible model for the unknown history of the parasitic order Strepsiptera. Females of the parasitic genus *Metoecus* rather unexpectedly lay their eggs on dead wood, although the larvae are parasitic on *Vespula*. Other rhipiphorids have larvae which are free-living in dead trees (Imms *et al*, 1957). So this one family spans the life styles of a normal beetle and a stylops, and the oviposition site mentioned suggests dead trees as the likely ancestral habitat. Secondly, consider the path to another type of parasitism, that of Cimicidae. *Cimex* is flattened and also wingless and we would expect some sign of these characters in a bug proposed as its closest non-parasitic relative. Flattening suggests a subcortical insect and sure enough both characters are to be found under bark in representatives of the predatory and wing polymorphic genus *Xylocoris*. Usinger (1966) finds *Xylocoris* to be the anthocorid genus with certain characters most suggestive of a primitive cimicid. Here flattening and winglessness suggest the direction of evolution. The genus *Anthocoris* itself suggests a bridge from subcortical life to phytophagy: in summer *Anthocoris* species are important predators

Table 10.1. Primitivity in insects connected with dead trees

INSECT GROUP ORDER SUBORDER	In dead or dying trees			
	Dead tree association exists in:	Primitive with respect to:	On or in live land plants (examples not mentioned)	Comments
Subclass APTERYGOTA				
Collembola				
Arthropleona	*Neanura*, etc	Arthropleona	None	c.f. *Rhyniella*
Symphypleona	*Allacma*, etc	–	Some	
Protura	Some			
Diplura	Some	Insecta	None	
Archaeognatha	Some			
Thysanura	Some			
Subclass PTERYGOTA				
Division Palaeoptera: 2 orders and 5 suborders, all aquatic unconnected with terrestrial plants.				
Division Neoptera				
Section Polyneoptera				
Dictyoptera				
Blattaria	Many	Dictyo-Isoptera	Some	Note *Cryptocercus*
Mantodea	A few	Neoptera	Most	
Isoptera	Most	Neoptera	None	Primitive termites in wood
Zoraptera	All	Neoptera	None	
Plecoptera	Some Archiperlaria?	Plecoptera	None	Aquatic
Grylloblattoidea	None?	Neoptera	None	Montane
Phasmida	*Eurycantha*, etc	Phasmida	Almost all	
Orthoptera				
Ensifera	*Deinacridia*, etc	Ensifera	Most	
Coelifera	None?	–	Most	
Embioptera	Some	Neoptera	Some (bark)	Soil and litter typical
Dermaptera	Many	Neoptera	A few	Note flattened subcortical Apachyoidea
Section Paraneoptera				
Psocoptera				
Trogiomorpha	*Lepinotus*, etc	Psocoptera	?	Principally
Troctomorpha	*Embidopsocus*, etc	Psocoptera	None?	under or on bark
Eupsocida	None?	–	Most	

Taxon				
Thysanoptera				
Terebrantia	Merothripidae	Terebrantia	Most	
Tubulifera	Many	–	Many	
Hemiptera				
Coleorrhyncha	None?	Hemiptera	All?	
Homoptera	Some Fulgoroidea	Homoptera	Most	Note also *Newsteadia*, *Orthezia* (see text)
Heteroptera	Dipsocorimorpha	Heteroptera	Most	
Section Oligoneoptera				
Neuroptera				
Megaloptera	Raphidioidea, *Chauliodes*	Neuroptera	None	Mainly aquatic
Planipennia	**	–	Most	
Mecoptera				
Protomecoptera	Unknown	Mecoptera	?	
Eumecoptera	None	Oligoneoptera	Many	
Trichoptera	Xiphocentron	(not primitive)	None	Almost all aquatic
Lepidoptera*				
Zeugloptera	None?	Lepidoptera	All?	On bryophytes
Monotrysia	Some Hepialidae, etc	Monotrysia	Most	
Ditrysia	Various	–	Most	
Diptera				
Nematocera	Some, in about $\frac{1}{2}$ of families	Diptera	Some	Aquatic species also common
Brachycera	Some, in majority of families	–	Some	
Cyclorrhapha	Some	–	Some	Leaf miners occur
Hymenoptera				
Symphyta	Siricoidea, Orussoidea	Hymenoptera	Most	***
Apocrita	Various	Apocrita	Many	
Coleoptera				
Archostemata	Cupesidae	Coleoptera	None	
Adephaga	Rhysodidae	Coleoptera	Some	
Polyphaga	Many, in almost all families	–	Many	

Table 10.1 shows a conservative classification of insects to the level of Sub-orders (or Orders, if Sub-orders are not defined), but with all wholly epizootic groups excluded. Column two indicates the degree of association of species with dead trees, and column three the extent to which these species are regarded as 'primitive' with respect their own suborder or to some more inclusive taxon (a dash indicates that primitivity is indecisive). Column four indicates the extent of attachment of each group to living plants: no attempt is made to mention specific groups and status within these, since this would greatly lengthen and complicate the table to reveal only a minor extent of primitivity.

*The earliest known Lepidopteran fossil is a larval head capsule in Cretaceous Amber embedded with webbing, frass and crumpled plant remains; this suggests a feeding site on or under bark or else in litter at ground level (McKay, 1970).

**Larvae of Psychopsidae and Berothidae under bark flakes – only on live trees? *Megalithone* (Ithonidae)?

***Species from living plants are mostly parasitoids of plant feeders; some are seed chalcids, gall wasps, etc.

of aphids, and feeding on the body fluids of aphids cannot be very different from imbibing phloem sap directly.

I do not know of any actually phytophagous Anthocoridae, but perhaps a parallel transition can be identified – my third example – in the Coccinellidae, where, along with the aphid-feeders, plant-feeding species also occur. Here, the divergence is more advanced and family Endomychidae is the proposed parallel to genus *Xylocoris* (although whether larval endomychids are predatory or mycetophagous I have not ascertained). Coccinellids have evolved preference for drier and quieter hibernacula than the sites under bark where one finds *Endomychus*; on the other hand *Anthocoris nemorum* can be found in winter under bark mixed with *Xylocoris* colonies and at such times is conceivably predatory there – in what is here suggested to be its ancestral home.

Examples of insects that have most probably *entered* dead trees from elsewhere appear to me much harder to find. This may be partly due to my current bias and I will be glad to be informed of counter-cases overlooked. Perhaps the best case noted so far is that of the genus Forcipomyia. Larvae of this genus are common under dead bark in Britain. They possess a closed tracheal system normally characteristic of aquatic larvae and in fact many related genera of the same family of flies (Ceratopogonidae) are aquatic as larvae (Saunders, 1924). It could be that the closed tracheal system was evolved for survival in species dwelling in bark habitats subject to frequent flooding and then later proved pre-adaptive for permanent life in water; but the opposite course of evolution seems at least equally likely. Outside the Insecta a rather similar example is provided by woodlice. These are isopod crustacea (Oniscoidea) and the majority of isopods are aquatic. On land some woodlice are, as the British common name implies, abundant under loose bark, but all are found widely in litter, compost and similar habitats as well. In Britain *Oniscus asellus* is, perhaps the most constant bark species and its slightly flattened form, so similar to that of some subcortical roaches, suggests bark as its ecological 'headquarters' (*sensu* Elton). It seems most likely that the ancestor of terrestrial Oniscoidea came ashore (probably in at least two invasions, Vandel, 1943) via litter and soil, as some isopods and amphipods and other primarily aquatic crustacean groups seem to be doing at the present day. However, isopods in *Limnoria* and *Chelura* suggest that a course from water directly into dead trees is at least possible.

Rather as woodlice seem recently to have gate-crashed the land fauna and, in consequence of a lack of groups adapted to exploit them, may obtain a kind of freedom of diverse habitats into which they can begin to radiate, so some *Drosophila* colonists which had the fortune to enter the Hawaiian Islands may have found a similar freedom, in this case probably mainly freedom from competitors. Superabundant speciation in the *Drosophila* in question gives their case a different complexion from the isopods, but the parallel to be pointed out here is that the colonist *Drosophila* stocks seem to have been particularly successful in invading the subcortical habitat (Carson & Kaneshiro, 1976). But again, an alternative possibility, that the invasion took place through species arriving in driftwood, should be borne in mind.

There are various enigmatic cases in Hemiptera which might be transferrals from phytophagy to mycetophagy. Most of them concern soil (Kuhnelt, 1961), but some refer to rotting logs, for example most of the Achilidae, or to rotting tree ferns, for example *Oliarus* in Cixiidae (Zimmerman, 1948). I do not know about the primitivity status of these log- and soil-associated groups within Fulgoroidea but this group itself is treated by Goodchild (1966) and other as the basal branch of Heteroptera. With Sternorrhyncha also the most primitive groups tend to be those most associated with soil and claustral habitats. Such a distribution of primitivity makes the mycetophagy which is, for example, at least sometimes indulged by *Orthezia* (Thorpe, 1968), lose weight as a counter-example. (A like argument also applies to what may be reversions to mycetophagy in scarabaeids – Bornemissza, 1971; Howden, 1955).

A more serious implication in Hemiptera comes from the Peloridiidae, usually considered the basal branch for the whole order. If the common ancestor which this family has with the rest did indeed such mesophyll, as peloridiids are thought to do, then, in Goodchild's phylogeny, there is an important reversion in passing from this ancestor to the supposed litter-dwelling non-phytophagous ancestor of the Heteroptera. The change in the articulation of the rostrum to the front of the head is certainly suggestive of a move under bark. Predatory enicocephalids and anthocorids make good use of forwardly-directable mouth-parts in this situation; yet achilids and mycetophagous thrips manage without them. The uncertain reversion indicated by peloridiids and also the general implication of their moss habitat will be touched on again below.

Rotten wood is favoured by very diverse groups as a site for hibernation (Elton, 1966) but transition into rotting wood in this sense is usually a trivial event from the point of view of breeding structure. Whether it carries an implication of 'ancestral familiarity' with dead wood is an intriguing question in view of the use of this site by *Anthocoris*, coccinellids, ichneumonids, etc (Larkin & Elbourn, 1964). Perhaps an equal list of cases with no such likely ancestral connection could be cited. *Vespa crabro*, a 'primitive' social vespine, uses dead wood for hibernation site, nest site and nest material; on the other hand *Vespula* queens seem to use equally readily any other hibernation site that offers suitable cover and microclimate. The same applies to *Bombus*. As regards wood-nesting in *Xylocopa*, Hurd (1958) claims that soil is the likely ancestral nest site, and Barrows (1973) equally claims this for log-nesting Halictinae. Wood or soil should make little differences to breeding structure in these very free-flying flower-visiting insects.

Some other cases of transition which might repay study are indicated in Table 10.1 and in the following additional list: Sminthuridae (Collembola); Liposcelidae (Psocoptera); Derbidae (Hemiptera); Chrysomeloidea, Scarabaeoidea (e.g. *Oryctes rhinocerus*), Curculionoidea, Scolytidae (e.g. *Hypothenemus*), Elateridae, Nitidulidae (Coleoptera); Hepialidae (Lepidoptera); Tipulidae, Lonchopteridae, Syrphidae, Phoridae (e.g. *Megaselia*), Dolichopodidae (Diptera). Possible trends in Scolytidae have already been outlined by Schedl (1958).

THE FOSSIL RECORD AND COMPARATIVE EVIDENCE

The record of arborescent land plants extends back as far as the record of insects but evidence directly connecting the two in the Palaeozoic is extremely scanty (Scott, 1977). This applies almost as much to evidence of attack on living plants, arborescent or not, where damage that could be attributable to insects should be relatively easy to recognise, as it does to evidence of inroads on dying and dead remains. As regards indirect evidence, the point can at once be made that the earliest of all known insects (using a broad view of the class), the collembolan *Rhyniella*, is remarkably like *Neanura muscorum* (Massoud, 1976), a species which occurs today under dead bark and in rotting wood. It is also to be noted that some forms of attack on tree detritus, especially the habit of feeding on the associated fungi, would leave little trace of any kind that could be fossilised, and that wet conditions such as prevailed where Palaeozoic plant fossils were formed are hostile even to the preservation of arthropod coprolites.

Standing dead lepidodendroid trunks of Carboniferous age on Arran, described by Williamson (1880) – hollow, and containing lodged within them twigs and 'leaves' and other fragments of various plants (and all finally buried in volcanic ash) – represent not only the type of habitat where one would confidently search for *Neanura* today, but also just that kind of food object, discrete, bulky yet cavernous, which our hypothesis has required. These trunks hint that we need not take too seriously the lack of evidence of bored insect galleries in pre-permian plant remains. In truth such evidence of boring in the Palaeozoic seems to be scanty and unconvincing: sinuous engravings on the cortex of *Sigillaria* in Germany (Geinitz, 1885) look to me as attributable to, say, impressions left by fungal rhizophores as to galleries of insects; and, in some other fragments, holes indubitably bored by some animal are of a size too minute to be probable work of insects (Williamson, 1880).

Borings, however, appear in fossil wood of the Permian and so does bark (Crowson, 1974). As Hinton (1948) has noted, there is a basic conflict between burrowing and the possession of wings and the most striking solution of this has been holometabolous development. Perhaps a move by arborescent plants to protect phloem beneath bark (Smart & Hughes, 1973) was the principal stimulus to holometaboly. In this essay we are mainly concerned with the bonanza of *dead* phloem but no doubt in all ages there have been saprophagous insects that pressed their attacks earlier or brought in parasitic fungi to prepare ground ahead, much as cerambycids, scolytids and siricids do today; bark needs to protect against these too. In this connection it is interesting to note the somewhat unexpected appearance of ovipositors in certain Palaeozoic groups (Carpenter, 1977). Among their many other uses ovipositors serve to put eggs in deep crevices and under bark: ovipositors occur, for example, in various stages of evolution, in many present-day insects whose larvae live under bark or in wood (*Helops* and *Lonchaea* show cases of incipience). Species that lay eggs from the outside of bark are usually outbreeders, and this according to our thesis counter-indicates the ancient groups with such appendages as being quite perfect images of the ancestors of major groups of the present day; correspondingly, of course, a connection of ovipositors with apparent evolutionary stagnation (as in raphidians, cupedids, siricids, orussids) is not surprising. Adults going in under bark, going in further through flattening or by development of horny forewings (as in roaches, beetles, and some psocoptera – and perhaps also in *Issus*, *Merope* and others) or by dealation (as in some Embioptera, Isoptera, Zoraptera, Psocoptera, Thysanoptera and Hymenoptera), going perhaps further still through an apterous adult morph or total wing-loss (Appendix I, *Cryptocercus*, etc.), sending on unencumbered larvae ahead as specialised burrowers, finally as the ultimate development in a few lines, allowing those larvae to breed for themselves without any need for more space (*Heteropeza*, *Micromalthus*) – this whole sketched evolutionary sequence for the increasingly rapid penetration and utilisation of dead phloem hidden beneath bark seems to me eminently possible. One step is perhaps currently reillustrated by the sluggish 'pupal' stage of subcortical tubuliferan thrips. In the main stock formation of the pupa produced the great insectan advance to holometaboly.

BAST AND WOOD

Bast – that is, phloem plus cambiums – offers a rich and well-balanced diet. On the death of a tree these layers are the most speedily consumed. Sapwood is consumed more slowly and by fewer species. Lignified heartwood and suberised bark, where these are present, go more slowly still.

The wide variety of taxa that can be found directly under the bark has already been emphasised, and the same habitat has had repeated mention in connection with other phenomena. Besides consumers of dead plant tissues there are many species that feed primarily on the bacteria, yeasts and fungi that are soon abundantly present, including on parts of those fungi that are primarily concerned with the decay of wood. There is also a surprisingly high diversity of predators. All these insects tend to be small, with a size range overlapping (e.g. *Ptinella*, 0.6 mm) that of typical Acarina, which group is itself abundantly and diversely represented. They also tend to have short generation times which, more than offsetting rather low total fecundities, gives them potential for high rates of increase. As our hypothesis requires, many species do indeed readily mate within the habitat and the inbreeding which this implies is reflected in biassed sex ratios which further improve the potential for colonisation and increase. Polymorphism in colonial species of this habitat is summarised in Appendix 1. It might be thought that winglessness in a morph would always be a further device for greater effective fecundity and increase, as it is in all other winged/wingless morph comparisons that have ever been made in non-social insects. No doubt this is often the case with the subcortical polymorphisms but one notable exception has recently appeared.

Studying the very marked polymorphism found in both bisexual *Ptinella aptera* and thelytokous *P. errabunda*, Taylor (1975) found that winged females were longer lived, were more fecund and had larger spermathecas than their wingless counterparts. This is suggestive of the beginning of a termite-like social development. In fact on the basis of this example plus the general absence of wing polymorphisms in xylophagous insects, it can be suggested (Hamilton and Taylor, unpublished) that social termites arose from their roach-like ancestors in the habitat of dead phloem, and that *Cryptocercus* is consequently connected to them as a parallel invasion of the wood, rather than as a 'wax-work' image of their ancient way of life as is often implied (e.g. Wilson, 1971).

In contrast to the subcortical insects, those of dead wood are usually larger (and include the largest of all, e.g. *Titanus giganteus*, up to 200 mm), and tend to develop much more slowly; generation time of 2 or 3 years is common in temperate latitudes. Consequently wood feeders are rather slow to increase in a habitat and for reasons perhaps connected with this but not fully understood they are also much less inclined to endogamy and to claustral continuance of the colony even if the wood provides bulk for it (primitive termites, *Micromalthus* and perhaps passalids are exceptions here). As expected if mating is outside the log, sex ratios are on the whole normal; both parthenogenesis and wing polymorphism are almost unknown (termites and *Micromalthus* again excepted).

Bearing in mind reduced faunal competition and slower generation turnover and bearing in mind also the more outbred breeding structure, it is particularly here that we would expect to find true relict insects. And possibly the concept can apply, for example, to *Cryptocercus*, Cupesidae and Siricidae.

At the same time the wood has many obviously advanced invaders, for example Cerambycidae. These are suspected to have moved inward from the phloem relatively recently. Some lines in Scolytidae appear to be so evolving at the present time (Schedl, 1958). Where beetles move in as 'ambrosia' feeders the diet remains rich and rapid endogamous breeding is sometimes retained. Here might be mentioned the large genus *Xyleborus* which also illustrates various cases of transition towards attack on the green parts of living plants. *Micromalthus*, combining arrhenotoky, thelytoky and paedogenesis, is much less successful but confined within rotten wood its trend is similar but more extreme. Certain cecidomyiids, moving in through shrinkage cracks without ambrosia but again feeding on fungi and not on wood, offer rough parallels to the ambrosia beetles – *Pezomyia* (or, probably better, the little known *Micropteromyia*) parallel to *Xyleborus*, and *Heteropeza* parallel to *Micromalthus*.

Primitivity in other habitats

Apart from habitats of the dead-tree complex, others especially frequently mentioned in connection with 'primitivity' are litter and soil, moss (and to some extent other primitive green plants), and finally fresh water. The living parts of higher plants tend to carry insects that taxonomists rate as relatively advanced, the main exception to this being, perhaps, an undue frequency of 'primitive' insects feeding on pollen and the pollen-producing organs of plants (see below). Seed insects, on the contrary, and also those of 'stored-product' type environments tend to be classed as advanced and so do parasites (probable routes to these various life-styles can be drawn out of dead trees via the hollows in which birds, lizards and mammals nest and keep their stores – for species suggestive of this see lists in Hicks, 1959).

Dung and carrion tend to carry types of rather intermediate primitivity. Carrion of large animals might at first thought seem very like the rich phloem masses of dead trees in size and nutritive value. Perhaps it is partly the quality of thinness that tends to reserve dead phloem for insects. Besides this carrion is even more scattered, even more nutritive, and less resistant to entry and destruction. Thus larger animals take a far larger share and tend to leave only dispersed fragments and bones – the latter a worse resource for insects even than the heartwood of trees. Altogether the attributes of carrion left available to insects almost necessitate that those that breed in it be good fliers and fly in every generation. This encourages panmixia. Excepting perhaps dried carcases, carrion is too ephemeral to support colonies.

MOSS

Plant litter and soil are habitats with no sharp separation from rotting tree trunks and branches. Moss too is intimately associated with all these, and like them provides sheltered and hidden spaces. The spaces in dense moss cushions may, like cavities in soil and wood (and like the similar spaces in dense grass tussocks), tend to encourage local inbred colonies. Yet, I believe – admittedly on no easily presented evidence – that feeding in living moss is more often derived from mycetophagy and detritophagy than vice versa. One example, that of Peloridiidae again, must serve to illustrate both moss primitivity and this bias of the writer. The family was largely lost as an illustration of my dead-tree theme when I read that some members had been shown to exist in *Sphagnum* and other mosses distant from trees (Evans, 1941); however, this has not been shown for all species and I preserve a small hope that at least *Peloridium hammoniorum* may prove mycetophagous in rotting wood: this species was first found under a rotten log, and it is perhaps significant that this is also the one species known to show wing polymorphism (China, 1962).

LITTER AND SOIL

Almost every group mentioned in connection with dead trees in Table 10.1 also has representatives in other kinds of plant litter and in soil. Soil and litter insects likewise tend to be more 'primitive' than collateral groups that feed (not necessarily as herbivores) on aerial parts of living plants. The special case made for dead trees is based partly on the *a priori* considerations already mentioned and partly on an impression that evolutionary novelty is really more common in dead-tree insects than in those of soil. What has soil to offer quite so odd as, for example, the intra-haemocoelic insemination of Cimicoidea and related bugs? There are, of course, abundant examples of ingenious adaptation for life in the soil and doubtless many routes to subaerial phytophagy do actually lie through soil-dwelling forms. But where such a path seems apparent the evidence is often somewhat equivocal. In the case of *Sminthurus viridis*, for example, derivation from the commoner litter habitat of other *Sminthurus* species is likely, but we note also the existence of the 'primitive' relative *Allacma fusca* which is associated with rotting wood and fungi. Similarly two rather unexpected pests of vines, *Lethrus* and *Vesperus* (Balachowsky & Mesnil, 1935), seem to have more connection with soil than rotting wood, but on surveying slightly more distant relatives this case too becomes more doubtful.

In a less dismissive approach to the competing claims of soil and litter a stand can be taken on the already mentioned weaker version of our thesis; that is, on the claim that, like dead trees, soil and litter forces subdivision and inbreeding on its inhabitants. Certainly a great deal of what has been said about necessities and adaptive responses of insects in dead trees applies to soil insects as well, and the evolutionary potential of these can be equally contrasted to that of the insects in more panmixial habitats on plants. It has to be insisted that neither in tree trunks nor in patchily distributed resources in the soil do we expect specially high rates of evolution *unless* some local inbreeding really occurs; if the resource, whatever its location, is used up in one generation and all offspring have to disperse and mix, then the species is certainly worse off than is, say, a coccid or a spider mite which is capable of local differentiation by drift through the perennation of its colonies and its restricted mobility. [Coccids have, in fact, very high rates of evolution at least for some traits (Nur, 1977); for tetranychid breeding structure see McEnroe (1969)]. Apart from soil insects this caution applies with force, as already mentioned, to a lot of dead-wood flies and beetles which have annual or longer life cycles. In being a more continuous habitat one might well expect soil and litter to enforce even less long-range dispersal than dead trees do; and the total loss of flying wings in many soil and litter insects of permanent habitats tends to confirm that this is so (see, e.g., den Boer, 1970). But unfortunately apart from the vague portent of such flightlessness we have little factual information about dispersal in environments presumed to be stable and uniform, either soil or any other. In the realm of theory too the problem remains little explored but models already developed make it clear that we cannot assume that existence of stable ongoing resource will imply minimal dispersal (Hamilton & May, 1977*).

Rather as in the field of sociobiology the factor of relatedness has been added to others previously adduced in order to improve our understanding of, for example, evolution of social insects (e.g. Hamilton, 1972; West-Eberhard, 1975; Trivers & Hare, 1976), so the present considerations of breeding structure must be considered additions to a continuously improving, general picture of insect evolution. I see no conflict between the present emphasis on dead trees and Ghilarov's (1956) emphasis on soil as mediating transition to drier habitats, or with Hinton's (1977) deduction that conditions of alternate dryness and flooding were important in the stem of endopterygote evolution. Dead trees in a showery climate, indeed, almost idealise the concepts of these writers – a wick through which insects can evaporate to the air.

*Recently Comins has found it possible to extend the models cited in such a way that less dispersal probabilities can be given for some of the classical breeding structures of population genetics (Comins, Hamilton and May, in prep.), so far, however, this development is more relevant to expected inbreeding and other properties for the 'stepping stone'-type distributions of dead tree insects than it is for the more purely 'viscous'-type distributions here supposed typical of soil insects.

WATER

Equally, trees fallen into water could encourage transition to fully aquatic life. The idea that insects as a whole originate in aquatic habitats is now almost abandoned; consequently even the modern palaeopteran orders have to be regarded as secondarily aquatic. I am not aware that these orders show any hint of a significant connection with dead trees or even with litter. (The terrestrial litter-dwelling larva of *Megalagrion oahuensis* in forest litter is obviously a re-transference to land.) Inability to flex the wings flat over the abdomen in these orders also strongly contraindicates any connection with bark (contrast the habitat and presence of this ability in Liposcelidae and Phylloxeridae with lack of it in related families); but then of course the arborescent plants had no substantial kind of bark at the time these groups originated.

For most of the rest of the numerous invasions of water by insects, a route through dead trees seems more possible but is not strongly indicated against routes through other kinds of wet litter. This particularly is true for the Diptera, and, as already indicated for Ceratopogonidae, there may be some good cases for transition from water into rotting trees on land rather than the reverse. However, a phyletic closeness of 'wood' and 'water' is also indicated by a surprising number of families and these are mostly at the 'primitive' end of the dipteran range; examples are Tipulidae, Chironomidae, Ceratopogonidae, Syrphidae, Stratiomyiidae, Tabanidae, Dolichopodidae.

In Trichoptera the species with exceptional terrestrial larvae are not considered primitive The larva of *Xiphocentron*, for example, lives on the outside of damp rotting trunks and that of *Enoicycla* lives in moss and litter. But both occur in woodland and so faintly suggest that tree trunks partly in water could have provided a bridge – to land in these cases, but also, perhaps, in an ancestor, to water. According to Ross (1967) the most primitive genera of Trichoptera are strongly convergent to a habitat of 'cool, moderately rapid small streams running through shaded woodland'. Trunks partly in water are particularly common in such places. With regard to the Megaloptera, such trunks at about the water level do in fact provide the headquarters of *Chauliodes* (Needham & Betten, 1901) a genus which well connects terrestrial subcortical raphidians with the more fully aquatic genera such as *Corydalus* and *Sialis*. Notwithstanding some claims (e.g. for Coleoptera, instancing *Corydalus*, Bradley, 1947) there seems no need to suppose that any major group of insects, excepting possibly Diptera, has arisen via an ancestral aquatic larva. However, the distribution of fresh water itself often imposes a pattern of small demes so that the evolutionary characteristics of this habitat may not be so very different from those of dead trees. In fact, ponds and streams (Riley, 1920; Parshley, 1922; Southwood, 1962) are indeed, like dead trees and grass tussocks (Luff, 1964), sites where wing polymorphism is common. And, corresponding to taxa of dead trees, aquatic taxa evidently also survive long, are often dubbed 'primitive' and at the same time manifest many very original adaptations: mention may be made here of labial masks of Odonata, subimagos of Ephemeroptera, the cases and feeding devices of Trichoptera. Paedogenesis has been achieved by Chironomidae. Male haploidy has not evolved in aquatic insects but Plecoptera seem to approach it, and outside the Arthropoda its only other full attainment is in an aquatic phylum, Rotifera.

POLLEN

Examples of primitivity in this habitat are few but quite striking. Here examples from three groups must suffice. Adult Micropterygidae feed on pollen but their larvae apparently on bryophytes – perhaps also on rotting materials. This family is regarded as the basal branch of the Lepidoptera, and by some as perhaps sister group to the Trichoptera. Three primitive genera of the Curculionoidea, *Cimberis* (Nemonychidae), *Allocorynus* (Oxycorynidae) and *Bruchela* (Anthribidae) breed, respectively, in male cones of *Pinus*, of *Zamia* and in the seed-capsules of *Reseda*. The last beetle and site merit mention here because the carpels of *Reseda* show an unusual primitive gymnospermous character – they are not closed. Such 'primitive' genera connected with sporangia of 'primitive' plants convey an obvious suggestion. However, they are but three genera in an immense group; and even with attention confined to these families plus the others that are primitive for the superfamily (e.g., Aglyceridae and Brenthidae), habitats in the dead-tree complex are far more abundantly represented than are attachments to microspores. A transference of diet from mycelium to fungal spores and thence to pollen is not difficult to imagine; Crowson (1974) suggests such a transference in *Micrambe* (Cryptophagidae). A close parallel to the situation of these curculionoids exists in the symphytan Hymenoptera; *Xyela* can be compared to *Cimberis*, Siricidae to Brenthidae, and the sawflies to the weevils. Malyshev (1969) thought that *Xyela* might betray the diet of the mecopteran-hymenopteran ancestor, i.e., microspores scattered on forest litter. In view of what has been said above a stem through mycetophagy with early branches to spore and pollen feeding seems at least equally likely. Wood, litter, moss, pollen and even water all cluster near to the faint phantom of the earliest hymenopteran. It is possible that, in the near future, uncovering the lives of the Meropeidae may slightly brighten this dim scene.

Pollen feeding is sometimes far from primitive. In connection with the unexpected pollen-feeding and other evolutionary novelties of heliconiine butterflies (Gilbert, 1972; this symposium) it is relevant to note that these insects seem unusually capable of creating a structure of small demes through their own social behaviour (Turner, 1971). In this they show a parallel tendency to polygynous eusocial insects (Hamilton, 1972) and to mammals (Wilson *et al*, 1975). Such animals can speed their evolution without need for the kind of *forced* deme structure which, as this essay supposes, dead-tree cavities have provided to so many juvenescent groups of insects.

APPENDIX: Insect wing polymorphisms from live and dead trees

Table 10.2 lists genera containing species which are both colonial inhabitants of cavities in dead trees and polymorphic for wings in the adults. For this table wing polymorphism is interpreted broadly as any natural co-occurrence among adults of some individuals capable of flight and others incapable: hence sexual flight dimorphisms are potentially included (e.g. *Xyleborus*, with flying wings always vestigial in the male and always functional in the female). The kind of correlation between flight and sex is indicated in the last column; a key to symbols used in this column is given below the table. In some cases a category has been assigned on the basis of pooled information regarding several species.

The last column of Table 10.2 shows that slightly more frequently than not the wing-sex correlation leans towards males being more commonly flightless than females. Flightlessness of males when females can fly is strongly indicative of inbreeding (Hamilton, 1967); likewise an opposite condition, flightless females combined with flying males tends to indicate outbreeding, although this second implication is less definite than the first. Very small winged males of such flightless-female species, as of cocceids, *Dusmetia*, some sciarids etc.,

Table 10.2. Flight-polymorphic insects living inside dead trees

ORDER Sub-group†	Habitat: Under dead bark	Habitat: In sound wood	Habitat: In rotted wood	Food	Wing-sex Association
ZORAPTERA					
*Zorotypus	x		x	Fungi? Mites?	–
PSOCOPTERA					
*Embidopsocus	x			Yeasts, etc	–
Psoquilla	x			Yeasts, etc	
THYSANOPTERA					
Megathrips	x			Spores	–
Cryptothrips	x			Spores	–
*Hoplothrips, etc.	x			Fungi	–
HEMIPTERA					
Aradidae	x			Fungi	–?
Henicocephalidae	x			Arthropods	+
*Xylocoris, etc.	x			Arthropods	–
COLEOPTERA					
Micromalthus			x	Wood, fungi?	x
*Ptinella	x			Fungi	–
Pteryx	x			?	–
Astatopteryx	x			?	?
*Xyleborus		x		Fungi	– –
DIPTERA					
*Heteropeza	x		x	Fungi	x
*Pezomyia	x		x	Fungi	+
Micropteromyia	?		?	Probably fungi	–
Plastosciara	?		?	Probably fungi	+
Coenosiara	?		?	Probably fungi	+
*Pnyxia	x			Probably fungi	+
Sciara semialata	?		?	Probably fungi	– –
HYMENOPTERA†					
Sycosoter	x			Beetle larvae	+
Theocolax		x		Beetle larvae	+
*Cephalonomia	x		x	Beetle larvae	+
ACARINA					
*Pygmephorus††	x		x	Fungi	–

*Groups personally observed by the author.

†Species where either (a) female lays eggs by ovipositor from outside of bark, or (b) male is both large and winged, are considered incapable of continuous colonial life inside dead trees and are excluded from the list. Notable exclusions by this rule are in Hymenoptera, for example *Eupelmella* by (a), and Thynnidae by (b). In Diptera *Chonocephalus* has relatively large males and phoretic copulation like Thynnidae and sometimes breeds under bark; absolute smallness of males and their agility argues for inclusion of this genus suggesting that continuous colonies would be possible, but the fact that rotting fruit is more typical habitat than bark argues for exclusion.

††This genus of mites shows phoretomorphs analogous to the winged morphs of subcortical insects (see Hamilton, *in press*; Moser & Cross, 1975).

– – All males flightless, all females capable of flight. – One or both sexes polymorphic such that males are more often flightless than females. ? Data on wing-sex correlation is lacking or inconclusive, or, in other columns that the main site of breeding is uncertain. + One or both sexes polymorphic such that females are more often flightless than males. x Species with thelytokous paedogenesis; outbreeding status unclear.

Table 10.3. Insect groups living externally on trees and showing some species with male winged and female flightless

ORDER Subgroup†	Habitat: Bark, Sometimes inside	Habitat: Bark, Outside	Leaves	Food
DICTYOPTERA				
Blattaria	×	×		Omnivorous
Perlamantinae	?	×		Arthropods
EMBIOPTERA		×		Lichens, etc
PSOCOPTERA		×		
HEMIPTERA				
Coccoidea		×	×	Sap
Microphysidae	?	×		Arthropods
COLEOPTERA				
Ptinidae	?	×		Lichens, etc
LEPIDOPTERA				
Psychidae		×	×	Leaves
Lymantridae			×	Leaves
Geometridae			×	Leaves

probably normally mate in their natal colony even though they might and sometimes do fly to enter other colonies. In insects as a whole such condition, with the female sex the more flightless, is far more common than the reverse. This fact underlines the hint of inbred colonies given by the correlations indicated in Table 10.2. In flightless-female species the sex ratio is usually about normal (1:1) whereas in flightless-male species (including all those known from dead trees) the sex ratio is female-biassed. This also has an understandable connection with inbreeding (Hamilton, 1967). If flightless-female species are relative outbreeders and flightless-male species relative inbreeders, the suggestion from Table 10.2 that subcortical habitats of dead trees force on some types of inhabitants a more than usual amount of inbreeding is further reinforced by noting that on the *exterior* of bark (and on the exterior of the living tree as a whole) we find a majority of all the flightless-female conditions that are known. Flightless-male conditions, on the contrary, are here far more rare and, when they occur, occur along with some kind of special cover that could confine the colony (e.g. *Archipsocus*, with colonies living under self-made webs – it must be admitted, however, that the contrast between polymorphism in *Archipsocus* and in *Reuterella* or bark-dwelling Embioptera, which also live in webs, remains puzzling). Table 10.3 summarises the well-known flightless-female groups that are known on trees. Trees are so favoured by flightless-female species because trees are particularly exposed to wind and this allows for passive dispersal, usually by young larvae (Ghilarov, 1966).

To avoid over-encumbering the reference list of this paper, supporting references for the statements of Tables 10.2 and 10.3 are not given: full bibliography will be included in a future publication. Groups for which the author has personal experience and data are marked with an asterisk.

Acknowledgements

Dead-tree insects, especially those manifesting wing polymorphism, were studied in Britain with assistance from a grant (GB3/1383) from the Natural Environment Research Council. For help with various facts, ideas and bibliographic items, thanks are also due to G. Borgia, M. Deyrup, L. A. Mound, U. Nur, V. A. Taylor and J. Waage.

References

Ashlock P. D. (1974) The uses of cladistics. *A. Rev. Ecol. Syst.* 5, 81–99.

Balachowsky A. & Mesnil L. (1935–36) *Les insectes nuisables aux plantes cultivées*, 2 vol. 1922 pp. Paris.

Barrows E. M. (1973) Soil nesting by wood-inhabiting Halictine bees, *Angochlora pura* and *Lasioglossum coeruleum* (Hymenoptera: Halictidae). *J. Kans. ent. Soc.* **46**, 496–99.

Beebe W. (1973) *Jungle days*. Garden City, New York.

Bedding R. A. (1967) Parasitic and free-living cycles in entomogenous nematodes of the genus *Deladenus*. *Nature, Lond.* **214**, 174–175.

Boer P. J. den (1970) On the significance of dispersal power from populations of carabid beetles (Coleoptera, Carabidae). *Oecologia (Berl.)* **4**, 1–28.

Bornemissza G. F. (1971) Mycetophagous breeding in the Australian dung beetle *Onthophagus dunningi*. *Pedobiologia* **11**, 133–42.

Borgia G. (1978) On the evolution of haplodiploid genetic systems. *Theoretical Population Biology* (in press).

Bournier A. (1961) Remarques au sujet du brachypterisme chez certaines espèces de thysanoptères. *Bull. Soc. ent. Fr.* **66**, 188–91.

Bradley J. C. (1947) Classification of Coleoptera. *Coleopts Bull.* **1**, 75–85.

Brothers D. J. (1975) Phylogeny and classification of the Aculeate Hymenoptera with special reference to Mutillidae. *Kans. Univ. Sci. Bull.* **50**, 483–648.

Bush G. L., Case S. M., Wilson A. C. & Patton J. L. (1977) Rapid speciation and chromosomal evolution in mammals. P.N.A.S.? *Proc. Natn. Acad. Sci. U.S.A.* **74**, 3942–3946.

Bush G. L. (1975) Modes of animal speciation. *A. Rev. Ecol. Syst.* **6**, 339–64.

Campos L. A. O., Velthuis-Kluppell F. M., & Velthuis H. H. W. (1975) Juvenile hormone and caste determination in a stingless bee. *Naturwissenschaften* **62**, 98–9.

Carpenter F. M. (1977) Geological history and the evolution of the insects. *Proc. XV int. Congr. Ent., Washington*. 63–70.

Carson H. L. & Kaneshiro K. Y. (1976) *Drosophila* of Hawaii: Systematics and ecological genetics. *A. Rev. Ecol. Syst.* 7, 311–65.

China W. E. (1962) South American Peloridiidae. *Trans. R. ent. Soc. Lond.* **114**, 131–61.

Crowson R. A. (1960) Observations on the beetle family Cupedidae with descriptions of two new fossil forms and a key to the recent genera. *Ann. Mag. nat. Hist. (13)* **5**, 147–57.

Crowson R. A. (1974) The evolutionary history of coleoptera as documented by fossil and comparative evidence. *Atti Congr. naz. ital. Ent., 10th, Sassari*, 47–90.

Elton C. S. (1966) *The Pattern of Animal Communities*. Methuen, London.

Evans J. W. (1941) Concerning the Peloridiidae. *Aust. J. Sci.* **4**, 95–97.

Evans J. W. (1967) The biological significance of the Peloridiidae (Homoptera, Coleorrhyncha) and a new species from Lord Howe Island. *Proc. R. Soc. Qd* **79**, 17–24.

Fager E. W. (1968) The community of invertebrates in decaying oak wood. *J. Anim. Ecol.* **37**, 121–42.

Fox L. R. (1975) Cannibalism in natural populations. *A. Rev. Ecol. Syst.* **6**, 87–102.

Geinitz H. B. (1885) *Die Versteinerungen der Steinkohlen-formation sachsen*. Leipzig.

Ghilarov M. S. (1956) The significance of the soil in the origin and evolution of insects (In Russian) *Ent. Obozr.* **35**, 487–94.

Ghilarov M. S. (1966) The evolution of insects during transition to passive dissemination and the principle of reverse links in phylogenetic development (In Russian) *Zool. Zh.* **45**, 3–23.

Gilbert L. E. (1972) Pollen feeding and reproductive biology of *Heliconius* butterflies. *Proc. Natn. Acad. Sci. U.S.A.* **69**, 1403–7.

Goodchild A. J. P. (1966) Evolution of the alimentary canal in the Hemiptera. *Biol. Rev.* **41**, 97–140.

Hamilton W. D. (1964) The genetical evolution of social behaviour I & II. *J Theoret. Biol.* 7, 1–16 & 17–52.

Hamilton W. D. (1967) Extraordinary sex ratios. *Science* **156**, 477–488.

Hamilton W. D. (1972) Altruism and related phenomena, mainly in social insects. *A. Rev. Ecol. Syst.* **3**, 193–232.

Hamilton W. D. (1974) Evolution sozialer Verhaltensweisen bei sozialen Insekten. In *Sozialpolymorphismus bei Insekten.* Ed., G. H. Schmidt, pp. 60–93. Wissenschaftliche Verlagsgesellschaft, Stuttgart.

Hamilton W. D. & May R. M. (1977) Dispersal in stable habitats. *Nature, Lond.* **269**, 575–581.

Hamilton W. D. (1978) Wingless and fighting males in fig wasps and other insects. In *Reproductive Competition and Selection in Insects*, Eds. M. S. Blum and N. A. Blum. Academic Press, New York.

Hicks E. A. (1959) *Check-List and Bibliography on the Occurrence of Insects in Birds Nests.* 681 pp. Iowa State College Press, Ames, Iowa.

Hinton H. E. (1948) On the origin and function of the pupal stage. *Trans. R. ent. Soc. Lond.* **99**, 395–409.

Hinton H. E. (1977) Enabling mechanisms. *Proc. XV int. Congr. Ent., Washington*, 71–83.

Hood J. D. (1940) The cause and significance of macropterism and brachypterism in certain Thysanoptera, with description of a new Mexican species. *An. Cienc. Biol. (Mexico)* **1**, 497–505.

Howden H. F. (1955) Biology and taxonomy of North American beetles of the subfamily Geotrupinae with revisions of the genera *Bolbocerosoma, Eucanthus, Geotrupes* and *Peltotrupes* (Scarabaeidae). *Proc. U.S. Natn. Mus.* **104**, 151–319.

Hurd P. (1958) Observations on the nesting habits of some new world carpenter bees with remarks on their importance in the problem of species formation. *Ann. ent. Soc. Am.* **51**, 365–75.

Hussey N. W., Read W. H. & Hesling J. J. (1969) *The Pests of Protected Cultivation.* Edward Arnold, London.

Imms A. D., Richards O. W. & Davies R. G. (1957) *A General Textbook of Entomology.* Methuen, London.

Kuhnelt W. (1961) *Soil Biology.* Translated by N. Walker. Faber & Faber, London.

Larkin P. A. & Elbourn C. A. (1964) Some observations on the fauna of dead wood in live oak trees. *Oikos* **15**, 1–92.

Levin D. A. (1976) The chemical defenses of plants to pathogens and herbivores. *A. Rev. Ecol. Syst.* 7, 121–59.

Luff M. L. (1964) The occurrence of some Coleoptera in grass tussocks, with special reference to microclimatic conditions. University of London, Ph.D. Thesis.

Malyshev D. I. (1969) *The Genesis of the Hymenoptera and the Phases of Their Evolution* (Translation from Russian). Methuen, London.

Mamaev B. M. (1971) The significance of dead wood as an environment in insect evolution. *Proc. 13th Int. Congr. Entomol. Moscow 1968* **1**, 269.

Mamaev B. M. (1975) *Evolution of Gall-forming Insects – Gall Midges* (translation from Russian by A. Crozy) The British Library.

Massoud Z. (1976) Essai de Synthese sur la phylogenie des Collemboles. *Revue Ecol. Biol. Sol.* **13**, 241–52.

Mayr E. (1963) *Animal species and evolution.* Harvard University Press, Cambridge.

McEnroe W. D. (1969) Spreading and inbreeding in the Spider Mite. *J. Hered.* **60**, 343–45.

McKay M. R. (1970) Lepidoptera in Cretaceous amber. *Science, N.Y.* **167**, 379–380.

Moser J. C. & Cross E. A. (1975) Phoretomorph: a new phoretic phase unique to the Pyemotidae (Acarina: Tarsonemoidea). *Ann. Ent. Soc. Amer.* **68**, 820–2.

Mound L. A. & O'Neill K. (1974) Taxonomy of the Merothripidae, with ecological and phylogenetic considerations (Thysanoptera). *J. Nat. Hist.* **8**, 481–509.

Needham J. G. & Betten D. (1901) Aquatic Insects of the Adirondacks. *Bull. N.Y. St. Mus.* **47**, 1–612.

New T. R. (1971a) An introduction to the natural history of the British Psocoptera. *Entomologist* **104**, 59–97.

New T. R. (1971*b*) A new species of *Belaphopsocus* Badonnel from Brazil with notes on its early stages and bionomics (Psocoptera). *Entomologist* **104**, 124–33.

Nur U. (1977) Electrophoretic comparisons of enzymes of sexual and parthenogenetic mealy-bugs (Homoptera: Coccoidea: Pseudococcidae). Virginia Polytechnic Institute and State University Research Division Bulletin **127**, 69–84.

Obreen H. C. (1967) Contributions to the biology and taxonomy of *Rhizoglyphus echinopus. Meded. Rijksfac. Landbouwwet. Gent.* **32**, 602–6.

Parshley H. M. (1922) A note on the migration of certain water-striders. *Bull Brooklyn ent. Soc.* **17**, 136–7.

Riley C. F. C. (1920) Migratory responses of water-striders during severe droughts. *Bull. Brooklyn ent. Soc.* **15**, 1–10.

Ross H. H. (1967) The evolution and past dispersal of the Trichoptera. *A. Rev. Ent.* **12**, 169–206.

Saunders L. G. (1924) On the life history and the anatomy of the early stages of *Forcipomyia (Diptera, Nematocera, Ceratopogonidae). Parasitology* **16**, 164–213.

Schedl K. E. (1958) Breeding habits of arbicole insects in Central Africa. *Proc. X int. Congr. Ent., Montreal* **1**, 183–97.

Scott A. C. (1977) Coprolites containing plant material from the Carboniferous of Britain. *Palaeontology* **20**, 59–68.

Smart J. & Hughes N. F. (1973) The insect and the plant: progressive palaeoecological integration. In Insect/Plant Relationships. *Symp. R. Ent. Soc. Lond.* **6**, 163–155.

Southwood T. R. E. (1961) The numbers of species of insect associated with various trees. *J. anim. Ecol.* **30**, 1–8.

Southwood T. R. E. (1962) Migration of terrestrial arthropods in relation to habitat. *Biol. Rev.* **37**, 171–214.

Southwood T. R. E. (1973) The insect/plant relationship – an evolutionary perspective. In *Insect/Plant Relationships. Symp. R. Ent. Soc. Lond.* **6**, 3–20.

Stannard L. J. (1968) The thrips or Thysanoptera of Illinois, *Bull. Ill. St. nat. Hist. Suv.* **29**, 1–552.

Steffan A. W. (1975) Morphological and behavioural polymorphism in *Plastosciara perniciosa* (Diptera: Sciaridae). *Proc. ent. Soc. Wash.* **77**, 1–14.

Stubbs A. E. (1972) *Wildlife Conservation and Dead Wood.* A Supplement to J. Devon Trust for Nature Conservation. Exeter, Devon, U.K.

Swain T. (1977) The effect of plant secondary products in insect co-evolution. *Proc. XV int. Congr. Ent., Washington*, 249–56.

Taylor V. A. T. (1975) *The biology of feather-winged beetles of the genus Ptinella with particular reference to coexistence and parthenogenesis.* Thesis for degree of Ph.D., University of London.

Thorpe W. H. (1968) *Orthezia cataphracta* (Shaw) (Hemiptera, Coccidae) feeding on a basidiomycete fungus, *Collybia* sp. *Entomologist's mon. Mag.* **103**, 155.

Trivers R. L. (1971) The evolution of reciprocal altruism. *Quart. Rev. Biol.* **46**, 35–57.

Trivers R. L. & Hare H. (1976) Haplodiploidy and the evolution of the social insects. *Science* **191**, 249–263.

Tuomikoski R. (1957) Beobachtungen uber einige Sciariden (Diptera), derem Larven in faulen Holz oder unter der Rinde abgestorbener. *Suomen hyönt Aikak.* **23**, 3–35.

Turček F. J. (1964) Beitrage Zur Ökologie der Kiefernrindenwanze *Aradus cinnamoneus. Panz. Biologia Bratisl.* **19**, 762–77.

Turk F. A. (1953) A new genus and species of pseudo-scorpion with some notes on its biology. *Proc. zool. Soc. Lond.* **122**, 951–4.

Turner J. R. G. (1971) Experiments on the demography of tropical butterflies. II. Longevity and home-range behavior in *Heliconius erato. Biotropica* **3**, 21–31.

Usinger R. L. (1966) *Monograph of Cimicidae*, Vol. 7. xi + 585 pp. The Thomas Say Foundation.

Usinger R. L. & Matsuda R. (1959) *Classification of the Aradidae (Hemiptera-Heteroptera).* British Museum, London.

Vandel A. (1943) Essai sur l'origine l'evolution et la classification des Oniscoidea (Isopodes terrestres). *Bull. biol. France et Belg., Suppl.* **30**, 1–136.

West-Eberhard M. J. (1975) The evolution of social behaviour by kin selection *Quart. Rev. Biol.* **50**, 1–33.

Williamson W. C. (1880) On the organisation of the fossil plants of the coal measures. X. Including an examination of the supposed radiolarians of Carboniferous rocks. *Phil. Trans. R. Soc.* **171**, 493–539.

Wilson A. C., Bush G. L., Case S. M. & King M. C. (1975) Social structuring of mammalian populations and rate of chromosomal evolution. *Proc. Natn. Acad. Sci. U.S.A.* **72**, 5061–65.

Wilson D. S. (1975) A theory of group selection. *Proc. Natn. Acad. Sci. U.S.A.* **72**, 143–46.

Wilson D. S. (1976) Evolution at the level of communities. *Science* **192**, 1358–60.

Wilson E. O. (1971) *The Insect Societies.* Harvard Univ. Press, Cambridge, Mass.

Wilson E. O. (1975) *Sociobiology*. Harvard Univ. Press, Cambridge, Mass.

Wright S. (1932) The roles of mutation, inbreeding, cross-breeding and selection in evolution. *Proc. 6th int. Congr. Genet.* **1**, 356–66.

Wright S. (1940) Breeding structure of populations in relation to speciation. *Am. Nat.* **74**, 232–248.

Wright S. (1969) *Evolution and the genetics of populations,* Vol. 2. University Chicago Press, Chicago.

Zimmerman E. C. (1948) Homoptera: Auchenorrhyncha. *Insects of Hawaii* **4**, 1–268.

11 • Constancy of insect species versus inconstancy of Quaternary environments

G. R. COOPE

Department of Geological Sciences,
University of Birmingham B15 2 TT England

No one can doubt that the Quaternary Ice Age with its rapid and numerous oscillations between glacial and interglacial conditions must have had a drastic effect upon insect diversity. Even in those parts of the world not directly afflicted by the repeated expansion and contraction of continental glacier ice, associated climatic changes must have been widespread and wrought havoc on the established order of ecosystems on a global scale. In the presumed absence of an adequate fossil record of Quaternary insects, the effects of these climatic changes have been to a large extent matters of speculation. It is my intention here to put this record straight.

It is certainly easy, from the comfort of the present interglacial, to view the effects of the ice age as if they were remote and of only marginal relevance to our understanding of present day ecology and evolution. Even in the most recent synthesis of evolutionary science (e.g. Dobzhanski *et al*, 1977; Grant, 1977) there is hardly a mention of the part played by ice ages and the environmental instability that accompanied them. And yet this is the arena in which the drama of our current ecology and evolution is set. It seems to me to be unrealistic to believe, or at least act as if we believe, that apart from minor variations on the scale of a few years, or at the most decades, the ecological circumstances and in particular the climatic zones have remained in the same place and retained more or less the same parameters throughout the relevant past. Since the geologic past is so remote and populated by such strange organisms it seemed safe to relegate its importance to macroevolutionary events.

But the Quaternary period can not so easily be dismissed. It covers the last two million years when much of the flora and fauna would have been familiar to the present day biologist and during which the present composition and geographical distributions of the modern biota became established. This is the area where interactions between biologists and earth scientists are proving most profitable, especially in botany and in vertebrate and molluscan zoology where there is a well-known Quaternary fossil record. What is still not generally recognised is that insects provide ample fossils from this period which in numbers and varieties outclass their more well-established rivals. Unfortunately, as in so many interdisciplinary fields, there is a difficulty of communication. Few of us can keep abreast of our own scientific literature let alone that of adjacent disciplines.

Thus at the outset I reveal my hand. I am by training a geologist though I would like to believe that I am a natural historian, a calling that has been greatly calumnised of recent years. I came to entomology, therefore, by a rather unconventional route, and view time in a way that must seem rather unfamiliar to the biologist (geologists are quite at home in this environment). It is my contention that insect diversity, linked inextricably with their evolution, cannot be adequately understood outside a historical context. If, in my efforts to justify this viewpoint, I display a mysterious ignorance of some of the most obvious of recent advances in ecological thinking may I retreat into the lines judiciously prepared for such an emergency at the end of the previous paragraph. For my part I will be more than satisfied if I provide entomologists with some insight into recent advances in our understanding of the geology and palaeontology of the Quaternary that may have been published in the 'wrong' place and thus have escaped their notice.

Quaternary (Ice Age) environments

For numerous reasons the term 'Ice Age' is unsatisfactory. Firstly it gives the impression of a single event in the past that has now, fortunately, come to an end; its termination being dated at about ten thousand years ago. Ever since the pioneer work of Penck and Bruckner (1909), who recognised four glacial periods in the northern foothills of the Alps, it has been recognised that the ice age was a multiple event. Their classification of glaciations into Gunz, Mindil, Riss and Würm has provided the framework upon which the glacial and interglacial periods of the whole of western Europe have been correlated, albeit with different names; a taxonomic situation that would not be unfamiliar to entomologists. This quadripartite subdivision has even been adopted in North America where the glacial episodes have been called the Nebraskan, Kansan, Illinoian and Wisconsinan. The term 'Ice Age' is also unsatisfactory because it was widely believed that the interglacial periods were much longer than the intervals of ice expansion. However, an ever increasing amount of evidence is now coming to light indicating that this belief is not well founded.

The seductively simple classification of glacials and interglacials has gradually become more and more difficult to apply. Without going into details it has become evident that there were several glacial episodes older than the Gunz and that each of the subsequent glacials was, in its turn, polyphase. Furthermore, it has become increasingly likely that there are more interglacials known than can be accommodated between the classical glacial stages. This rather unsatisfactory situation arises for two reasons. Firstly, the stratigraphy of terrestrial Quaternary deposits is immensely complex and the traditional lithological and palaeontological criteria for interpreting sequences of strata can only be applied to a very limited extent. Secondly, there is no satisfactory way of absolute dating of deposits beyond about 50 000 years, the acceptable limit of ^{14}C dating techniques. Because of the difficulties in the interpretation of terrestrial sequences, quaternary geologists have turned increasingly to deep sea deposits where more continuous successions are to be expected.

The record of environmental change is preserved in these abyssal sediments by a number of parameters whose variation can be measured at different depths in cores recovered from the sea bed. Only two of these variables will be briefly discussed here, the changes in oxygen isotope ratios in the shells of foraminifera and the changes in specific composition of foraminiferal assemblages. Both give a measure of frequency and intensity of climatic changes, the first on a global scale and the second of more local significance.

ISOTOPIC RECORD FROM DEEP SEA CORES

The proportions in which the two stable isotopes of oxygen, ^{18}O and ^{16}O, are incorporated into the calcium carbonate shells of foraminifera depend on two environmental factors; the isotopic composition of the ocean water and the temperature at which the shell is secreted. The ratio of $^{18}O/^{16}O$ decreases with rising temperature. At first it was believed that the isotopic composition of the ocean could be taken as constant and thus the oxygen isotope ratio of the shells would provide a measure of past temperature changes in the ocean water (Emiliani, 1955). However, the expansion of continental ice sheets deprives the ocean of the lighter isotope of oxygen, since water vapour derived by evaporation from the ocean surface contains relatively more ^{16}O than the water that remains. When this vapour condenses and falls as snow precipitation over the ice caps it is there stored as glacier ice and does not return to the ocean as surface runoff. During the times of great ice accumulation on land the ocean waters had a relatively high proportion of the heavier isotope of oxygen. The assumption that the ocean water has remained isotopically constant is therefore unsound, and as Shackleton (1967) has shown, the isotopic difference between glacial and interglacial ocean water is significant.

Since the foraminiferal $^{18}O/^{16}O$ ratio changes in the same direction as temperatures are lowered and as land ice increases, the two effects reinforce one another and could distort the record by amplifying this change. To avoid this potential source of error Shackleton and Opdyke (1973, 1976) selected only benthonic foraminifera whose living conditions might be expected to be largely insulated from changes in surface temperature. The resulting curve of variation in the $^{18}O/^{16}O$ ratio down the core thus reflects the waxing and waning of global land ice and can thus be taken as an integrated measure of global climate.

To turn this record into history, core depth must be calibrated with time. This is a tricky problem that is gradually yielding to concerted attacks from a number of directions. One or two courses of action may be mentioned here. Ocean bottom deposits are far from sources of sediment and were never, in the time we are considering, raised up above sea level by earth movements. They thus represent slow and continuous accumulations whose sedimentation rate may be nearly constant. Two fixed points in time may be recognised, the upper levels may be dated by ^{14}C in the marine shells, and lower down the sudden reversal in the earth's magnetic field at the boundary between the Brunhes and the Matuyama magnetic epochs can be dated at 690 000 years ago. Other less reliable time control points can also be used, such as the times of apparent extinction of certain foraminifera or the first appearance of others. The calibration of numerous isotopic records from all the oceans of the world is now underway and remarkably consistent results are being obtained.

Figure 11.1 shows the isotopic changes in ocean water during the last half million years, representing chiefly the changes in Nothern Hemisphere ice volume (Hays *et al,* 1976). It must be emphasised here that this represents only the terminal third of the Quaternary period and that similar oscillations occurred at least as far back as the Pliocene several million years ago. There are a number of interesting features about this curve. Firstly, climatic changes do not divide themselves naturally into glacial and interglacial units; rather there are numerous oscillations of varying intensity, although which of these qualify for glacial, interglacial or interstadial status is a semantic argument that need not detain us here. Secondly, the slopes of the sides of the oscillations are not symmetrical but resemble saw teeth in which the gradients representing ice destruction are much steeper than those for ice accumulation. In other words it is much easier to remove the continental ice sheets than to establish them. Thirdly, if the extreme left of the curve indicates the global ice volume at the

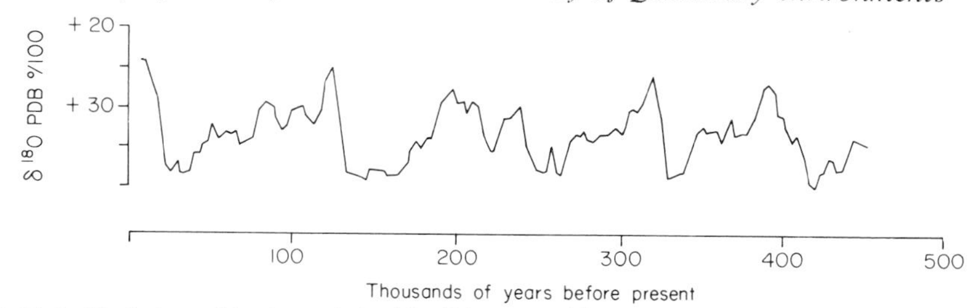

FIG. 11.1. Variation with time of the $^{18}O/^{16}O$ ratio (^{18}O PDB °/100) in foraminiferal shells from a core taken from the bottom of the subantarctic Indian Ocean. This curve gives a measure of changing world ice mass. Troughs indicate periods of maximum ice accumulation (Hays *et al*, 1976).

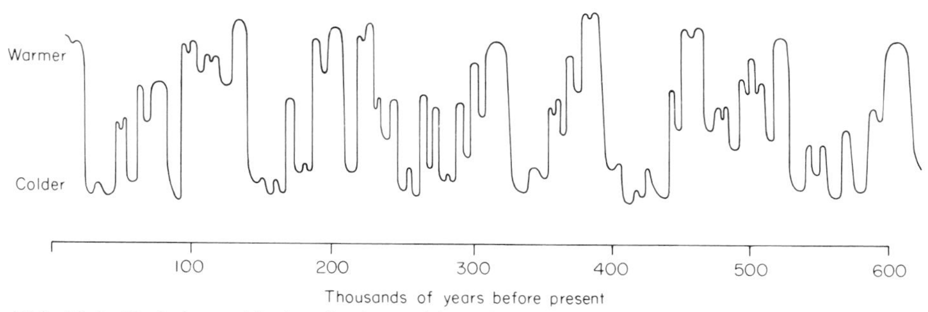

FIG. 11.2. Variations with time in the position of the polar front in the North Atlantic. This curve gives some measure of ocean temperatures and thus of the changing climates on adjacent land. (Ruddiman & McIntyre, 1976).

present day, then it is clear that during the last half million years periods with as little world ice as the present day are remarkably rare and short lived. This is a complete reversal of the earlier view that the interglacials were the predominant climatic mode of the Quaternary punctuated by brief onslaughts of glacial conditions. (I am sure that this latter view stemmed rather from wishful thinking than from any scientific data.) Lastly this curve must not be exactly equated with climate, though for sure there is a broad correlation, since ice build-up requires time and there must be a lag in the ice accumulation curve compared with the climatic change that set the process going.

FORAMINIFERAL FAUNAS AS INDICATORS OF QUATERNARY CLIMATIC CHANGE

Whereas the oxygen isotopic data provide us with a global picture of changing climate, placing its major emphasis on the Nothern Hemisphere because that was where the greatest changes of ice volume occurred, the record of changing foraminiferal faunas gives a picture of fluctuations in ocean surface temperatures on a very much more local scale. It is fortunate, therefore, that the north Atlantic has been the object of an intense research effort (Ruddiman & McIntyre, 1976); fortunate, not only because it is near home, but because the changes in the upper part of the cores can be compared with changes in fossil insect faunas from adjacent areas of western Europe.

Briefly the procedure is as follows. Foraminiferal species used in this study are identical with present day forms and are assumed to have the same ecological requirements. This assumption seems justified by the orderly procession of ecologically homogeneous faunas down the length of the cores. Using the present day requirements of the species as a data base, the species are allocated to a series of water mass bodies; polar, subpolar, transitional, subtropical northern and southern and finally a cosmopolitan group of species. The varying fortunes of these groups can then be plotted down the core and thus, in a manner similar to that outlined earlier for the isotopic record, set against a time scale. Since each core will reflect the sequence of changes in the water mass ecology over the sampling site, it is possible from a widely scattered series of boreholes to plot the variations in time of the geographical distributions of the water masses.

Ruddiman and McIntyre (1976, p. 111) summarise their findings in the following way. 'In the subpolar Atlantic Ocean during the Quaternary Period, water mass environments have migrated across more than 20° of latitude, which is equivalent to temperature oscillations of the ocean surface of at least 12°C. The migrations have occurred along a northwest trending axis at mean rates of approximately 100m/yr sustained over intervals of several centuries. During peak glaciations polar water moved south to Lat. 42°N, where an abrupt frontal system separated the cyclonic subpolar gyre from the anti-cyclonic subtropical gyre. Seven complete climatic cycles have occurred in the past 600 000 yrs within which at least 11 separate major southward advances of the polar front have occurred'. Only during those periods when the polar front lay to the north of Britain did the warm ocean current of the north Atlantic drift wash along our western seaboard, giving us the warm moist oceanic climate to which we are accustomed today in these islands.

Figure 11.2 shows the movement of the polar front across the north Atlantic during the last 600 000 years. As might be expected it is a more complex curve than that of Fig. 11.1 because the shift in ocean current systems is a more excitable variable than the accumulation and destruction of continental ice sheets. Again this curve shows how unusual our present warm climate is compared with the major part of this period. In fact, it might safely be said that the usual environment for Britain and most of North-west Europe during the last half million years was in many aspects analogous to the polar tundras of present day. The temperature gradient across the polar front was about 7°C, so the passage of the front northwards or southwards must have been associated with drastic changes in the thermal environment of the adjacent continents. The rates at which the polar front could sweep across the North Atlantic ocean are indeed dramatic, even though the measurements available at the moment only provide us with minimum values and the actual speed of movement may have been even higher than this. For example, 14 000 years ago, the front stood at Lat. 46°N, that is in a position westwards from the northern tip of Portugal. By 11 000 years ago the front had been pressed back against the southern tip of Greenland and was oriented in a more north-south direction. Furthermore, by 10 200 years ago the front had once again come southwards, to stand off the South-west coast of Ireland (data from Ruddiman *et al*, 1977.) Although this period cannot be considered to be typical of the changing patterns of north Atlantic circulation, it provides us with some understanding of the actual rates of changes that are involved and a view of sudden shifts rather than gradual ones that we have, by tradition, come to expect. Climatic changes in North-west Europe match these circulatory changes in their timing, duration and intensity (Pennington, 1977; Coope, 1977), and are reflected in marked alterations in the vegetational record and in rapid switches in the specific composition of the insect faunas of the times.

SUMMARY OF QUATERNARY ENVIRONMENTAL CHANGES

The picture that is now emerging from these recent advances in earth science is much more complex than the older classification of glacials and interglacials. Major variations in climate have been more numerous and the shifts from one extreme to the other have been more rapid than was hitherto expected. The frequency of these climatic oscillations may be measured in tens of thousands of years, or less in some cases, though the actual period of transition may be very much less than this and may have taken place in only a few centuries.

Though the effects of these climatic changes are likely to be most pronounced in the Northern Hemisphere, where large continental masses permitted greater accumulations of ice than in the Southern Hemisphere, it is now clear that no part of the earth escaped without considerable environmental changes. Thus in South America the forest repeatedly became fragmented and in part replaced by open savannah. Alternating periods of heavy rainfall and drought, on a far greater scale than those experienced in historic time, expanded and contracted African lakes such as Chad and Victoria, or the extent and area of the Caspian Sea, with consequent changes in the flora and fauna. Sea level alterations, associated with the accumulation of ice on land leading to the abstraction of ocean water, meant that at episodes of fully glacial conditions the sea level was as much as 130 m lower than it is today.

Thus in our efforts to understand diversity in animal and plant communities we cannot assume that environmental parameters were constant throughout the relevant past, unless we believe that the species and their interactions in ecosystems are almost entirely the result of the last ten thousand years of adaptation. The Quaternary fossil insect record shows that this belief is untenable.

Effects of Quaternary environments on insect diversity

It would not be easy, in the absence of a good fossil record, to determine whether the intense environmental instability that characterised the Quaternary period should have led to an increase or decrease in insect diversity. If by diversity is meant the numbers and varieties (using this in its English rather than taxonomic sense) of species, then it might be expected that diversity should increase as rates of speciation increased and thus be related to evolutionary rate. But did the oscillating climate of the Quaternary increase or decrease the speed of evolution? Did the succession of major environmental changes increase the number of extinctions so that, regardless of the rate of recruitment of new species, the net effect was one of faunal impoverishment? Such questions would be impossible to answer were it not for the good fortune that insect cuticle is so durable, and that Quaternary sedimentary environments are so often favourable to its preservation. Thus insect fossils make up the most abundant, diverse and identifiable component of Quaternary terrestrial fossil assemblages.

Before discussing the significance of this fossil record, it would perhaps be helpful to consider briefly some of the inferences that might have been, or indeed have been, made on the effects of Quaternary climatic events on insect evolution in the absence of this fossil information. I must emphasise at the outset, however, that the hypothesis outlined below merely sets the scene of our expectations without the constraints on our imagination imposed by the fossil record. In actual fact, these expectations are given little support by Quaternary insect fossils. The setting up of what is essentially a straw hypothesis seems necessary, however, because statements are still being made about the likely consequences of recent Ice Ages without any reference whatsoever to the fossil remains of the insects that actually lived at the time. These are, after all, the ultimate arbiters in our discussions about past events.

The result of climatic deterioration (in the sense of cooling) on the insect fauna of the times is not just the symmetrical opposite of climatic amelioration. This is because, with the onset of glacial conditions, vast areas of the northern continents were reduced to polar desert or covered with ice sheets so that insect life must have been impossible in many places at such times. Interglacials do not involve the same expansion of uninhabitable areas of the world. For this reason the effects of climatic deterioration have to be dealt with separately from those of climatic amelioration.

During periods of ice advance, huge tracts of tundra-like environments existed in Europe south of the ice sheet. In North America the tundra zone was much more limited and the ice, in places, actually pushed down coniferous forests. During such periods thermophilous insect species would have been confined to limited geographical areas where their populations would have been greatly reduced. Intense selection pressures, induced by the severity of the climate, would lead to rapid genetic change in these populations. It would also be likely that many populations would be eliminated, and thus widespread extinctions might be expected to occur at such times. Since these cold periods were frequent and prolonged, these effects would be intensified at each glacial/interglacial cycle.

During periods of climatic amelioration great areas of previously uninhabitable land would become available for colonisation, probably very rapidly indeed. A scramble ecology would ensue with the most efficient pioneers, both animal and plant, moving first and furthest into the new territories in which, at least initially, competition would be low. Amongst these pioneers would be many thermophilous species released from their refuges in the south. This early phase of colonisation, with numerous isolated populations becoming established, might have been expected to lead to speciation by genetic drift and the fixation of novel genetic combinations that would have been swamped by outcrossing in larger parent populations. This is the ideal situation for 'quantum evolution' (Simpson, 1944) or speciation by genetic revolution (Mayr, 1954), and was the mechanism postulated for rapid speciation of Hawaiian *Drosophila* species (Carson, 1970). Of course many of these new populations would fail, either because the novel genetic combinations were inadaptive, or because of competition from late comers or the ecological development of the habitats born of time and changing edaphic factors. This boom and bust speciation could have been characteristic, not only of episodes of climatic warming after glacial periods, but it may also have occurred when northern species moved southward during periods of deteriorating climate to occupy areas rendered uninhabitable to thermophilous species. However, in the latter case they would not have been presented with a *carte blanche* since thermally less fastidious species would have survived.

This sequence of events seems entirely in keeping with the readiness with which present day insect species form geographical races on both a local and regional scale. Also, the existence of sibling and sister species that are exceedingly close genetically seems to imply recent speciation. Although the term 'recent' is scientifically inadequate, it is tempting to attribute the origin of such species to the last glacial period. Yet the northern areas that were subject to the most intense glaciation at such times are poor in specific diversity. If the process outlined above had led to extensive speciation then there must have been complementary massive extinctions to redress the balance. Thus Downes (1964) and Dunbar (1968) point out that the arctic environment is not intrinsically lacking in ecological diversity, but, being near the physiological tolerance limit for most organisms, the development of slowly diversifying ecosystems must have been set back at each glacial episode by widespread extinctions.

This attractive and internally consistent picture of the effects of glacial/interglacial cycles on the evolution and diversity of insect faunas would lead us to expect that the record of Quaternary insect fossils should present evidence of speciation and extinction on a grand scale. Certainly, a brief survey of some early accounts of these insect fossils might lead one to believe that this was indeed the case, since a great many fossils, even from the youngest Quaternary deposits, were credited with new specific names. Some of these new names showed clearly that their author was aware of their close relationship with extant species (Lomnicki, 1894), others adopted names that betrayed a certain lack of sympathy with the fossil material (Scudder, 1895, 1900). However, this was the easy way out of the problem, and where a re-examination of the fossils has been possible it has become apparent that there was inadequate justification for this taxonomic timidity (Coope, 1968; Angus, 1973).

Quaternary insect fossils

Sadly, the Quaternary fossil insect record does not live up to the expectations of it outlined above. I say sadly, because in so many ways the visions of earlier workers on the effects of Ice Ages on insect evolution have often been so elegant. Yet an empirical view of the fossil record shows that there is no evidence of any morphological evolution during the last half million years at least. In other words insect species have remained immutable regardless of the vicissitudes of the Quaternary climate. Certainly for the past few glacial/interglacial cycles, back as far at least as the Cromerian interglacial, almost all our fossil insects match their modern counterparts with extraordinary exactness even down to the intimate internal sclerites of the genitalia. Though the vast majority of insect fossils are Coleoptera, because the robustness of their exoskeleton lends itself to easy preservation, a broad spectrum of insect orders is represented in these fossil assemblages although structural frailty makes identification in many other groups a rather rare occurrence. The ever growing literature on this abundant fossil record has been reviewed by Coope (1970), and since then insect faunas from numerous other localities have been investigated (Coope & Brophy, 1972; Ashworth, 1972; Morgan, 1973; Osborne, 1972).

Of outstanding importance in this context is the work of Matthews on the insect fossils from early Pleistocene and late Tertiary deposits of Alaska and North-west Canada. Thus in early Pleistocene deposits at Cape Deceit, Alaska (Matthews, 1974) numerous extant species were recorded, some articulated specimens complete with their genitalia. Fossils of *Tachinus apterus*, however, showed a very slight statistical difference from modern specimens and may suggest the first hints of phyletic change. Even more exciting are preliminary results of work currently being carried out by Matthews (1976*a, b*) on insects from late Miocene deposits on Meighen Island and at Lava camp, Alaska. The deposits at the latter site can be dated at just over five and a half million years ago. The fossils from these localities fall outside the range of variation of modern species, but it is clear that many of them are exceedingly close morphologically to living forms. These discoveries cast specific longevity in quite a new light, showing that stability is in fact the norm rather than the exception, and seemingly that most of the upper Tertiary fossils represent the phyletic predecessors of living species rather than cladistically related sister species.

The part played by extinctions in the structuring of Quaternary insect faunas is a rather more intractable problem. The main reason for this difficulty is philosophical and more a matter of theory than of practice. It is not merely the problem of proving a negative, because circumstantial evidence can accumulate so as to leave no reasonable doubt that many organisms are now extinct. But since absence of evidence is not evidence of absence, we are on much more shaky ground when faced with the problem of assessing the part played by extinction in this context. Here, then, I must rely on my personal beliefs. The more experience I have of Quaternary fossil insects the more convinced I become that extinction played a very minor role in the development of our present day insect fauna, or if it did play an important part then the timing of these extinctions must be pressed back into the Tertiary period.

This belief is founded on a number of facts. Firstly, if we consider the insect fossils that are complex enough to provide an adequate basis for comparison with modern specimens, a very high proportion (more than 90%) can be matched precisely with modern species. There are no extensive residues left over that defy identification except for nondescript fragments such as for example disarticulated bits of *Atheta* species. This is particularly true of arctic faunas where the proportion of diagnosable fragments may be even higher, but where, if mass extinction was at all widespread its effects might have been expected to be most intense. Secondly, when we have felt assured that some highly distinctive specimen must at last be an extinct form, we have found, usually through the kind offices of some taxonomic specialist, that the species concerned is in fact well known after all. Thirdly, there is always the possibility that the few problem species remaining are not yet 'known to science'. Such species are, however, few and even if they have died out in recent time they in no way qualify as mass extinctions.

It is easy to see why earlier workers on Quaternary fossil insects frequently believed that they were looking at extinct species. Firstly, there was the general expectation that, since fossils were usually the remains of extinct animals, it was reasonable to infer that Quaternary fossils should also represent extinct forms. Furthermore, there was the very strange impression created by the disarticulated skeletons that made keys to identification almost impossible to use and the body shape difficult to appreciate. Occasionally post-mortem distortion added to the problem (though I should add here, for fear that some aspiring Quaternary entomologist might be deterred by this last point, that most of this distortion occurs when the specimens are dried in the laboratory and can be avoided if the fossils are preserved in spirit). Finally, it is now becoming clear that these earlier workers were indeed often faced with unfamiliar species many of which live today hundreds or even thousands of kilometres from the places where their fossil remains were found (see Coope, 1973, for a Tibetan species of *Aphodius*, and Ullrich and Coope, 1974, for a Siberian and North American species of *Tachinus*, both abundant in British deposits of the Last Glaciation).

These large scale changes in geographical distributions of insect species, even within the limited time of the last glacial/interglacial cycle, are one of the major features of this investigation into Quaternary insect fossils. Lindroth (1948) was one of the first to recognise that species of the present day tundra had moved southward as far as central Sweden under the influence of climatic cooling. Later Coope (1962) showed that many of these arctic species occurred in the English Midlands during the Last Glaciation, namely 38 000 years ago. After this, numerous publications have documented the ebb and flow of northern and temperate species in western Europe in sympathy with the to and fro of climatic changes (Coope, 1970, 1975*a*, *b*, 1977; Morgan, 1973). In North America, similar specific mobility has been documented by Ashworth (1977) and Morgan (1972) where again insect species can be shown to have undergone large scale alterations of their geographical distributions.

Whereas morphological constancy and the large scale changes in the geographical distributions of species during the Quaternary period are now demonstrable facts, physiological stability can only be inferred indirectly. If we assume that the physiological characteristics of a species are reflected in its environmental requirements and preferences, then it should be possible to obtain some measure of physiological evolution by investigating the ecological homogeneity of our fossil assemblages using present day ecologies of the species concerned as our yardstick. In fact, species kept much the same company in the past as they do today as the numerous studies on Quaternary insect assemblages clearly show.

There are, however, exceptions to this generalisation, showing that in the past environmental combinations may have occurred that strictly have no analogue at the present time. Some species so persistently offend against ecological conformity (Coope, 1977, p. 315–6) that they may display tolerance limits that are different from their modern relatives and may thus be cases of physiological evolution without any equivalent morphological change. It must be emphasised, however, that such cases are rare, and for the bulk of insects it seems likely that the demonstrable morphological stability during the Quaternary period is matched by a corresponding degree of physiological constancy.

Conclusions

It is now possible to draw together the known facts and reasonable inferences that can be made from the Quaternary fossil record, and to relate these to present day insect diversity. The facts are simple enough, even though they were unexpected to one brought up to see incipient speciation in the intricacies of geographical variation and, in the recurrent ice ages, a powerful evolutionary agent. But the fossil record provides ample evidence of specific constancy for the last half million years (generations?) at least, and for much further back in time than this if we bear in mind the dramatic results of Matthew's work on the lower Pleistocene and late Tertiary insect faunas. However, it cannot be argued that the fossil record precludes any *possibility* of recent speciation, but it does provide concrete evidence for the geological longevity of an ever increasing list of species and confirms our belief that this long-term stability is the usual state of affairs for insect species throughout the Quaternary period. Furthermore, there is now ample circumstantial evidence that morphological stability entailed a similar degree of physiological constancy.

The fossil record clearly shows that insect species altered their geographical ranges on an enormous scale as the climatic zones shifted, so that, for example, northern tundra communities came south to occupy the lowlands of central Europe during the cold phases to be replaced by suites of southern European species during some of the warmer interglacials. Although it is still early in the investigation of Quaternary insect faunas, there are hints that groups of species moved *en bloc* reflecting the migration of whole ecosystems or at least major components of ecosystems. Be that as it may, it is evident that even the most soil-bound and flightless species had adequate mobility to enable them to keep pace with the changing geographical locations of suitable climates. Thus the environment in which a species actually lived remained more or less constant throughout the perturbations of the Quaternary climate; it was the geographical locations of suitable environments that changed. Here, then, we remove one of the main reasons for believing that successive glacial/interglacial cycles must have involved large scale extinctions and the consequent cuts-back in insect diversity. If the peripatetic ecosystems were not to be pruned in this way there would have been little opportunity for any new growth, especially if they have survived for hundreds of thousands of years and have their environmental niches more or less saturated.

Of course specific constancy itself could have its origin in the instability of the Quaternary climate (Coope, 1970; 1977). Certainly, the forced marches that many species were compelled to undertake, according to the dictates of Quaternary climatic changes, must have repeatedly led to the breaking down of the geographical separation of populations. Repeated outbreeding must have resulted from this geographical shuffling and the gene pools kept well stirred. The rapidity and intensity of the climatic changes can be readily appreciated from Figs. 11.1 and 11.2 and, wherever the fossil record has been investigated, these climatic changes can be matched by corresponding comings and goings of insect assemblages. It is difficult to see how, in such effervescent circumstances, any species was left in peace long enough for genetic isolation to become established.

A corollary of this argument is that for rapid speciation environmental stability in a constant geographical location is required. In the present phase of climatic instability it is difficult to see where we might find such a place, for even the tropics were not immune to fluctuations in the physical environment. However, the intensity of the forces tending to shuffle populations might be expected to be diminished in equatorial latitudes, and thus opportunities for speciation might be correspondingly greater there than in the temperate latitudes. Is this a contributing factor to the greatly enhanced diversity of insect faunas in the tropics?

The recognition that the option of movement is the one usually taken up by populations of insects under stress from environmental change has important consequences for our understanding of evolution in trap situations from which emigration is denied. We might site caves, oceanic islands and equatorial mountain tops as examples of inescapable environmental prisons. Here, with option of movement no longer available, any environmental change must be endured on the spot so that the choices available are now limited to adaptation to fit the new conditions or extinction. Under such circumstances a rapid rate of evolutionary change is to be expected and, may be, more frequent extinctions; the balance between gain and loss in this way will determine specific diversity. Here again it would be the fossil record that could show to what extent these speculations are justified. Though remote islands, or similar evolutionary traps, provide excellent opportunities for the study of evolutionary processes, nonetheless the rates of evolution in such localities are likely to be atypically high compared with the bulk of evolutionary changes taking place on the continental masses.

Finally I would like to point out that the science of Quaternary entomology is still in its infancy (or at least I hope it is!) so that what I have said here should be seen in this light. Almost all the fossil insects so far recognised are Coleoptera but we have high hopes that many other orders of insect will eventually play their part in unravelling the complicated interplay of recent insect history. Also the areas of the world from which Quaternary fossil insects have been investigated are relatively small, chiefly North-west Europe and the British Isles in particular, and there is a growing contribution from North America and the U.S.S.R. In principle there is no reason why Quaternary insect fossils should not be found in any part of the world wherever water-logged sediments of terrestrial origin are found.

References

Angus R. B. (1973) Pleistocene *Helophorus* (Coleoptera, Hydrophilidae) from Borislav and Starunia in the Western Ukraine, with a reinterpretation of M. Lomnicki's species, description of a new Siberian species and comparison with British Weichselian faunas. *Phil. Trans. R. Soc.* (B) **265**, 299–326.

Ashworth A. C. (1972) A Late-glacial insect fauna from Red Moss, Lancashire, England. *Ent. scand.* **3**, 211–224.

Ashworth A. C. (1977) A Late Wisconsin Coleopterous assemblage from southern Ontario and its environmental significance. *Can. J. Earth. Sci.* **14**, 625–1634.

Carson H. L. (1970) Chromosome tracers of the origin of species. *Science N.Y.* **168**, 1414–1418.

Coope G. R. (1962) A Pleistocene coleopterous fauna with arctic affinities from Fladbury, Worcestershire. *Q. J. geol. Soc. Lond.* **118**, 103–123.

Coope G. R. (1968) Coleoptera from the 'Arctic Bed' at Barnwell Station, Cambridge. *Geol. Mag.* **105**, 482–486.

Coope G. R. (1970) Interpretations of Quaternary Insect fossils. *A. Rev. Ent.* **15**, 97–120.

Coope G. R. (1973) Tibetan species of dung beetle from Late Pleistocene deposits in England. *Nature, Lond.* **245**, 335–336.

Coope G. R. (1975*a*) Climatic fluctuations in north west Europe since the Last Interglacial, indicated by fossil assemblages of Coleoptera (*in 'Ice Ages Ancient and Modern'*, Wright, A. E. and Moseley, F. eds.) *Geol. J.* Special Issue No. **6**, 153–168.

Coope G. R. (1975*b*) Mid-Weichselian Climatic Changes in Western Europe, Re-interpreted from Coleopteran Assemblages. In *Quaternary Studies*, Ed. R. P. Suggate & M. M. Cresswell, pp. 101–108. Royal Society of New Zealand, Wellington.

Coope G. R. (1977) Fossil coleopteran assemblages as sensitive indicators of climatic changes during the Devensian (Last) cold stage. *Phil. Trans. R. Soc.* (B) **280**, 313–340.

Coope G. R. & Brophy J. A. (1972) Late Glacial environmental changes indicated by a coleopteran succession from North Wales. *Boreas* **1**, 97–142.

Dobzhansky T., Ayala F. J., Stebbins G. L. & Valentine J. W. (1977) *Evolution*, Freeman & Co. San Francisco.

Downes J. A. (1964) Arctic Insects and their Environment. *Can. Ent.* **96**, 279–307.

Dunbar M. J. (1968) *Ecological Development in Polar Regions, a study in evolution.* Prentice Hall inc., New Jersey.

Emiliani C. (1955) Pleistocene temperatures. *J. Geol.* **63**, 538–578.

Grant V. (1977) *Organismic Evolution.* Freeman & Co San Francisco.

Hays J. D., Imbrie J. & Shackleton N. J. (1976) Variations in the Earth's Orbit: Pacemaker of the Ice Ages. *Science N.Y.* **194**, 1121–1132.

Howden H. F. (1969) Effects of the Pleistocene on North American insects. *A. Rev. Ent.* **14**, 39–56.

Lindroth C. H. (1948) Interglacial insect remains from Sweden. *Arsbok Sveriges Geol. Undersokn.* (C) **42**, 1–29.

Lomnicki A. M. (1894) Pleistocenskie owady z Boryslawia (Fauna Pleistocenica Insectorum Boryslaviensium). *Wydaw. Muz. Dzieduszyck.* **4**, 3–116.

Matthews J. V. (1974) Quaternary Environments at Cape Deceit (Seward Peninsula, Alaska): Evolution of a Tundra Ecosystem. *Bull. Geol. Soc. Am.* **85**, 1353–1384.

Matthews J. V. (1976*a*) Insect fossils from the Beafort formation: Geological and biological significance. *Geol. Surv. Pap. Can.* **76-1B**, 217–227.

Matthews J. V. (1976*b*) Evolution of the subgenus *Cyphelophorus* (Genus *Helophorus*, Hydrophilidae, Coleoptera) description of two new fossil species and discussion of *Helophorus tuberculatus* Gyll. *Can. J. Zool.* **54**, 652–673.

Mayr E. (1954) Change of genetic environment and evolution. In *Evolution as a Process*, Eds. J. Huxley, A. C. Hardy & E. B. Ford. George Allen & Unwin, London.

Morgan A. (1972) The fossil occurrence of *Helophorus arcticus* Brown (Coleoptera, Hydrophilidae) in Pleistocene deposits of the Scarborough Bluffs, Ontario. *Can. J. Zool.* **50**, 555–558.

Morgan A. (1973) The Pleistocene environmental changes indicated by fossil insect faunas of the English Midlands. *Boreas* **2**, 173–212.

Osborne P. J. (1972) Insect faunas of Late Devensian and Flandrian age from Church Stretton, Shropshire. *Phil. Trans. R. Soc.* (B)**263**, 327–367.

Penck A. & Brückner E. (1909) *Die Alpen im Eiszeitalter*, Leipzig.

Pennington W. (Mrs T. G. Tutus) (1977) The Late Devensian flora and vegetation of Britain. *Phil. Trans. R. Soc.* (B) **280**, 247–271.

Ruddiman W. F. & McIntyre A. (1976) Northeast Atlantic palaeoclimatic changes over the last 600 000 years. *Geol. Soc. Am. Mem.* **145**, 111–146.

Ruddiman W. F., Sancetta C. D., & McIntyre A. (1977) Glacial/Interglacial response rate of subpolar North Atlantic waters to climatic change: the record in oceanic sediments. *Phil. Trans. R. Soc.* (B) **280**, 119–142.

Scudder S. H. (1895) The Coleoptera hitherto found fossil in Canada. *Contr. Can. Palaeont.* **2**, 27–56.

Scudder S. H. (1900) Additions to the Coleopterous fauna of the interglacial clays of the Toronto district. *Contr. Can. Palaeont.* **2**, 67–92.

Shackleton N. J. (1967) Oxygen isotope analyses and Pleistocene temperatures re-assessed. *Nature, Lond.* **215**, 15–17.

Shackleton N. J. & Opdyke N. D. (1973) Oxygen isotope and palaeomagnetic stratigraphy of Equatorial Pacific core V28.238: Oxygen isotope temperatures and ice volumes on a 10^5 year and 10^6 year scale. *Quaternary Res. N.Y.* **3**, 39–55.

Shackleton N. J. & Opdyke N. D. (1976) Oxygen isotope and palaeomagnetic stratigraphy of Equatorial Pacific core V28.239 Late Pliocene to Latest Pleistocene. In *Investigation of late Quaternary palaeoceanography and palaeoclimatology*, Eds. R. M. Cline & J. D. Hays. *Mem. Geol. Soc. Am.* **145**, 449–464.

Simpson G. G. (1944) *Tempo and Mode in Evolution*. Columbia University Press, New York.

Ullrich W. G. & Coope G. R. (1974) Occurrence of the east palaearctic beetle *Tachin us jacutus* Poppius (Col. Staphylinidae) in deposits of the Last glacial period in England. *J. Ent.* (B) **42**, 207–212.

12 • The dynamics and diversity of insect faunas

ROBERT M. MAY

Biology Department, Princeton University, Princeton, N.J., 08540

and

Imperial College Field Station, Silwood Park, Ascot, Berks.

> 'Entomologists can speak with fervor of the intricacies of taxonomic investigation. Among all biologists, they seem to have made the worst mess of it, characterizing so many families, genera, and species that they have far outstripped the whole field of their taxonomist brethren. This is not really their fault; it is merely a feeble attempt to sort out the avalanche of insects that Nature has lavished on the Earth.'
>
> Brues (1946, p. viii)

Several contributors to the present symposium have already remarked, with pardonable entomological chauvinism, that most species of living things are insects: insects account for something like 50–60% of all currently denumerated species (plants and animals), and around 90% of terrestrial animal species.

Any attempt to understand this diversity of insect faunas must ultimately deal both with absolute diversity (why are there around 10^6 species of organisms, rather than 10^8 or 10^4?), and with relative diversity (why are so many of these species insects?). My paper is addressed almost exclusively to the question of relative diversity.

The first part of the paper aims to review various of the dynamical and evolutionary factors that influence species diversity: these include single-species dynamics, competition, predation, food web structure, numerical abundance of the species, short- and long-term aspects of evolutionary rates, and evolutionary genetics. Some of these have been treated in greater depth earlier in the symposium. For each such factor, my emphasis is on determining in what way, if at all, it favours insects over other animals. The gist of the discussion is that, although all these factors help mould absolute diversity, most of them have little bearing on the diversity of insects relative to other organisms. It is argued that the main reason for the relative diversity of insects is their small size; this, coupled with their dispersal ability, enables them to carve the world into niches that are smaller (in both space and time), and thus more numerous. The point has, of course, been made by previous authors, and it is discussed more fully in Southwood's paper.

This conclusion leads to the second part of my paper, which is a preliminary attempt to determine the empirical relation between numbers of species of terrestrial animals and their size (length or weight). I think this is a fundamental relationship, and that it must be documented and understood before we can hope for any quantitative explanation of the relative diversity of insect faunas.

1: Factors influencing diversity

In his classic *Homage to Santa Rosalia*, Hutchinson (1959) asked 'why are there so many kinds of animals?' In what follows, we focus rather on the question 'why are there so many more kinds of insects than of other animals?'

1.1: SINGLE-SPECIES DYNAMICS

The past few years have seen significant advances in our understanding of the extraordinary richness of dynamical behaviour that is latent within the simplest of non-linear deterministic models (Li & Yorke, 1975; May, 1974*a*, 1976*a*); earlier work had tended to be confined to the analysis of small disturbances about equilibrium values. In particular, essentially all the first-order difference equations that have been propounded in the biological literature (as deterministic models for the behaviour of single-species populations with non-overlapping generations) have been shown to exhibit stable points, or sustained stable cyclic oscillations, or apparently chaotic fluctuations, depending on the values of intrinsic growth rates and other biological parameters (May & Oster, 1976). This work has innate mathematical interest. More importantly for biologists, the fact that simple deterministic models can lead to dynamical trajectories indistinguishable from the sample function of a random process holds disturbing implications both for the analysis and interpretation of data, and for hopes of long-term population forecasting.

Various people (Hassell *et al*, 1976; Stubbs, 1977) have attempted to estimate the parameter values in such models, by fitting them to population data. The parameter values thus estimated for field populations tend to lie in the domain corresponding to a stable equilibrium. Although this is a comforting conclusion, and one that is perhaps plausible on evolutionary grounds, it is not final. Quite apart from technical difficulties in estimating the relevant parameters, there are no truly single-species situations in the real world; once one has a multi-species situation, chaotic dynamical behaviour is likely to be more common (Guckenheimer *et al*, 1977), with all its attendant problems for data analysis and long-term predictions.

Despite these complications, it is often possible to estimate the population parameters for particular organisms, and to show how they correlate with life history strategies (for good reviews, see Southwood, 1976, 1977*a*). This permits the synthesis of a great deal of natural history information, and also paves the way for codifying the types of control strategies that are appropriate to particular kinds of insect pests (Southwood, 1977*b*; Conway, 1976; May, 1976*b*).

Reviewing this whole body of work, we see that the range of dynamical behaviour manifested by insect populations is not strikingly different from that of other organisms. Instances of, e.g., stable cycles can be drawn evenhandedly from insects or mammals: witness the almost physics-like fit between Nicholson's blowfly data and a simple model, and the use of the same model to give a somewhat more general explanation of the 3-to-4 year cycles of mice, voles and lemmings (May, 1976*c*, ch. 2). Examples of systems with multiple stable states are similarly ecumenical (May, 1977).

Insofar as insect dynamics differ from the dynamics of larger animals, it is that their comparatively small size and short generation time endow them with the capacity for relatively high intrinsic rates of population growth, *r*. This makes it easier for them to escape from superior competitors by accepting the boom-and-bust economy of the 'fugitive species' (Hutchinson, 1951; Southwood, 1977*a*). We will return to this below.

1.2: COMPETITION

In the 1920s and 1930s, the theoretical work of Lotka and Volterra, and the experiments of Gause, Park and others led to the enunciation of the competitive exclusion principle: species that make their livings in identical ways cannot coexist. More recently, Hutchinson, MacArthur and others have posed the more contentful question of how similar can species be, yet coexist? What are the limits to similarity, the limits to niche overlap and species packing?

Although most competitive situations are too complicated to unravel, insight can be gained from those special situations where a set of competitors sort themselves out mainly along one resource axis (such as food size, or foraging height). Hutchinson (1959) observed that there are many examples, including both vertebrates and invertebrates, of sequences of competing species in which each is roughly twice as massive as the next. This leads to ratios of around 1.3 in the linear dimensions of successive species. Many other examples that conform to the 1.3 ratio have since been given: e.g., for birds by MacArthur (1972), Diamond (1975), Cody (1974), for spiders by Uetz (1977); other examples are mentioned by Southwood and by Halkka (this symposium). Indeed Dyar (1890) long ago noted the closely related fact that successive larval instars of many insects have weight ratios of 2 and linear ratios of 1.3. This provoked much discussion (which seems to have been forgotten), and even speculation that the underlying mechanism was a doubling of the number of cells between instars (for a review, see Bodenheimer, 1933).

This empirical relation is still not understood. There is a growing body of theory (for a review, see May, 1976*c*, ch. 8) which suggests that, with regard to the size of food items, the average difference between two species should not be significantly less than the characteristic range of food sizes utilised by either species. This, however, does not explain why the average intraspecific range of food items typically spans a weight ratio of around 2.

It is worth noting that the Dyar–Hutchinson rule also holds for instars of children's bicycles, for sets of kitchen skillets, for ensembles of recorders, and for consorts of viols (Horn & May, 1977). The rule may well have more to do with assembling sets of tools, than with anything directly biological.

As far as the Dyar–Hutchinson rule goes, insects show no significant differences from vertebrates or other invertebrates. Although the limits to similarity and niche overlap are clearly central in determining absolute diversity, they appear to have little to do with the relative diversity of insect faunas.

1.3: PREDATION

As many writers have stressed (e.g., Connell, 1975), predation can profoundly modify the outcome of competition among the prey, thus increasing diversity.

A generalist predator, acting impartially on prey species that all have roughly equal intrinsic growth rates, makes no difference to competitive coexistence among the prey. Most vertebrate predators, however, concentrate their attacks disproportionately upon the prey species that happens to be most abundant at any one time. This 'switching' behaviour, which commonly derives from the way vertebrate predators form a 'search image', can enable the coexistence of prey species, of which some would otherwise be competitively excluded (Holling, 1959; Murdoch, 1969; Roughgarden & Feldman, 1975). Empirical and theoretical work has recently shown that many invertebrate predators also exhibit attack patterns that result in the predation rate on any one prey species being of the sigmoidal 'Type III' or 'vertebrate' form (Murdoch, 1977, and references therein; Hassell *et al*, 1977; Hassell, 1978). One pervasive mechanism whereby this can come about is if the spatial distribution of prey is patchy (either for a single prey species, or for a mixture of several prey species), and if the predators have searching behaviour that leads to differential aggregation in patches of high prey density; in a multi-prey situation, the upshot is tantamount to predator switching. (Incidentally, Beddington *et al* (1978) have made a convincing case that this combination of spatial heterogeneity and predator aggregation is the key mechanism in maintaining those arthropod prey-predator or host-parasitoid systems in which the prey population is well below the level set simply by the environmental carrying capacity.)

There is a second important way in which predation can enhance diversity, even in the absence of 'switching' effects. Generalised predation can promote coexistence if, among the prey, the inferior competitors have higher intrinsic growth rates.

We noted earlier that insects and other small animals can have relatively large values of r, which facilitates their playing the role of fugitive species. In addition, we now see that large r-values enable insects to exploit the second (non-switching) mechanism whereby predation can enhance diversity. These two factors, either separately or in conjunction, thus contribute to the relative diversity of insect faunas. It is my view, however, that this contribution is a comparatively minor one.

1.4: FOOD WEB STRUCTURE

As observed by Hutchinson (1959), trophic structure as such contributes little to species diversity. There are but few levels in even the longest food chains.

In this context, it should be noted that overall dynamical stability may be the main factor limiting trophic complexity (see, e.g., Pimm & Lawton, 1977, and references therein); this contrasts with the conventional explanation of the number of trophic levels, which sees them as set by considerations of energy flow. A corollary of this view is that complex ecosystems are likely often to have evolved as loosely coupled assemblies of simpler subsystems. This notion, which was originally advanced on abstractly theoretical grounds (May, 1974*b*), has gained some support from the empirical studies of Gilbert (1975 and this symposium), Lawton (this symposium) and others, and from D. S. Wilson's (1978) studies of models for the evolution of ecosystems. The notion is appealing, because it carries the implication that studies of interacting populations may have direct relevance at the ecosystem level (Lawton, 1976).

None of these broad aspects of food web topology, however, make appreciable contributions to the diversity of insects relative to other animals. Among the exceptions to this sweeping statement are the insect parasitoids. The biochemistry and life history of parasitoids are closely matched to their hosts, so that, from the standpoint of many of the host's predators, parasitised and unparasitised hosts are indistinguishable; the parasitoids have, in a sense, succeeded in slipping in an extra trophic level. But I think this is a tactical detail, contributing relatively little to overall insect diversity.

1.5: SPECIES VERSUS NUMBER OF INDIVIDUALS

The number of individual insects is, by virtue of their small size, vastly greater than the number of individuals of larger animals. May this numerical abundance, of itself, account for the larger number of insect species?

A crude answer may be given by borrowing the methods used in deriving the Preston–MacArthur–Wilson species-area relation. This theoretical relation is obtained by: (i) assuming a particular distribution of species relative abundance, (see, e.g., Taylor, this symposium), which then gives a relation between the total number of individuals, N, and the number of species, S; and (ii) assuming that N is linearly proportional to area A. Here we need only the first, and less dubious, of these assumptions. Preston (1962) and MacArthur and Wilson (1967) assumed a special ('canonical') member of the 1-parameter class of lognormal distributions of species relative abundance. This gives a complicated relation between N and S, which for large values of $S(S > 30$ or so) reduces to

$$S \sim N^z \tag{1}$$

with $z = 0.25$. More generally, relinquishing the 'canonical' hypothesis, reasonable lognormal distributions give eq (1), with z having values in the range $z \simeq 0.20$ to 0.35 (for a much more full discussion, see May, 1975). In short, the number of species does depend on the total number of individuals in a given taxonomic class, but only weakly (typically as the 1/4 power).

By making some further very crude generalisations, we can relate the total number of individuals, N, to their typical linear dimension, L. First, we recall Odum's (1968; see also Van Valen, 1973) observation that, within a given trophic level, the net productivity, P, is very roughly the same for organisms spanning a wide range of sizes (bacteria to deer): $P \sim$ constant. Second, biomass B may be roughly approximated as $B \sim PT_g$, where T_g is of the order of the generation time of the organism. Third, numerical abundance, N, scales as biomass divided by the mass of an individual, and hence, again roughly, $N \sim B/L^3$. Finally, noting that T_g scales approximately as L (Bonner, 1965), we arrive at $N \sim (\text{constant})/L^2$. Eq. (1) now takes the form

$$S \sim L^{-x} \tag{2}$$

with x somewhere in the neighbourhood of $x \simeq 0.5$.

I strongly emphasise that the species-size relation (2) pays regard *only* to statistical generalities about the distribution of relative abundance of individuals among species. It *ignores* ecological aspects of species' size (which is dealt with separately, below). Eq (2) suggests that a 100-fold decrease in the characteristic length of a group of organisms (from, say, 30 cm to 3 mm) will, by virtue of these statistical generalities alone, produce something of the order of a 10-fold increase in the number of species. As is shown in detail below (Figs. 12.1 to 12.7), this is nowhere near enough to explain the observed increase in species number with decreasing size.

1.6: EVOLUTIONARY RATES: A GRAND ARGUMENT

Compared with larger creatures, insects tend to have short generation times and high mortality rates. These two factors can accelerate evolutionary processes. As pointed out at different times by Mayr (1976), evolutionary rates may also be faster in very large populations or, alternatively, in species with many scattered subpopulations; both of these phenomena are preeminently exhibited by insects.

In short, the evolutionary clock ticks faster for insects and other small organisms. This is clearly seen in disease and pest systems, where significant evolutionary changes have taken place in decades, or even years, in response to antibiotics and insecticides.

If the living world showed a pattern of yet increasing and unsaturated diversification of organic forms, the faster 'evolutionary clock' of insects could account for their relatively great diversity. But the evidence assembled by Simpson (1953, 1969), and more recently reviewed by Raup (1977 and references therein; see also Coope, this symposium), shows a pattern of saturation which, for terrestrial animals, reaches back (with major and minor fluctuations, and with much relay and replacement) to the Permian, and possibly beyond. Indeed for insects Carpenter (1977, p. 69) has gone so far as to write 'In terms of diversity of form and the association of generalized and specialized species, the fauna of the Permian was probably the most diverse in the history of the Insecta'.

If it is accepted that the diversity of terrestrial faunas is roughly in an equilibrium or saturated state on a geological time scale, then insects' potential for faster evolutionary rates cannot be invoked as a direct explanation of their relative diversity.

1.7: EVOLUTIONARY RATES: A MORPHOLOGICAL ARGUMENT

A more detailed variation on the above theme is that insects' potential for fast evolutionary changes permits them to indulge in a greater amount of morphological 'fine-tuning' than is possible for taxa of bigger animals. The great diversification of insect mouth-parts, which contrasts with the evolutionary conservatism of the vertebrate mouth, could for example be attributed to this mechanism.

However, I think this puts the cart before the horse. The diversity of insect mouth-parts is more likely to be a secondary effect, deriving from many more niches being available to creatures of small size (see section 1.9 below). It seems unlikely that the tourist experience on the Serengeti plains would have been enriched by the presence of many more species, had the evolution of the vertebrate jaw manifested less conservatism; when the occasion has arisen, vertebrate mouth-parts have shown considerable adaptability (witness the baleen whales). It can be contended that vertebrate feet show as much functional diversity as insect feet.

This whole question is a complicated and unresolved one. It has, for example, been argued that the remarkable adaptive radiation and diversity of cichlid fishes stems from an 'evolutionary breakthrough' in the morphology of mouth-parts (see, e.g., Fryer & Iles, 1972, and references therein; for further discussion, see Sage & Selander, 1975). These issues are related to those treated in the next section.

1.8: EVOLUTION AND POLYMORPHISM

In a most interesting paper, Selander and Kaufman (1973) have compiled data from electrophoretic studies of protein polymorphisms in a variety of animals. They show that the average invertebrate individual is heterozygous at 15% of its loci, whereas the mean level of heterozygosity for vertebrates is 5%. Selander and Kaufman review possible causes and implications of their findings (see also Southwood, this symposium).

For our present purposes, it is sufficient to note that such intraspecific variability is the raw stuff with which evolution works. If this difference in polymorphism levels between vertebrates and invertebrates is taken at face value, it reinforces the arguments outlined above, to the effect that insects have the capacity for relatively rapid evolutionary responses. But there is an important caveat. Electrophoretic techniques test for qualitative differences between proteins coded by structural genes; they tell nothing about quantitative differences at the molecular level due to variation in regulatory genes (see, e.g., Feldman, 1978). This point is underlined by King and Wilson's (1974) demonstration that chimpanzees and humans differ to about the same degree as sibling species of *Drosophila*, if the difference is measured by structural genes (as revealed by gel electrophoresis). On the other hand, chimpanzees and humans do indeed differ morphologically at the taxonomic level conventionally attributed to them. This example supports the contention that in advanced groups, such as vertebrates in general and primates in particular, evolution primarily involves regulatory rather than structural genes.

Our current inability to assess variability at the genetic loci that code for timing and development means that we cannot say whether the overall genetic variability of insects is greater, less, or much the same as that of vertebrates. This unresolved problem lies at the heart of evolutionary genetics.

1.9: THE SIZE OF ANIMALS AND THE NUMBER OF NICHES

Each of the above factors has been argued to play little or no role in explaining the *relative* diversity of insects. I think most of the explanation lies simply in the small size of insects, which allows them to divide their environment much more finely than larger animals can; a given species of tree may be only part of an elephant's resource base, whereas it can be subdivided into a myriad of niches for tiny animals. This notion has of course been propounded by many people, including MacArthur and Levins (1964; see also MacArthur, 1971, and Levins, 1968), who introduced the concept of 'grain size'. Southwood (this symposium; see also Lawton, this symposium) has developed some of these ideas in a more quantitative way.

There have, however, been very few attempts to pursue these notions to get some estimate of the relative numbers of species of animals in given size classes. The main thing I know of is a difficult paper by Hutchinson and MacArthur (1959): within a given biotype, their model implies 'rapid increase in number of species up to a modal size and a slow decline in number, ideally asymptotic to unity, as the size increases'. At the large end of the spectrum, the number of species, S, falls off with the characteristic linear dimension of an animal, L, roughly as

$$S \sim L^{-2} \tag{3}$$

Such a relation implies that a 10-fold reduction in length (e.g., 3 cm to 3 mm) will see a 100-fold increase in the number of species. Hutchinson and MacArthur give some data, mainly from North American mammals, which tend to be in support. Other discussions of species-size relations have been largely confined to particular groups, and are reviewed below.

In essence, eq (3) comes from the assumption that animals see their environment as a 2-dimensional mosaic (e.g., the total surface of a tree) to be partitioned up, on a scale set by their perceived grain size (which goes as L^2). More generally, one could argue that it is volumes rather than surfaces that are divided, and that time is also a resource axis to be divided on a scale set by generation times (which tend to scale as L); thus a generalised version of eq (3) is

$$S \sim L^{-y} \tag{4}$$

with y somewhere in the neighbourhood of 2 to 3.

These fanciful speculations should not be pursued in advance of the facts. What we need is more empirical information than is currently available concerning the number of species as a function of size. Some such information is presented below. In each of the figures I have indicated (by a light, dashed line) the form of the theoretical relationship (3), $S \sim L^{-2}$; I did this more for entertainment than from any sense of conviction.

2: Number of species versus size

This part begins with some fairly exact results for the number of species versus size in particular groups of animals, and ends with a very crude estimate of the overall species-size relation obtained by pooling all terrestrial animals. For any one group (e.g., beetles), ecological aspects of the species-size relation tend to be masked by the group blending into ecologically similar, but taxonomically different, groups at both low and high ends of its size range.

2.1: PREVIOUS WORK

Van Valen (1973) has pulled together a vast amount of data, to discuss the relation between number of species (and genera) of mammals, birds and flowering plants and their sizes. Van Valen breaks up the groups in various ways, and discusses the implications of his results: this paper should be read. Figs. 12.1 and 12.2 are based on his results.

Schoener and Janzen (1968; see also Janzen and Schoener, 1968) have gathered data, based on some 10^4 sweep net samples, on the number of species, number of individuals and sizes of insects in several tropical (Costa Rica) and temperate (Massachusetts) locations. Their main aim was to understand how the faunal composition, and size, of insects varied among these different regions. Basing their analysis on Hemmingson's (1934) conclusion that, for a given taxon, the size distribution of species in a given region often approximates a lognormal, Schoener and Janzen give the mean and variance of the length of insects in various orders (weighted according to the number of individuals, or to the number of species). The lognormal, however, provides a Procrustean fit to the data, and the authors end up using a '3-parameter lognormal' to accommodate to the facts that the species-size distributions suggest a 'minimum size limit or threshold', and have 'relatively long tails to the right'. Janzen (1973*a,b*; 1977) has subsequently presented further results, based on a similarly heroic set of 800 sweeps at each of 25 sites in Costa Rica and on Caribbean islands. For beetles, bugs, and a lumped class of all other arthropods, he catalogues the number of individuals (but not the number of species) according to size classes. It would be nice to see all this, or similar, data reanalysed to give species-size distribution patterns of the kind shown below (rather than only means and diversity indices). One problem here is that sweep net samples may not give an unbiassed estimate of the species-size distribution in a given biotype; if relatively large-sized species range more widely (Southwood, this symposium), they may tend to be over-represented in such samples.

2.2: MAMMALS

Fig. 12.1 shows a log-log plot of the total number of terrestrial mammals (excluding bats) as a function of the animal's weight. The figure is obtained by aggregating Van Valen's (1973) results. Fig. 12.1 also gives the analogous histogram for all British mammals (again excluding bats), based on Southern (1964).

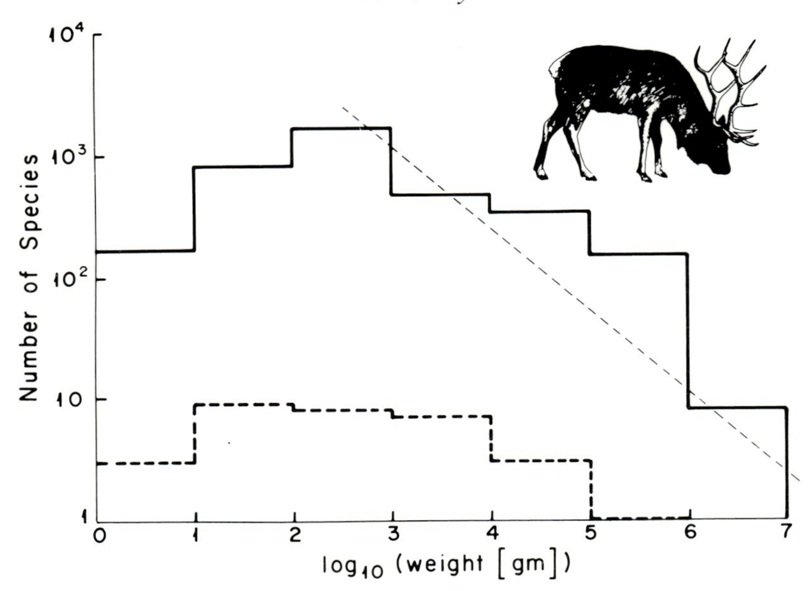

FIG. 12.1. Number of species of all terrestrial mammals (solid histogram) and of British mammals (dashed histogram), excluding bats, according to weight classes. Note the doubly logarithmic scale. The thin dashed line illustrates the shape of the relation (3), $S \sim L^{-2}$ (assuming weight scales as L^3 : see McMahon, 1973, and references therein).

Notice that the species-size distribution for British mammals is not very different from that for all terrestrial mammals, uniformly scaled down to allow for the much smaller number of species.

[Marine mammals have been omitted. So have bats, partly because their species-size distribution is more like that for passerine birds (Van Valen, 1973). The distribution for British bats is again close to a scaled-down version of the global distribution (not shown here). But Britain contains a larger fraction of the world's species of bats than of other mammals, and so a figure that combined all terrestrial mammals (including bats) would show the average British mammal to be noticeably smaller than the global average.]

2.3: BIRDS

Fig. 12.2 is also based on Van Valen's (1973) work, and shows the number of species of terrestrial birds according to weight classes. Birds whose way of life is mainly aquatic have been omitted; such birds are systematically heavier than most, and although their inclusion in Fig. 12.2 would run counter to the spirit of the exercise, it would give a much nicer fit to the dashed ($S \sim L^{-2}$) line. I have resisted this temptation.

2.4: BRITISH COLEOPTERA

As we move from large and aesthetically appealing animals down to small ones, we encounter the problem that collectors and systematists have tended to give more attention to larger insects. Thus a current count of Coleopteran species stands around 350 000, but it has been suggested the true count may be over one million.

One way of minimising this difficulty is to confine attention to relatively well-studied

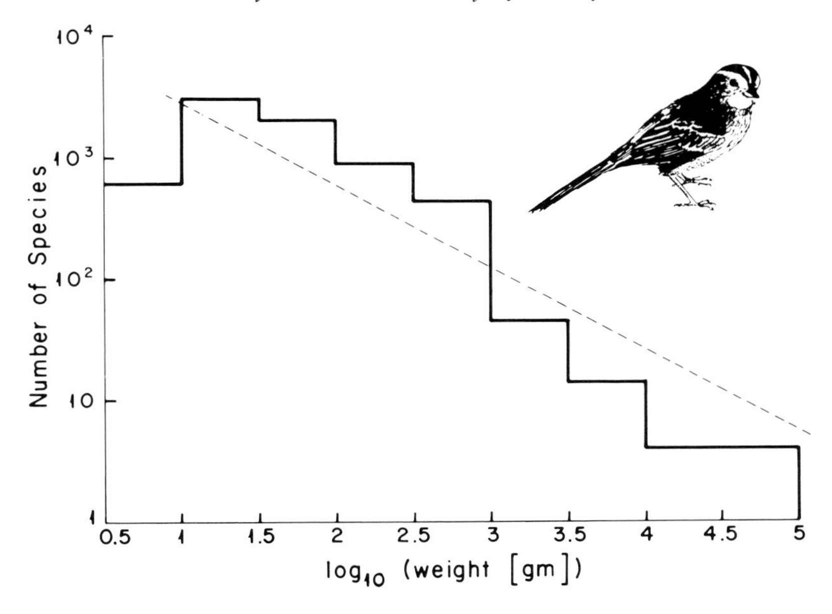

FIG. 12.2. The number of species of non-aquatic birds, classified according to weight. The dashed line is as in Fig. 12.1. (Data compiled by Van Valen, 1973).

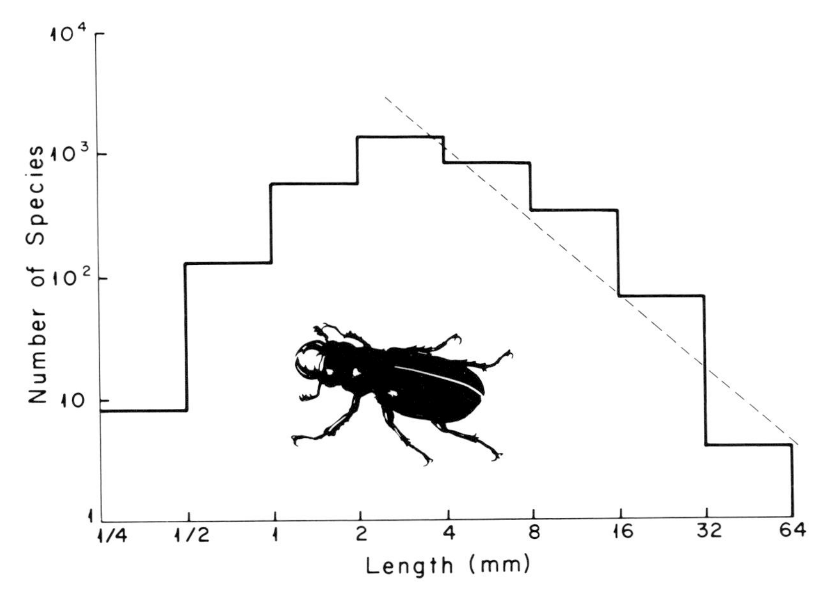

FIG. 12.3. The number of species of British Coleoptera, classified according to length. (Information taken mainly from Fowler, 1887.) The dashed line illustrates the shape of the relation (3), $S \sim L^{-2}$.

groups of insects. British Coleoptera are one such group, comprising some 3500 species. Fig. 12.3 is a log-log plot of British beetle species according to length classes, which I have compiled from Fowler's (1887) monograph. This figure invites comparison with Fig. 12.1 (all terrestrial mammals): the numbers of species on which the two figures are based are roughly equal; they span similar ranges (over two orders of magnitude in length classes); and they manifest similar shapes.

The purpose of Fig. 12.4 is to show that patterns are harder to discern if one gives a linear plot of species against size.

2.5: BRITISH LEPIDOPTERA

I have derived Fig. 12.5 mainly from Meyrick's (1927) monograph on the British Lepidoptera, another well-studied group. The figure displays the 2200-odd species according to length classes, using the linear dimension conventionally employed to characterise these creatures (the wing-spread or 'expanse of wings', defined by Meyrick as twice 'the distance from the tip of the forewing to the centre of the thorax'). This group shows a smaller size range (less than a factor of 100) than the British beetles.

2.6: BUTTERFLIES

Moving from Britain into the more biologically exciting tropical world, I have fastened on butterflies as likely to be the best-studied group of insects. Fig. 12.6 shows the species-size relation for butterflies (superfamily Papilionoidea, not including the Hesperioidea) in the Australian geographical realm, compiled with ruler and patience from D'Abrera (1971: even here the systematics of the smaller species is still in a state of flux, particularly among the Lycaenidae, which comprise more than one third of the butterfly species in the region).

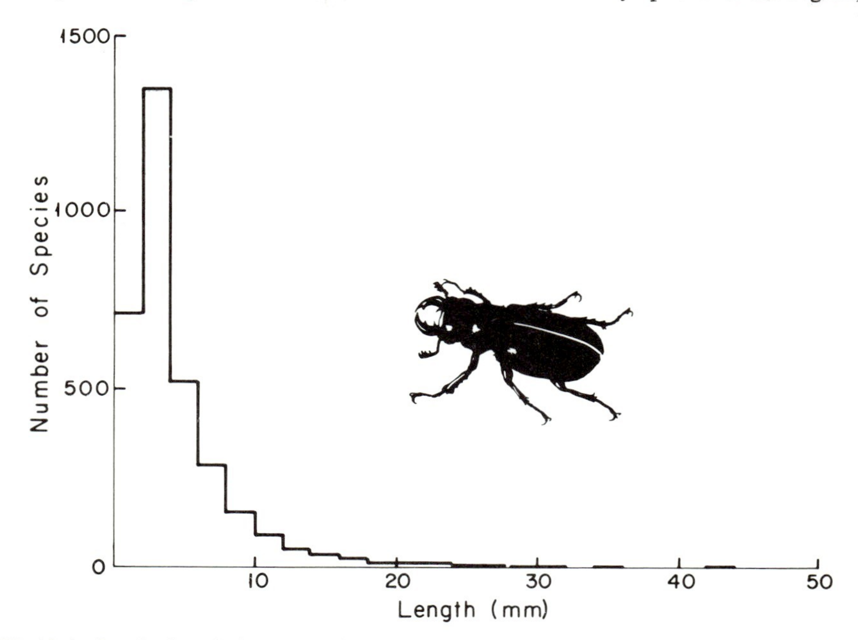

FIG. 12.4. Species-length histogram for British Coleoptera, as in Fig. 12.3, except here linear axes are used instead of logarithmic ones.

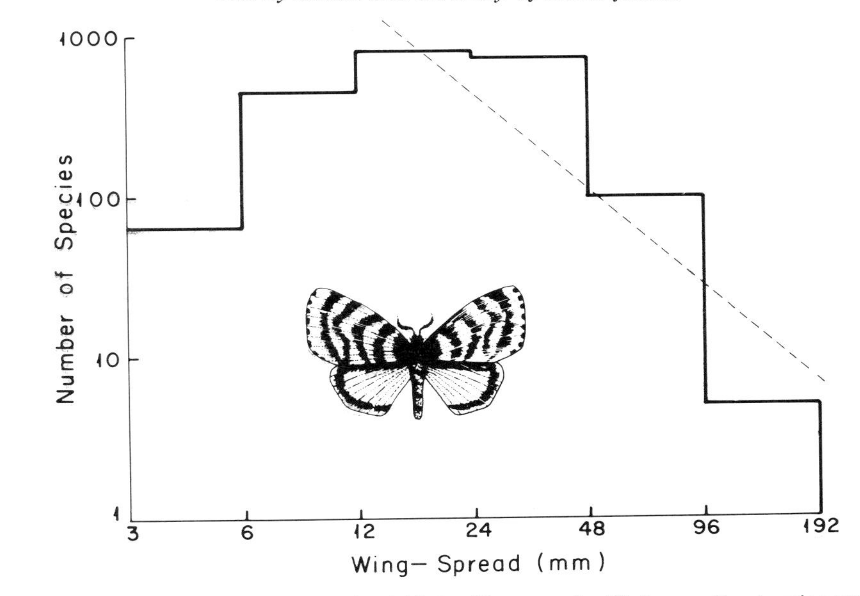

FIG. 12.5. The number of species of British Lepidoptera, classified according to wing-spread, as defined in the text. (Information taken mainly from Meyrick, 1927.)

The size distribution of the 60 species of British butterflies is shown for comparison. As in Fig. 12.1, the British distribution is not very different from a uniformly scaled-down version of that for the Australian realm. From such a naive scaling, the numbers of British species in the ascending size classes of Fig. 12.6 would be 2.6, 30.3, 23.5, 3.1, 0.5, which is to be compared with the actual numbers of 1, 31, 27, 1, 0.

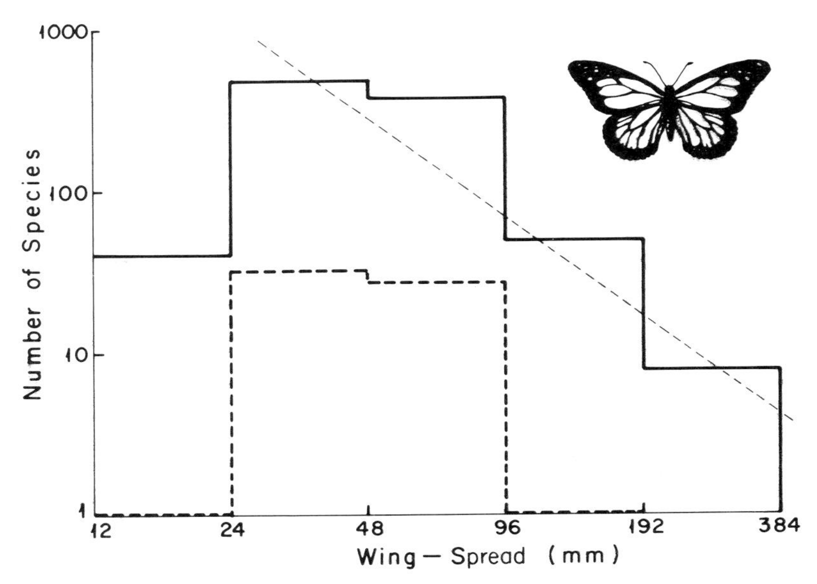

FIG. 12.6. The number of species of butterflies in the Australian geographical realm (solid histogram) and in Britain (dashed histogram), classified according to wing-spread. For discussion, see the text.

2.7: SIZE DIFFERENCES BETWEEN TROPICAL AND TEMPERATE INSECTS

Figs. 12.1 and 12.6 suggest the species-size distribution of British mammals and butterflies may be uniformly scaled-down versions of the global or tropical distributions, so that the mean sizes are the same. This is contrary to many people's intuition, which is that within most taxonomic groups British species are systematically smaller than tropical ones (e.g., Humphries, 1972). Such an intuition may derive in part from collections and monographs that give more thorough attention to large species of tropical insects than to small ones; some collections and writings treat races of larger species as carefully as they treat the smaller species themselves.

There is, however (to the contrary of Figs. 12.1 and 12.6), data to support the view that the typical species of tropical insect is larger than its temperate counterpart. Waloff (1954) has noted that, among the Acridoidea (a superfamily of Orthopterans), the average length among 31 temperate species is 25 mm, while for 60 tropical species it is 34 mm; this is only a fraction of the tropical species, and the sample could be biassed toward larger ones. The above-mentioned work of Janzen and Schoener gives a detailed discussion of the average sizes of several groups of insects in a variety of tropical and temperate environments. These results are very crudely summarised in Table 12.1. They exhibit a tendency for species to be bigger in the tropics. Janzen (1973*a,b*; 1975; 1977; see also Elton, 1973 and Enders, 1975) has advanced arguments to explain why this is so. But, as was noted above, sweep net samples could well be biassed toward a more complete representation of larger insect species, especially in the tropics. Such samples are not exhaustive lists in the way Figs. 12.1–6 are.

In short, it is not clear whether there are systematic average differences in sizes between tropical and temperate species of insects, nor whether such differences are present in some taxa and some regions and not in others. These questions deserve further exploration, in the field and in the museum.

2.8: SPECIES VERSUS SIZE FOR TERRESTRIAL ANIMALS

As was mentioned above, if we want a species-size relation in which the ecological aspects are not hidden by details of taxonomy and classification, we need to combine data for all

Table 12.1. The average species length (weighted according to number of species, not number of individuals) for various taxa of insects in tropical versus temperate habitats†

Taxon	Average tropical length (mm)	Average temperate length (mm)
Coleoptera	3.0	3.7
Formicidae	5.1	3.8
Other Hymenoptera	3.9	2.7
Homoptera	4.9	3.5
Hemiptera	9.3	6.3
Diptera	3.0	3.1
All insects	5.1	3.5

†I have obtained the average tropical length by the dubious procedure of taking an unweighted average over Schoener and Janzen's (1968) tropical Areas I, II and III, and the average temperate length by an unweighted average over their numbers for the months June–October.

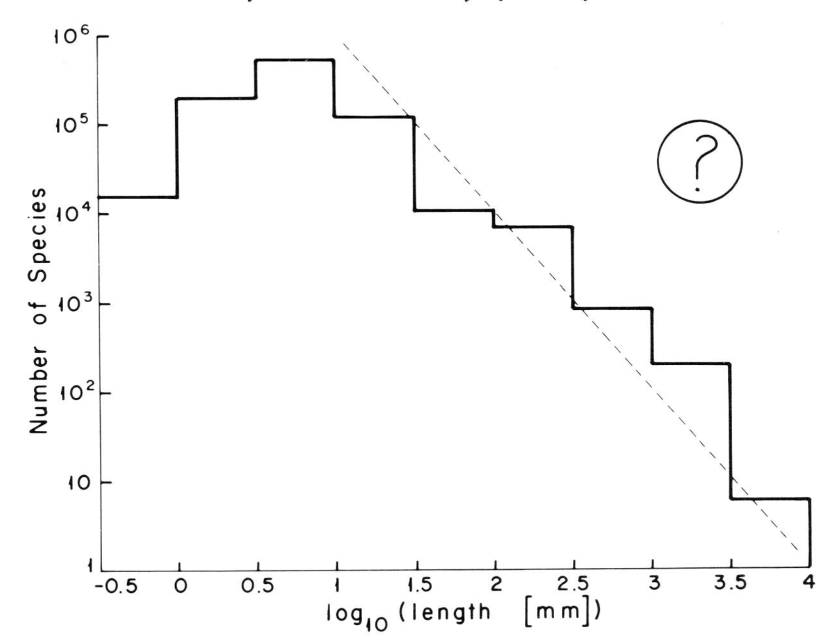

FIG. 12.7. The number of species of all terrestrial animals, classified according to length. The question mark serves to emphasise that this is a tentative figure, based on very crude approximations. The dashed line illustrates the shape of the relation (3), $S \sim L^{-2}$.

terrestrial animal species. Such information is hard to come by. Fig. 12.7 offers a first attempt at such a synoptic species-size distribution. This figure is based on crude approximations and outright guesses; it may be regarded as a cartoon, both in the older sense of a preliminary sketch, and in the contemporary sense of something risible.

Some of the main problems underlying Fig. 12.7 are as follows.

Bacteria, viruses, etc.,

As one drops below 1 mm and beyond, the number of species (protozoans, bacteria, viruses) decreases markedly. This could be a real phenomenon, and Janzen (1977) has given an interesting explanation of how it could come about. An alternative explanation is that conventional taxonomy begins to break down. The way bacteria exchange genetic information across species and generic lines, and the way R-factors and the like are incorporated in the bacterial chromosome, suggest that the concept of species is less well defined for bacteria than for larger animals. One could argue that at the bacteria-virus level one has a continuum of organic forms rather than distinct species, or alternatively one could argue that rates of evolution are so fast that the species concept is a loose one. In either event, it could be that the decrease in numbers of species at very small sizes is more apparent than real.

Small animals

It is widely acknowledged that the systematics of small arthropods and other invertebrates is, in most instances, in a rudimentary state. As Mayr (1969, p. 13) has written 'We must take it for granted that a large part of the mite fauna of the world will remain unsampled, unnamed, and unclassified [not to mention unwept, unhonoured, and unsung] for decades to come.' This could mean the size classes below 1 cm in Fig. 12.7 are underestimated by a factor 2 or more.

Other worries

The shape of Fig. 12.7 is determined almost entirely by the mammals, birds, reptiles and arthropods. For all except the five largest arthropod orders, I simply assigned a characteristic size to the order. For the five orders with the most species, I made a crude partitioning into size classes along lines guessed with the help of Figs. 12.3–6. (Detailed notes on the construction of Fig. 12.7 are available.) The saving grace in all this is that Fig. 12.7 is painted with a broad brush; a change of a factor of 2 in the number of species in any one size class would hardly alter the overall shape.

The relation sought in Fig. 12.7 is of fundamental biological importance. The pattern appears indeed to be very roughly described by eq (4), $S \sim L^{-y}$, with y a little less than 2. Although it would be silly to try to specify a precise y-value, it can be said that the figure is inconsistent with y-values as small as 1 or as large as 3.

The next step, I think, is to try for a more accurate version of Fig. 12.7. This should be possible, at least for the terrestrial animals in well-studied regions, such as Britain or (maybe) North America. Then the stage would be set for an explanation of the underlying ecological mechanism, and thence for an explanation of the relative diversity of insect faunas.

Although some people will find it dissatisfying, I find it pleasing that this symposium ends with several large questions, rather than with answers.

Acknowledgements

I have been helped by more people than I can list. Preeminent were M. P. Hassell, C. Istock, J. H. Lawton, S. McNeill, O. W. Richards, R. K. Selander, T. R. E. Southwood, D. W. Tonkyn, G. K. Waage, and N. Waloff. This work was supported in part by the US National Science Foundation, under grant DEB 77-01565.

References

Beddington J. R., Free C. A. & Lawton J. H. (1978) Modelling biological control: on the characteristics of successful natural enemies. *Nature, Lond.* **273**, 513–519.

Bodenheimer F. S. (1933) The progression factor in insect growth. *Q. Rev. Biol.* 8, 92–95.

Bonner J. T. (1965) *Size and Cycle: an Essay on the Structure of Biology*. Princeton University Press, Princeton, N.J.

Brues C. T. (1946) *Insect Dietary* xxvi + 466 pp. Harvard University Press, Cambridge, Mass.

Carpenter F. M. (1977) Geological history and evolution of the insects. In, *Proc. XVth Int. Congr. Ent. Washington D.C.* 1976: pp. 63–70.

Cody M. L. (1974) *Competition and the structure of bird communities.* Princeton University Press, Princeton, N.J.

Connell J. H. (1975) Some mechanisms producing structure in natural communities. In *Ecology and Evolution of Communities,* Ed. M. L. Cody & J. M. Diamond, pp. 460–490. Harvard University Press. Cambridge, Mass.

Conway G. R. (1976) Man versus pests. In *Theoretical Ecology: Principles and Applications*, Ed. R. M. May, pp. 257–281. Blackwell Scientific Publications, Oxford.

D'Abrera B. (1971) *Butterflies of the Australian Region* 415 pp. Lansdowne Press, Melbourne.

Diamond, J. M. (1975) Assembly of species communities. In *Ecology and Evolution of Communities*, Ed. M. L. Cody & J. M. Diamond, pp. 342–444. Harvard University Press, Cambridge, Mass.

Dyar H. G. (1890) The number of moults of Lepidopterous larvae. *Psyche, Camb.* **5**, 420–422.

Elton C. S. (1973) The structure of invertebrate populations inside neotropical rain forests. *J. anim. Ecol.* **42**, 55–104.

Enders F. (1975) The influence of hunting manner on prey size, particularly in spiders with long attack distances. *Am. Nat.* **109**: 737–763.

Feldman M. W. (1978) *Genetic Variation in Natural Populations.* (in preparation).

Fowler W. W. (1887) *The Coleoptera of the British Isles* (6 Vols.). Reeve and Co., London.

Fryer G. & Iles T. D. (1972) *The Cichlid Fishes of the Great Lakes of Africa.* T. F. H. Publications, Neptune City, N.J.

Gilbert L. E. (1975) Ecological consequences of a coevolved mutualism between butterflies and plants. In *Coevolution of Animals and Plants*, Ed. L. E. Gilbert & P. H. Raven. University of Texas Press, Austin, Texas.

Guckenheimer J., Oster G. F. & Ipaktchi A. (1977) The dynamics of density dependent population models. *J. Math. Biol.* **4**, 101–147.

Hassell M. P. (1978) *Arthropod Predator-Prey System*. Princeton University Press, Princeton, N.J.

Hassell M. P., Lawton J. H. & May R. M. (1976) Patterns of dynamical behaviour in single-species populations. *J. anim. Ecol.* **45**, 471–486.

Hassell M. P., Lawton J. H. & Beddington J. R. (1977) Sigmoid functional responses by invertebrate predators and parasitoids. *J. anim. Ecol.* **46**, 249–262.

Hemmingsen A. M. (1934) A statistical analysis of the differences in body size of related species. *Vidensk. Meddr. dansk naturh. Foren.* **98**, 125–160.

Holling C. S. (1959) The components of predation as revealed by a study of small-mammal predation of the European pine sawfly. *Can. Ent.* **91**, 293–320.

Horn H. S. & May R. M. (1977) Limits to similarity among coexisting competitors. *Nature, Lond.* **270**, 660–661.

Humphries B. (1972) *The Wonderful World of Barry McKenzie.* Deutsch, London.

Hutchinson G. E. (1951) Copepodology for the ornithologist. *Ecology* **32**, 571–577.

Hutchinson G. E. (1959) Homage to Santa Rosalia, or why are there so many kinds of animals? *Am. Nat.* **93**, 145–159.

Hutchinson G. E. & MacArthur R. H. (1959) A theoretical ecological model of size distributions among species of animals. *Am. Nat.* **93**, 117–125.

Janzen D. H. (1973*a*) Sweep samples of tropical foliage insects: description of study sites, with data on species abundance and size distributions. *Ecology* **54**, 659–686.

Janzen D. H. (1973*b*) Sweep samples of tropical foliage insects: effects of seasons, vegetation types, elevation, time of day and insularity. *Ecology* **54**, 687–708.

Janzen D. H. (1975) *Ecology of Plants in the Tropics.* E. Arnold, London.

Janzen D. H. (1977) Why are there so many species of insects? *In, Proc. XV Int. Congr. Ent. Washington D.C.* pp. 84–94.

Janzen D. H. & Schoener T. W. (1968) Differences in insect abundance and diversity between wetter and drier sites during a tropical dry season. *Ecology* **49**, 96–110.

King M.-C. & Wilson A. C. (1974) Evolution at two levels in humans and chimpanzees. *Science* **188**, 107–116.

Lawton J. H. (1976) Mathematical models in ecology (book review). *Nature, Lond.* **264**, 138–139.

Levins R. (1968) *Evolution in Changing Environments*. Princeton University Press, Princeton, N.J.

Li T-Y. & Yorke J. A. (1975) Period three implies chaos. *Am. Math. Monthly* **82**, 985–992.

MacArthur R. H. (1971) Patterns of Terrestrial Bird Communities. In *Avian Biology, Vol. I.* pp. 189–221. Academic Press, New York.

MacArthur R. H. (1972) *Geographical Ecology*. Harper and Row, New York, N.Y.

MacArthur R. H. & Levins R. (1964) Competition, habitat selection, and character displacement in a patchy environment. *Proc. natn. Acad. Sci. U.S.A.* **51**, 1207–1210.

MacArthur R. H. & Wilson E. O. (1967) *The Theory of Island Biogeography*. Princeton University Press, Princeton, N.J.

McMahon T. (1973) Size and shape in biology. *Science* **179**, 1201–1204.

May R. M. (1974*a*) Biological populations with nonoverlapping generations: stable points, stable cycles, and chaos. *Science* **186**, 645–647.

May R. M. (1974*b*) *Stability and Complexity in Model Ecosystems*. Princeton University Press, Princeton, N.J.

May R. M. (1975) Patterns of species abundance and diversity. In *Ecology and Evolution of Communities*, Ed. M. L. Cody & J. M. Diamond, pp. 81–120. Harvard University Press, Cambridge, Mass.

May R. R. (1976*a*) Simple mathematical models with very complicated dynamics. *Nature, Lond.* **261**, 459–467.

May R. M. (1976*b*) Coexistence with insect pests. *Nature, Lond.* **264**, 211–212.

May R. M. (ed.) (1976*c*) *Theoretical Ecology: Principles and Applications.* Blackwell Scientific Publications, Oxford.

May R. M. (1977) Thresholds and breakpoints: ecosystems with a multiplicity of stable states. *Nature, Lond.* **269**, 471–477.

May R. M. & Oster G. F. (1976) Bifurcations and dynamic complexity in simple ecological models. *Am. Nat.* **110**, 573–599.

Mayr E. (1969) *Principles of Systematic Zoology*. McGraw-Hill, New York, N.Y.

Mayr E. (1976) *Evolution and the Diversity of Life: Selected Essays*. Harvard University Press, Cambridge, Mass.

Meyrick E. (1927) *A Revised Handbook of British Lepidoptera*. Watkins and Doncaster, London.

Murdoch W. W. (1969) Switching in general predators: experiments on predator specificity and stability of prey populations. *Ecol. Monogr.* **39**, 335–354.

Murdoch W. W. (1977) Stabilizing effects of spatial heterogeneity in predator-prey systems. *Theor. Pop. Biol.* **11**, 252–273.

Odum E. P. (1968) Energy flow in ecosystems: a historical review. *Am. Zool.* 8, 11–18.

Pimm S. L. & Lawton J. H. (1977) Number of trophic levels in ecological communities. *Nature, Lond.* **268**, 329–331.

Preston F. W. (1962) The canonical distribution of commonness and rarity. *Ecology* **43**, 185–215 and 410–432.

Raup D. M. (1977) Probabilistic models in evolutionary biology. *Am. Sci. ent.* **65**, 50–57.

Roughgarden J. & Feldman M. (1975) Species packing and predation pressure. *Ecology* **56**, 489–492.

Sage R. D. & Selander R. K. (1975) Trophic radiation through polymorphism in cichlid fishes. *Proc. natn. Acad. Sci. U.S.A.* 72, 4669–4673.

Schoener T. W. & Janzen D. H. (1968) Notes on environmental determinants of tropical versus temperate insect size patterns. *Am. Nat.* **101**, 207–224.

Selander R. K. & Kaufman D. W. (1973) Genic variability and strategies of adaptation in animals. *Proc. natn. Acad. Sci. U.S.A.* **70**, 1875–1877.

Simpson G. G. (1953) *Evolution and Geography: an Essay on Historical Biogeography with Special Reference to Mammals*. Oregon Univ. Press, Eugene, Oregon.

Simpson G. G. (1969) The first three billion years of community evolution. In *Diversity and Stability in Ecological Systems*, pp. 162–177. U.S. Department of Commerce, Springfield, Va.

Southern H. N. (ed.) (1964) *The Handbook of British Mammals*. Blackwell Scientific Publications, Oxford.

Southwood T. R. E. (1976) Bionomic strategies and population parameters. In *Theoretical Ecology: Principles and Applications*, Ed. R. M. May, pp. 26–48. Blackwell Scientific Publications, Oxford.

Southwood T. R. E. (1977*a*) Habitat, the templet for ecological strategies? *J. anim. Ecol.* **46**, 337–366.

Southwood T. R. E. (1977*b*) Entomology and mankind (opening address). In *Proc. XV Int. Congr. Ent. Washington D.C.* pp. 36–51.

Stubbs M. (1977) Density dependence in the life-cycles of animals and its importance in K- and r-strategies. *J. anim. Ecol.* **46**, 677–688.

Uetz G. W. (1977) Coexistence in a guild of wandering spiders. *J. anim. Ecol.* **46**, 531–542.

Van Valen L. (1973) Body size and numbers of plants and animals. *Evolution* **27**, 27–35.

Waloff N. (1954) The number and development of ovarioles of some Acridoidea (Orthoptera) in relation to climate. *Physiol. Comp. Oecol.* **3**, 370–390.

Wilson D. S. (1978) *Evolution on the Level of Populations and Communities.* (in preparation).